The Complete Idiot's Reference C...

Everything You Need to Know Abo...

- ➤ Atoms are the smallest particles of an element.
- ➤ Molecules are groups of two or more atoms that are bonded together.
- ➤ DNA (deoxyribonucleic acid) is the molecule that carries genetic information for most living things.
- ➤ Stretches of DNA are called genes.
- ➤ Genes carry the instructions for making specific proteins.
- ➤ Humans have about 80,000 genes.
- ➤ Genes are located on specific chromosomes.
- ➤ Chromosomes are the rod-shaped structures found in the nucleus of a cell. Humans have 46 chromosomes, half of which are inherited from their mothers; the other half is inherited from their fathers.
- ➤ The nucleus of a cell contains the chromosomes.
- ➤ A cell is the smallest structure of a living thing that can grow and reproduce independently.

Biotech Terms

Biotechnology refers to the use of living things to serve the wants and needs of humankind.

Cloning refers to making an identical copy of all the DNA in an organism, instead of just changing a gene or two as in genetic engineering. The method used to produce Dolly the sheep is called nuclear transfer.

DNA ligase is a special chemical that acts like molecular glue to paste together a cut-out gene from one organism into another organism's DNA.

Genetic engineering (also called recombinant technology, rDNA, gene cloning, or gene splicing) refers to a technology in which researchers put one or more genes from one organism into another living thing. Using restriction enzymes, scientists can cut genes and then paste them into another organism's DNA.

Nuclear transfer is a way of cloning an organism. This is the way Dolly the sheep was produced. Ian Wilmut took an egg from another ewe and removed its DNA. Next, he took the DNA from the cells of an adult ewe. He put the adult ewe's DNA inside the "empty" egg, and when it started to divide, he put this embryo into a sheep that later gave birth to Dolly.

Restriction enzymes are special chemicals that act like molecular scissors to cut DNA in very specific places. Scientists use these enzymes to cut and paste genes from one organism into another.

alpha
books

D0478821

Genetic Timeline

1865	Mendel discovers the Laws of Heredity
1900	Correns, de Vries, and von Tschermak rediscover Mendel's laws
1902	Sutton's chromosome theory
1920s	Morgan shows that genes lie in a row on chromosomes
1941	Beadle and Tatum's one gene/one enzyme theory
1944	Avery discovers that genes are made of DNA
1951	Rosalind Franklin starts work on X-ray diffraction technology to learn more about the structure of the DNA molecule
1952	Hershey and Chase prove that DNA is the hereditary material
1953	Watson and Crick discover the structure of DNA
1966	Nirenberg and Ochoa crack the genetic code by figuring out DNA's three-letter "words" for 20 amino acids
1973	Cohen and Boyer put a toad gene into bacteria
1980	U.S. Patent and Trademark Office grants a patent on a genetically engineered form of bacteria
1988	Patent granted for the first genetically engineered mammal, the Oncomouse
1990	Anderson performs the first approved gene therapy on Ashanti DeSilva
1990	The Human Genome Project, an international effort to map and sequence human DNA, begins
1993	Hall and Stillman clone early-stage human embryos
1994	Calgene's FLAVR SAVR™ tomato is the first genetically engineered food on the market
1996	Wilmut clones Dolly the sheep from the DNA of an adult ewe
1997	Wakayama and Yanagimachi clone 50 mice
1998	Japanese researchers at Kinki University clone eight calves from the DNA of a cow
1998	Researchers at the Kyunghee University Hospital in South Korea put a woman's DNA into her eggs, and embryos start to grow

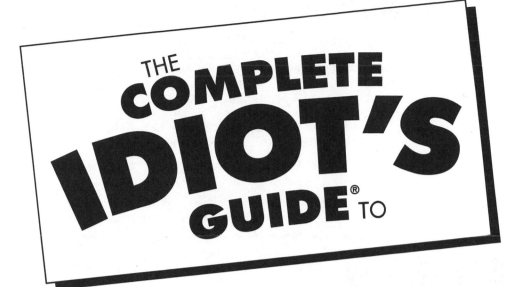

THE
COMPLETE
IDIOT'S
GUIDE® TO

Decoding
Your Genes

*by Linda Tagliaferro
and Mark V. Bloom, Ph.D*

alpha
books

A Division of Macmillan General Reference
A Pearson Education Macmillan Company
1633 Broadway, New York, NY 10019-6785

Contents

Part 7: Playing God? Ethical Issues   293

25 Patents on Life   295

26 Food for Thought: Are Biofoods Safe?   307

Appendices

Foreword

Genes are like a big zipper. And they are like a foreign language. Human DNA looks like a spiral staircase and operates like a giant Erector Set. If you look closely enough at our DNA you can see echoes of millions of years of genetic predecessors. Genes are but little drops of acid, yet they can form water, air, and dirt into a living, breathing human baby. You use them when you grow, and when you fight the flu, and when you make babies. You may not know much about your genes, but your genes know a lot about you.

We are entering the century of biotechnology. Soon, everyone will use genetic testing and before the first decade is over you will have a chip in your wallet that contains your complete genetic code. Cracking the human genetic code has become a trillion dollar industry and is the subject of the revolutionary Human Genome Project. It doesn't take much imagination to see a future where our children, our medicine, and even our careers and hobbies are the subject of genetic engineering. Already there are more than 800 genetic tests. The only question left to answer is: How do we figure out how to put our genes to good use?

The Complete Idiot's Guide to Decoding Your Genes is essential reading for, well, everybody. We all have dangerous genes, as well as other strengths and weaknesses that come to us from our parents. It is important to get a handle on what a gene is, how it works, and what it can and can't do. Learning about your DNA means taking responsibility for a world of new choices: Will you have genetic tests? Will you use a genetic therapy? How do you feel about human and animal cloning? What kind of child do you want, and what sorts of things should be off-limits for human medicine? Congratulations on your first step toward an exciting and dangerous journey into the world of DNA.

—Glenn McGee, Ph.D., professor of Bioethics, University of Pennsylvania

Glenn McGee, Ph.D., is a bioethicist at the University of Pennsylvania Health System. He is one of the world's leading experts on bioethics, genetics, reproduction, and health policy. A professor in Penn's School of Medicine, he is also a Commissioner on the new Federal Advisory Board on Clinical and Molecular Genetics and an Atlantic Fellow in Public Policy of the Commonwealth Foundation and British Government. He is Senior Editor of the MIT Press Basic Bioethics book series.

Dr. McGee's books include *The Perfect Baby: A Pragmatic Approach to Genetics* (Rowman & Littlefield, 1997), *Pragmatic Bioethics* (Vanderbilt Press, 1999), and *The Human Cloning Debate* (Berkeley Books, 1998). He has authored dozens of articles and chapters and his comments have appeared in virtually every world newspaper and in his MSNBC Online column. He has discussed cloning, genetics, and bioethics on national news programs such as *20/20, 60 Minutes*, CNN, *Charlie Rose*, and the *Jim Lehrer News Hour*. Professor McGee has given hundreds of named lectureships, public talks, and grand rounds in the continental U.S., U.K., Israel, Brazil, Japan, Norway, Denmark, Finland, Sweden, and Puerto Rico.

Dr. McGee is a Senior Fellow of the Leonard Davis Institute for Health Economics, a Senior Research Fellow of the Kennedy Institute for Ethics at Georgetown University and an Associate of Dartmouth College Ethics Institute. He has recently served as a member of the CDC Working Group on Genetic Testing and Public Health, the Coordinating Group on Bioethics of the American Bar Association, the American Association of Health Plans task force on ethical issues in managed care, and the Primary Working Group of the National Advisory Board on Ethics in Reproduction.

He is on the editorial board of several journals. He is the Director of bioethics.net and Senior Editor of "Breaking Bioethics" for *Cambridge Quarterly in Healthcare Ethics*, and Editor of Penn *Bioethics*.

Dr. McGee received his B.A. in Philosophy at Baylor and his M.A. and Ph.D. at Vanderbilt.

Introduction

The study of genes and DNA is fascinating, but it can seem pretty complicated to a beginner. Since the days of Gregor Mendel, an Austrian monk who puttered in a monastery garden, to Watson and Crick, who figured out the structure of the DNA molecule, to Ian Wilmut, who cloned Dolly the sheep from an adult ewe's DNA, a tremendous number of things have happened in the field of genetics in a relatively short period of time. Just check today's newspaper, and you'll probably read about a new gene that some scientist has discovered.

We need to know more about how our genes work so we can understand more about ourselves. But because so much is going on in biology these days, we first need to learn the basic principles of genetics. This book will make all that hard stuff easy. It organizes everything for you and includes everything you need to know about this important field that is affecting all of us.

This book is arranged to give you the basics of genetics in a fun and easy way. Here's what's covered:

Part 1, "Welcome to the Age of Genetics," gives you quick-and-easy explanations about early theories of inheritance. You'll read about Mendel, the Father of Genetics. Then you'll learn about the scientists who built upon Mendel's findings to figure out how characteristics are inherited. You'll also understand how cells contain DNA and genes.

In **Part 2, "The ABCs of DNA,"** you'll learn about how Watson and Crick discovered the structure of DNA. The reader-friendly text will explain how DNA and genes work and what all of this means to your health and well-being.

Part 3, "Faulty Blueprints: Mutations and Genetic Diseases," tells you what happens when genes don't function properly. You'll learn about sex (at least how gender is determined—nothing too racy here) and how changes in genes can be either good or bad news for you.

Part 4, "Manipulative Scientists: Genetic Engineering and Cloning," tells you about today's hottest genetic topics. Everyone's talking about how scientists are putting genes from animals into plants or human genes into animals. And there's the possibility of making "carbon copies" of humans in the future. In simple language, this part of the book examines the science behind all these developments.

Part 5, "DNA Detectives," tells you all about DNA tests, both in court and in the labs. You'll learn how scientists can find out whether you carry genes that may cause diseases, the science behind trials such as O.J. Simpson's, and more.

Part 6, "Mending Your Genes: Gene Therapy and New Medicines," describes the Human Genome Project, a large effort to catalog all our genes. You'll read about experiments with gene therapy, in which scientists try to fix problems inside a person's genes. There's also information about the new types of medicine that are coming out because of genetic technology and new ways for infertile couples to become parents through high-tech solutions.

Part 7, "Playing God? Ethical Issues," discusses the hottest scientific controversies today: whether scientists should manipulate genes, patent living organisms, or clone animals or even human beings.

At the back of this book, you'll find appendices that will help you to learn even more about these genetic topics. There's a list of resources that includes Web sites, organizations with newsletters, and companies that offer DNA testing. There's also a glossary, so you can quickly look up any genetic term that you're not familiar with. A selected bibliography lists some good books that will help you delve deeper into DNA.

Extras

In addition to simple and clear explanations of some very complicated topics, this book also includes some features to give you quick facts that are interesting, informative, strange, or outright provocative. There will be some trivia tidbits, definitions of unfamiliar terms, explanations of popular misconceptions, and anecdotes that will give you some genetic food for thought.

DNA Data

This feature provides you with interesting anecdotes about genes, DNA, scientists, and everyday people who've had a close encounter of the genetic kind.

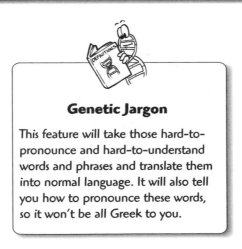

Genetic Jargon

This feature will take those hard-to-pronounce and hard-to-understand words and phrases and translate them into normal language. It will also tell you how to pronounce these words, so it won't be all Greek to you.

Mutant Misconceptions

There are lots of misunderstandings in the world of genes, and this feature will keep you from making a genetic faux pas.

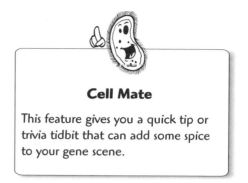

Cell Mate

This feature gives you a quick tip or trivia tidbit that can add some spice to your gene scene.

Part 1
Welcome to the Age of Genetics

A lot of excitement is surrounding your genes these days. To learn how scientists from the 1970s on found out so much about DNA and genes, first you need to understand the history of genetics, the science that studies heredity.

This section explains how early scientists helped lay the foundation for the genetic science of today. From Gregor Mendel and his experiments with pea plants, to later scientists such as Griffith, Avery, and Beadle and Tatum, you'll learn about the foundations of today's understanding of how traits are passed from one generation to the next.

Chapter 1

Join the Genetic Revolution

In This Chapter

➤ What is genetics anyway?

➤ Early ideas about heredity

➤ New developments in genetics

➤ Why you should bother to learn about this stuff

Did you ever wonder why you look like a human and not a hamster? Why cats don't give birth to canaries? Why your brother doesn't look like a chimpanzee? Or why he acts like one? It's probably soul-searching questions like these that led people in the past to ponder the mysteries of *heredity*, the study of how traits are passed from one generation to the next. This in turn led to the development of *genetics*, which is the scientific study of heredity.

In the past century, huge advances have been made in this field. Scientists discovered something called DNA, which has been called the code of life. It's contained in the cells of all living organisms, from petunias to pigs to people. Scientists found out that long stretches of DNA called *genes* are responsible for passing traits from one generation to the next. So, it's important for you to learn about your genes. By understanding how genes work, you can decode some of life's greatest secrets.

Genetics may seem like a confusing and intimidating subject, reserved only for folks who have lots of letters after their names. But by taking on the information bit by bit, you can learn how to decipher the code of life that's inside every one of your own cells

Genetic Jargon

The way traits are passed from parents to their offspring is called **heredity**. The scientific study of heredity is called **genetics**. **Genes** are the basic units of heredity, and they are found in the cells of living things.

right now. Even if you don't know a gene pool from a cesspool, or the difference between RNA and R&B, when you finish this book you will know enough about what goes on inside your genes to have a basic understanding of how they work and how they affect your life.

In this chapter, you'll get a basic overview of what genetics is, some common questions that are answered by genetics, some of the theories that scientists had before they began to truly understand how heredity works, and a little of what's going on in genetics today.

Everybody into the Gene Pool!

Do people tell you that you have your father's eyes? Your mother's nose? Your grandfather's smile? Most people hear that they resemble one or more of their relatives.

We all know that children usually resemble one or both of their parents. What's particularly fascinating is the combination of traits that can come from the same mother and father. If a husband has wavy black hair and brown eyes and stands 6'1", and his wife has baby blue eyes, straight blonde hair, and is 5'1" tall, they might have three children who look very different from one another, yet somehow still bear a resemblance to their mother and their father. The oldest daughter might have black hair and blue eyes and measure 5'2", while the middle child, a son, could have blond hair and brown eyes and stand 6' tall. The youngest child, another boy, might be blond, too, with blue eyes, and tower over his sister at 6'2".

People can also resemble their relatives in other ways. You might hear that you have your uncle's mannerisms, or your grandmother's talent for playing the piano, or your father's interest in reading *Complete Idiot's Guides*. Look into the mirror, and you might

Mutant Misconceptions

Mixing politics with science can have disastrous results. For example, Soviet dictator Joseph Stalin (1879–1953) thought that the idea that genes contribute to heredity was a self-supporting capitalist myth. The impact on Soviet agriculture was devastating.

notice that you have the same broad shoulders and rippling arm muscles as your brother. While eating at a restaurant with your family, you may realize that you walk with the same quick pace as your sister and that you and your mother both like chocolate ice cream with mounds of whipped cream on top.

Throughout the ages, people have wondered just how it is that children can resemble one or both of their parents and yet be very distinct individuals. Even though a set of siblings may have totally different personalities and appearances, there are usually ways in which they are very similar. Scientists have also done research to find out just how much of a person's physical, mental, and emotional makeup is inherited and how much is controlled by forces other than genes.

Nature Versus Nurture

Are your genes more important than the physical surroundings you grew up in? Does what you eat and where you live and what kind of lifestyle you lead contribute more, less, or the same as the inherited traits that were encoded in your genes before you were born? Scientists are still hotly debating this question of nature versus nurture. Some feel that genes are everything, others say that the environment is more important, and still others argue that who a person is results from a complex combination of genes and environment.

Questions such as these have fascinated scientists for centuries, but the current era is the one that's sometimes referred to as the Age of Genetics, because so many momentous discoveries about genes have been made in this century. The code of life has been deciphered, and the secrets of DNA are unraveling.

Why You Need to Know About Your Genes

Here are just a few of the reasons why you should understand how genes work:

➤ You were born with them, and you're stuck with them. So far, it's been nearly impossible to trade them in for new designer models. (In Chapter 22, "What Is Gene Therapy" you'll learn how some scientists are working on trying to tailor people's genes.)

➤ You're living in the Age of Genetics. People are also calling it the Genetic Revolution. Scientists are rapidly reporting the discovery of more and more genes that they believe contribute to different traits and diseases.

➤ You can discover what the genes you were born with mean to your physical appearance, personality, and well-being.

➤ You can also understand what your genes *can't* tell you about your health, personality, and appearance.

➤ New genetic technologies will impact every aspect of your life, including what you will eat, the new medicines you will take, and the way diseases will be treated. These technologies may even redefine what it means to be a human.

Inherited Misconceptions

Not surprisingly, some of the earliest theories about heredity were a little bit off the mark. The Greeks, Romans, and others came up with some pretty imaginative—but incorrect—ideas about the way traits are passed from one generation to the next. Here's a historical sampling.

It's All Greek to Me

Many ancient Greek philosophers and scientists thought that children were created from a combination of different elements from each parent. Theories like this were

Genetic Jargon

Most ancient Greek philosophers and scientists had what are called parti-culate theories about how heredity works. A **particulate theory** states that people inherit traits from their parents when some kind of physical particles from Mom and Dad mix together to form their children.

Mutant Misconceptions

The ancient Romans didn't know a whole lot about heredity and repro-duction either. They thought that female horses became pregnant as a result of being fertilized by the wind.

Genetic Jargon

A **preformation theory** holds that people (and other organisms) start off fully formed in the egg or sperm and simply grow larger until they are born. Now, of course, we know that this is not the case.

called *particulate theories*, because they stated that certain particles from the parents' bodies got mixed up together and formed a new person or animal.

This theory also held that some unusual creatures were the result of the mating of two very different types of animals. For instance, some people thought that if a camel and a leopard became enamored of each other, their love child would be a giraffe. (There's no account-ing for taste in love, is there?)

Aristotle, a Greek philosopher who lived from 384 to 322 B.C., theorized that particles of blood were passed from one generation to the next through semen, which he believed was made of the thickest parts of human blood. This theory, long proven untrue, is the basis of expressions we still use today, such as bloodline and blood relative.

Aristotle also believed that mothers' and fathers' responses to events in their environment could cause their future children to have distinct appearances. For instance, if a father received a serious cut on his arm that left a scar, his future son would be born with a replica of that scar in the same place on his body. However, Aristotle believed that the son's scar might be less pronounced than that of his father.

Another Greek philosopher, Empedocles, who lived in the fifth century B.C., told the story of a pregnant woman who gazed admiringly at a group of lovely Greek statues for a long time. According to Empedocles, when the woman later gave birth, her children were as classi-cally beautiful as the statues she had stared at while pregnant.

Pre-Fab Humans

Another early type of theory about heredity was the *preformation theory,* which states that a human being starts out in miniature form inside the sperm or the egg. One opponent of this theory was British doctor William Harvey (1578–1657). He discovered that a child forms from an egg inside his or her mother. Harvey concluded that this is true for all animals. He somehow skipped over the part that the male of the species plays. He con-tended that the egg was a self-contained unit capable of developing without any outside help. As late as the eighteenth century, an Italian scientist named Marcello Malpighi (not to be confused with actor Marcello Mastroianni) was convinced that you could find an entire miniature person-to-be inside each of a woman's egg cells.

During the eighteenth century, a similar theory evolved that sure struck a blow to women's lib. It held that all the makings of a new organism were contained not in the egg, but in the sperm. This theory stated that something called a *homunculus*, or a teeny, tiny man, was all curled up and crammed inside the sperm, ready to develop into an embryo and eventually a full-grown human being.

From Spontaneous Generation to Generation X

Another early theory about how life forms was called *spontaneous generation*. It held that living creatures could come into being automatically by arising from dead matter. For instance, people believed that maggots, which are worm-like baby flies, sprang to life from decaying meat. This theory was disproved when an Italian scientist named Francesco Redi (1626–1697) conducted an experiment in which he filled two jars with meat and then covered one and left the other uncovered. When maggots formed in the open jar but not in the covered jar, the spontaneous generation theory spontaneously disappeared.

In the 1700s, a final death blow to the theory of spontaneous generation was delivered by another Italian scientist, Lazzaro Spallanzani (1729–1799), who basically repeated Redi's experiment with water instead of meat. To disprove the idea that living things could spring forth directly from stagnant pond water, Spallanzani took two jars of water, covered one, and left the other open. When airborne bacteria wound up having a pool party in the uncovered jar of water, the covered water remained pristine.

A Theory Slightly Off Lamarck

Later in the eighteenth century, French biologist Jean Baptiste Lamarck (1744–1829) developed his

Cell Mate

Charles Darwin (1809–1882), the British naturalist famous for his theories on evolution, believed in the theory of pangenesis, which was first described by Hippocrates. The theory says that each part of the body reproduces itself. Darwin believed that the units of reproduction were called gemmules. According to this theory, all parts of the body contained gemmules and united with each other.

Genetic Jargon

A **homunculus**, in one obsolete theory of heredity, is a tiny person crammed into the sperm cell, ready to develop into an embryo and eventually into a fully grown person.

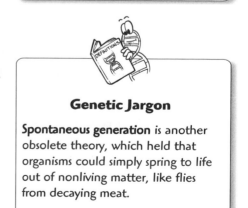

Genetic Jargon

Spontaneous generation is another obsolete theory, which held that organisms could simply spring to life out of nonliving matter, like flies from decaying meat.

theory of the inheritance of acquired characteristics. According to this idea, if a plant or animal changes during its lifetime, its future offspring will reflect this change. For instance, Lamarck theorized that when early giraffes strained to reach for tasty leaves that sat on high tree branches, this eventually led to more and more baby giraffes being born with longer necks. This theory became known, surprisingly enough, as *Lamarckism*.

Genetic Jargon

Lamarckism, the theory of inheritance of acquired characteristics, says that adaptations to events in the outside world cause changes in plants and animals and that these changes can be inherited by their offspring. This theory is considered obsolete these days.

Sex: A Cooperative Effort

After all these false fits and starts, scientists finally got it right. We now know that most living things make more of their own kind by sexual reproduction, in which the female provides the egg and the male provides the sperm. This discovery later led to the development of X-rated movies.

Scientists eventually learned that living things are made of cells, that cells contain chromosomes, that chromosomes contain genes, and that genes are made up of DNA. You'll read a lot more about cells, genes, chromosomes, and DNA in later chapters, especially in Chapter 5, "Nature's Blueprint," and you'll see how all of them function and interact.

Inside the dark spot of the cell, the nucleus, there are chromosomes, and inside chromosomes are genes. Genes are stretches of DNA, the Code of Life.

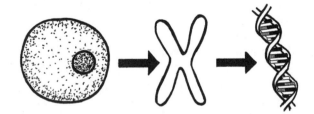

Mendel Was No Pea Brain

From the mid-nineteenth century through the present, scientists have made great strides in understanding how traits are passed from parent to child. Gregor Mendel (1822–1884), an Austrian monk, played around with garden pea plants and came up with some interesting findings that are the foundation of modern genetics. He asserted that some "factors" were responsible for the passing of characteristics to the plant's offspring—but he never discovered exactly what these factors were, exactly where they were located in the parent plant, or how they functioned in that plant. He did, however, lay down the laws of heredity and therefore is considered "The Father of Modern Genetics." We'll look at Mendel's life and work in detail in Chapter 2, "Mendel: The Father of Genetics."

Although no one in Mendel's time understood the significance of his momentous work, his findings were eventually rediscovered in the early 1900s. Building upon Mendel's research, a number of scientists in the twentieth century made contributions that ultimately led to the discovery of the roles that genes and DNA play in the inheritance of traits.

While studying the chromosome, the part of the cell that contains genes, American biologists Walter Sutton (1877–1916) and Nettie Stevens began to see the relationship between the way chromosomes function and the hereditary laws that Mendel had articulated. Other scientists, such as Frederick Griffith (1877–1941) in England in the 1920s and Oswald Avery (1877–1955) in the United States in the 1940s, worked hard to pinpoint what it is that carries information about traits from parents to children. Eventually, scientists realized it was something called DNA, which is divided into genes. (You'll learn about this in depth in Chapter 2 and Chapter 3, "Mendel's Successors.") From Mendel working with peas in his Austrian monastery all the way to James Watson (born 1928) and Francis Crick (born 1916) discovering the structure of DNA, it's been a long and winding road to our present understanding of genetics.

Cell Mate

Recombinant DNA recombines genetic material from two separate organisms. Such genetic manipulation is also known as genetic engineering. This area of science has the potential to affect everything we do, from the foods we eat, to the medicines we take, to the traits our children will have.

Altering Your Genes

After Watson and Crick's landmark discovery of the structure of DNA in the 1950s, scientists had a solid base of information to better understand the workings of heredity. This base led to even more momentous discoveries. In the 1970s, scientists figured out how they could cut and paste genes from one organism to another. This process is called recombinant DNA, or genetic engineering.

Scientists are now hoping to find cures for genetic diseases by adding to or deleting from the genes we start off with. Researchers are trying to improve upon nature by creating plants that provide superior nutrition and vegetables that can grow in inclement conditions or fend off insect pests. Animals are being genetically manipulated to produce rare human hormones in their bodies—hormones that would otherwise be in short supply for ailing people who need them. New medicines and vaccines are being created by splicing a gene or

Genetic Jargon

Cloning refers to making an exact copy of all the genetic material in a plant or animal. Technically, identical twins are clones because they have all the same genes. Identical twins occur when one fertilized egg splits into two separate embryos. Scientists have found ways to artificially induce cloning in different organisms.

two into plants or animals. Researchers even hope to someday be able to administer vaccines through genetically altered food such as bananas.

So far, understanding our genes has led scientists to the development of test tube babies and the *cloning* of human embryos. In 1997, we said hello to Dolly, the cloned sheep, and then to other cloned animals, like monkeys and cattle. (We'll go into cloning in detail in Chapter 16, "Send in the Clones.")

The Possibilities Are Endless

Some people anticipate using newfound knowledge about our genes to lower the risk of life-threatening diseases such as cancer and heart disease. By determining a person's genetic tendencies toward certain conditions, that person can be made aware of his or her predisposition and can avoid the occurrence of the disease by regular screening, a pharmaceutical regimen, or the like. The future may even hold the possibility of producing "designer babies" with great intelligence, physical beauty, and strength. Some people think it's conceivable that children could be programmed before birth to be born with the potential for better health.

As these developments continue to unfold, some people look forward to seeing this technology used for the betterment of humankind. On the other hand, it frightens others with its potential for the sticky ethical issues genetic engineering brings with it. Because this science is so new, they worry that we have no idea what the long-term effects of genetic manipulation may entail. They caution that this powerful technology could be misused if it winds up in the wrong hands. We will discuss these issues in Part 7, "Playing God? Ethical Issues."

Cell Mate

A transgenic organism is produced by mixing and matching genes. Using new genetic technologies, scientists can mix the genes of different types of plants and animals, including humans, in ways never done before. A few genes from a human can be put into a pig, or fish genes can go into a tomato.

Brave New World

Some people worry about the possibility of genetic discrimination, and the idea that a genetic underclass could develop if people in the future are judged by their genetic makeup. Some fear that it will lead to more abortions. They argue that parents who don't like what the genetic tests on their unborn children reveal will be tempted to abort and try again.

There are concerns that *transgenic* plants, which are genetically altered plants, could inadvertently produce a crop of superweeds. And if animals are genetically manipulated to produce more uniform varieties, they could become more susceptible to diseases than the wider variety of creatures that would evolve naturally. These ethical concerns will be discussed later in the book, in Chapters 26, "Food for Thought: Are Biofoods Safe?" and 27, "Creature Concerns."

The Human Genome Project

One reason why Mendel was successful in his search for the laws of heredity was because he chose to work with a relatively simple life form: the garden pea plant. Had he chosen to work with humans or even mice, he would never have accomplished so much. It isn't always easy to make predictions about heredity in humans because they are infinitely more complex. Humans have many obvious traits that are governed by several genes, instead of just one.

In addition, humans don't make great laboratory animals. If you were a scientist who wanted to study several generations of people, you'd have to wait nine months for each generation to be born in order to see the results of your experiment. Then, of course, there would be the ethical issues of using people to study the laws of genetics. It's much easier to work with simpler life forms like pea plants or, better yet, bacteria or fruit flies, which reproduce quickly, allowing geneticists to study many generations in a short period of time.

But there *is* one way that scientists are learning more and more about human heredity, and it doesn't involve telling anyone to marry a person with interesting genes. In 1989, the Human Genome Project was launched in the United States to probe even deeper into the mysteries of genes. Other countries, including England, France, and Japan, have started similar programs.

A Genetic "Road Map"

In essence, the Human Genome Project entails making a map of the human *genome*, which is the totality of all the genes that humans can have in their cells. The project is currently in full swing, and some scientists have high hopes that this knowledge will enable them to unlock even more secrets of how life functions. You'll read lots more about this in Chapter 21, "The Human Genome Project."

Scientists anticipate that knowing the location of different genes and the composition of each one may lead to cures for dreaded and incurable conditions. By trying to understand how genes function—and also how they malfunction—scientists hope to eventually be in a position to treat diseases with a genetic component, possibly by stopping them before they occur.

Scientists working on the project set the year 2005 as their goal for the completion of this task of mapping and delving further into every human gene in the entire human gene pool. So far, opinions are divided. Some scientists think that the Human Genome Project will succeed by that target date or even finish before then. One current

Genetic Jargon

The word **genome** means all of the genes, or units of heredity, that an organism can have. The human genome refers, of course, to all human genes, but every type of plant and animal has its own genome.

estimate predicts that it will be finished ahead of schedule in 2001. Others are not as optimistic, because they point out that much of the most challenging work will have to be done in the last five years of the project, and it could conceivably take much longer than estimated.

DNA Data

About one-third of the Human Genome Project is being handled by the United States Department of Energy; much of the remaining work is being done by the National Institutes of Health. The project is projected to be finished in the year 2001, unless the Y2K problem kicks in.

Touchy Science

There are also ethical issues involved in the Human Genome Project. The National Institutes of Health, the government organization responsible for the greater part of the research done on the Human Genome Project, has allocated five percent of the project's budget for studying ethical issues such as genetic privacy, genetic discrimination, and the right to patent genes. (Much more on this topic is coming in Chapter 28, "People Problems.")

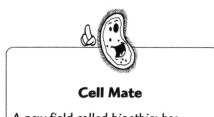

Cell Mate

A new field called bioethics has come into being because of touchy issues about genetic technology, reproductive technology, and other medical issues such as euthanasia (mercy killing) and abortion.

As the Human Genome Project draws nearer to completion, scientists are learning amazing things about our genetic past, present, and possible future. In the pages that follow, you will, too. You'll learn how early thinkers and later scientists discovered some of the basic laws of genetics, the science that explains how each generation inherits characteristics from its parents, be they plant or animal. You'll also learn about the mysteries of DNA, the molecule of life—you'll find out that DNA wasn't just something they invented for the O.J. Simpson trial or to prove who's the real father in a paternity suit.

The Least You Need to Know

➤ Studying genetics teaches you how characteristics are passed on from parents to children.

➤ This is the Age of Genetics, and scientists are learning more and more about how genes work.

➤ DNA, the code of life, underlies heredity.

➤ Genes are stretches of DNA.

➤ Scientists are learning how to switch genes around.

➤ The Human Genome Project is studying each and every gene human beings have to find out where it is and how it works.

Chapter 2

Mendel: The Father of Genetics

> **In This Chapter**
>
> ➤ A substitute science teacher gets some revolutionary ideas
>
> ➤ Mendel stops his experiments with mice and gives peas a chance
>
> ➤ Mathematics plus biology equals important genetic discoveries
>
> ➤ Phenotypes, heterozygous, and other fancy terms

Some people have original ideas that are considered too revolutionary in their time. If their ideas turn out to be right, fame comes for such people long after they have passed away, if it comes at all. Such was the case for Gregor Johann Mendel, an Austrian monk who was the first researcher to figure out how traits are passed from one generation to the next. In this chapter, you'll learn about "The Father of Genetics" and see how he laid the groundwork for numerous momentous discoveries that followed in the scientific study of inherited characteristics.

Minding His Peas and Pews

In 1822, when Gregor Johann Mendel was born in what was then Austria but is now the Czech Republic, most people believed that children were formed by a blending of their parents' traits. According to this theory, if a 6' tall man married a woman who was 5' tall, their children would all measure 5'6". If a man had a long nose, and his wife had a short one, people believed their children would inherit medium-sized noses. Dark-skinned men and light-skinned women were expected to produce children with skin tones in the middle, and so on. Things didn't always turn out exactly as expected, but the theory of the blending of traits explained things reasonably well to folks at the time.

The Linnaeus Catalog

When European explorers were bringing back exotic, "new" plants and animals from around the globe, it seemed like a good idea to catalog all these previously unknown life forms. So in 1735, a Swedish scientist named Carl Linnaeus (1707–1778) came up with a classification system to keep track of it all.

One of Linnaeus's contributions was the term *species*. You probably learned in biology class that two organisms belong to the same species if they can mate with each other and produce fertile offspring. In Mendel's time, people believed that each species represented the descendants of the original pair of that animal created by God, because this is what they read in the Bible. People felt that because God created organisms the way they were, they could not change.

Genetic Jargon

A **species** is a classification of organisms that have common characteristics distinguishing them from other species and can mate to produce fertile offspring. A **hybrid** is an offspring of two parents of different varieties of the same organism.

Room to Breed

Mendel grew up on a farm, and he must have learned a lot about plant breeding from watching his father go about his daily chores. He learned even more when he attended the village school. Mendel was fortunate enough to study with a teacher named Thomas Makitta, who taught his students about hybrids. A *hybrid* is produced by breeding plants or animals that have different variations. For instance, if two varieties of sheep mate, they will produce offspring that has inherited traits from each variety represented by the parents.

DNA Data

One thing we have to remember about scientific beliefs in Mendel's time is that the distinction between species and hybrid wasn't clear. In Mendel's day, crossing any organisms that were different—whether a species or a variety—was called a hybrid. But nowadays, we would say that Mendel crossed different varieties. (We say that crossing plants produces varieties, and crossing animals produces breeds.) Nowadays, we call the offspring of two different species a hybrid. In fact, when Mendel finally concluded his experiments and wrote a paper on his findings, he called it "Elements in Plant Hybridization," despite the fact that all the peas he studied belonged to the same species.

In Mendel's time, as now, many people were interested in breeding plants and animals of specific varieties that they considered more desirable than others. Even though they didn't understand the mechanisms of heredity behind this type of breeding, they were usually more than satisfied with the hybrid plants and animals that resulted.

Back to the Garden

In 1843, Mendel, a devout Catholic, decided to join the order of Augustinian monks, and he entered the monastery at Brünn (now Brno in the Czech Republic). There he met an elderly monk, Father Aurelius Thaler, who had tended the monastery's peaceful, secluded gardens for many years. When the elder monk passed away, Mendel happily became the new caretaker of the gardens. He also took on duties as a substitute science teacher in a high school near the monastery.

Mendel's greatest desire was to become a high school science teacher, but he failed the teachers' exam several times. Unable to fulfill his dream, Mendel went back to his beloved gardens in the monastery and decided to study plant breeding in a small section of the grounds. Ironically, Mendel's thwarted desire to become a teacher helped lead him toward the work that would immortalize him. He would inspire theories and set events into motion that would revolutionize the way people understood heredity and would eventually lead to a truly scientific understanding of genetics. Ultimately, Mendel's investigations paved the way for research into genetic engineering, cloning, and gene therapy.

Cell Mate

One of the keys to success in scientific inquiry is keeping it simple. One of the reasons Mendel's experiments were so successful was that he used pea plants, which have a relatively simple genetic structure. When he first tried the experiment on mice, it didn't work out.

The Seven Traits of Highly Effective Pea Plants

Mendel was fascinated by the fact that living things come in almost infinite varieties. He saw that even in one species of plant, there could be any number of colors, shapes, and sizes. He wanted to understand how these varieties occurred and how the parents passed some traits and not others to their offspring.

He started out by breeding mice, but decided this approach wasn't working out. (We can only speculate whether this might have led to a book entitled *Of Mice and Mendel*.) Instead, he decided that the easiest way to approach his study would be to breed different varieties of the common garden pea plant. Mendel decided on pea plants because:

➤ They weren't moving out of town any time soon.

➤ He didn't need a signed statement from them saying he wouldn't be held responsible for any ill effects from the experiments.

➤ Each generation matured quickly, so Mendel didn't have to wait very long to see how each generation would turn out.

➤ Mendel could easily decide which plants to mate with others, and none of them would protest.

➤ The different characteristics, such as color and height, were fairly easy to observe.

Genetic Jargon

The **pistil** of a plant is the seed-bearing reproductive organ of a flower. The pistil has parts called the stigma, the style, and the ovary. The **stamen** is the part that produces **pollen**, which is the plant's sperm cells. The **ovule** is the part of a plant that contains the egg cell. It is usually at the base of the flower. When pollen reaches the egg cell, the egg is fertilized, and new seeds begin to form.

Mendel decided to observe several generations of pea plants and meticulously note which traits were transferred from one generation to the next and which ones weren't. To keep things simple, he analyzed just one trait at a time.

Pistil-Packing Pea Plants

One reason it was so easy for Mendel to control his plant breeding experiments was that the garden pea's sex cells, like those of most plants, are in its flowers. When pollen—a plant's version of sperm cells—unites with a plant's egg cell, the egg is fertilized, and seeds develop. These seeds, of course, can be planted to produce another generation of plants.

The egg cell in the pea plant is found inside a part called the *ovule* (O-vyool). This is usually the bulgy green part at the base of a flower. Those thin little pieces that stick up from the center of a flower are known as *stamens* (STAY-mens), and they produce *pollen*, the powdery stuff that gets all over your nose if you're not careful when you sniff a flower.

Cross-section of a flower.

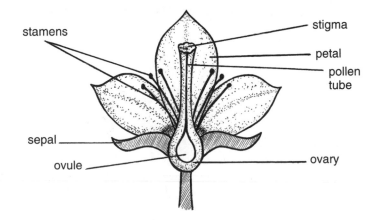

Some plants have male and female varieties. In other words, eggs are produced in one type of flower, and pollen is produced in another type. In such cases, pollen from a male plant can be carried by an outside agent, such as the wind or an insect like a busy bee, and transferred to the *pistil* on a female plant. A small tube leads from the top of the pistil to the ovule, which contains an egg. Fertilization occurs here, and seeds are eventually formed. When pollen from one plant fertilizes the egg from another plant, this is called *cross-fertilization.*

Some plants have different sexual preferences. The pea plant, for example, has it all. Eggs and pollen are contained in the same plant, so it can simultaneously be both the mother and the father of a new generation of seeds. This is called *self-fertilization.* To facilitate this process, pea plant flowers are closed during fertilization so pollen from other plants can't get in to fertilize the eggs.

So Mendel had his work cut out for him to get the pollen from one selected pea plant into another. To begin with, he had to open each of the closed flowers to dust them with foreign pollen. He even cut the stamens off of some plants so they couldn't form pollen for self-fertilization. To doubly ensure that no other pollen except the kind he chose would find its way to the female cells, Mendel tied little cloth bags over the flowers.

Genetic Jargon

Some plants have both male and female parts. The male parts produce pollen, the plant equivalent of sperm. If this pollen comes in contact with the plant's female parts and fertilizes the plant's own eggs, this process is known as **self-fertilization**. On the other hand, if one plant's pollen fertilizes the egg of another plant, this is known as **cross-fertilization**.

Peas and Quiet

Life must have been tranquil and simple in the monastery in Brünn. For eight years, Mendel intently focused on these experiments. He narrowed his investigation down to seven specific traits that pea plants could only exhibit in one of two ways.

The Seven Traits of Pea Plants That Mendel Studied

Trait	Varieties
Seed color	Yellow or green
Seed shape	Smooth or wrinkled
Flower color	Purple or white
Stem length	Tall or short
Shape when ripe	Inflated or constricted pods
Color of unripe pods	Green or yellow
Position of flowers	Either up and down the stem, or only on the top of the stem

Before Mendel started his experiments, he spent two years producing pea plant varieties that bred true for each of the seven traits he was going to study. This means that tall plants always produced tall plants, plants with green seeds always produced other plants with green seeds, and so on. He wanted to make sure that the plants he would later cross were stable varieties.

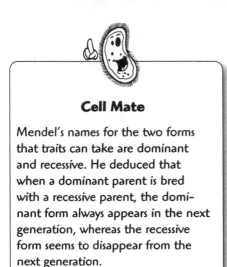

Cell Mate

Mendel's names for the two forms that traits can take are dominant and recessive. He deduced that when a dominant parent is bred with a recessive parent, the dominant form always appears in the next generation, whereas the recessive form seems to disappear from the next generation.

One trait Mendel investigated was height. Some pea plants spring up to six feet tall; others stop growing at only one foot. Mendel noted that when tall plants are crossed with other tall plants, they always produce tall offspring. Similarly, when short plants are pollinated by other short plants, the seeds they produce always grow into small plants.

But when Mendel crossed a tall plant with a short plant, something unexpected happened. Remember that the accepted theory of heredity at the time stated that there would be a blending of the male and female parent plants' traits, and therefore the next generation was expected to be medium in height. But this didn't occur. When Mendel crossed a tall plant with a short one, seeds for the next generation all yielded tall plants. When he crossed plants that had yellow pods with plants that had green pods, the harvested seeds always grew into plants with green pods instead of the expected yellow-green blend.

DNA Data

Mendel even tested Lamarck's theory of acquired characteristics. When he found a rare variety of ornamental plant that grew elsewhere in the garden, he planted it next to the usual variety. He even planted their offspring next to one another to see if any traits were passed on simply by being close to other varieties of plants. When nothing changed, he concluded that Lamarck's theories were off the mark.

So Mendel came to the conclusion that for the seven traits he was studying, one of the two possible forms was always stronger than the other in being passed to the next generation. Mendel called this the *dominant* form. The other trait, which seemed to disappear in the next generation, Mendel called *recessive*. We still use these terms today.

Double-Crossed Genes

As time went by, Mendel harvested even more surprises from his crop of garden peas. (Not to mention that the fruits of his experiments made a tasty side dish for the monks.) Now he was ready to cross the second-generation plants that had grown from the seeds of the first plants he worked with.

This time, he was astonished at the results. For example, when he crossed tall plants whose fathers were tall and whose mothers were short plants, three out of four of the offspring were tall, and the last was short. This type of result occurred whether the father plant was tall and the mother plant was short, or if the mother was tall and the father was short. So, Mendel assumed that the "factor" that controlled a plant's height could be inherited from either the "mother" or the "father" plant.

One pea plant fell short.

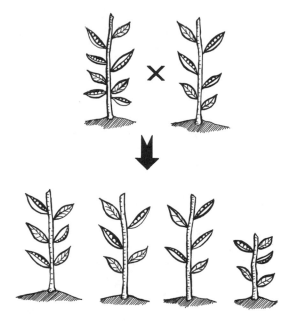

Factors You Can Count On

Mendel repeatedly performed these experiments, focusing on how the seven traits appeared in succeeding generations. What made Mendel's work significantly different from anything that had been done before was the fact that he meticulously counted and analyzed exactly how many plants exhibited each trait. This application of mathematics to biology was highly innovative for Mendel's time.

After his lengthy experiments were completed, Mendel came to some conclusions about heredity. He determined that something that made the traits appear the way they did was being passed from generation to generation. He called these mysterious entities factors. Today, we call Mendel's factors *genes*.

DNA Data

Through the years, some suspicious biologists have wondered about Mendel's mathematical data. They've gone over it with a fine-toothed comb and even accused him of fudging the results, because they feel his experimental findings are too good to be true.

Independent Traits

Mendel also concluded that each trait was inherited independently of the others. For example, when he crossed a tall plant that had smooth seeds with a short plant that had wrinkled seeds, their offspring would all be tall with smooth seeds because tallness and smoothness are the dominant forms of their traits.

However, things would change in the next generation. The offspring of the hybrid tall, smooth-seeded plants could be tall and smooth-seeded, but there was also a chance that they could be tall and have wrinkled seeds or be short and have smooth seeds. They could even be short and have wrinkled seeds. In other words, tallness and smooth seeds don't necessarily go together, and neither do shortness and wrinkled seeds. This is what Mendel meant when he said that every trait is inherited independently of the others.

Here is a summary of Mendel's conclusions:

➤ Each physical characteristic corresponds to one "factor." (Today, we call those factors genes.)

➤ "Factors" come in pairs.

➤ Only one of each pair is passed on to the next generation.

➤ It's equally probable that one "factor" or the other will be passed on.

➤ Some "factors" are dominant, and others are recessive.

Looking for Recognition

In 1865, hoping his discoveries would be recognized by the scientific community, Mendel presented his findings to a local organization, the Brünn Society for the Study of Natural Science. Unfortunately, no one seemed impressed or even particularly interested. Nevertheless, Mendel was allowed to publish his paper, "Experiments in Plant Hybridization," in the Brünn Society's journal.

DNA Data

In his quest for recognition of his work, Mendel even sent a copy of his paper to Charles Darwin, the scientist best known for his theory of evolution. The British naturalist who wrote *On the Origin of Species* never replied, so it's possible that he either disagreed with Mendel's conclusions or that he never even read the paper.

Mendel still hoped for recognition and continued to send his paper to numerous biologists. None replied. He even tried getting the attention of a noted Swiss botanist, Karl Wilhelm von Nägeli (fon NAY-ga-lee) (1817–1891), but he, too, failed to recognize the brilliance of the monk's thorough and groundbreaking research.

When Mendel became the abbot of the monastery in Brünn in 1868, his administrative duties kept him so busy that he had little time to continue his research on heredity. When the gentle monk died in 1884, no one could have suspected that his pioneering work would be rediscovered in 1900, and that he would come to be hailed as "The Father of Modern Genetics."

We know today that Mendel's original work was way ahead of the thinking and understanding of his peers. Nonetheless, it paved the way for many later scientists who built upon his pioneering work. We've also come to realize that Mendel was either extremely clever, extremely lucky, or both when he picked the garden pea as the subject of his research. The seven traits he chose to study were coincidentally determined by only one gene, and each was located on a different chromosome.

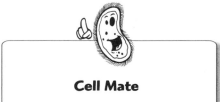

Cell Mate

In the course of Mendel's research, he studied approximately 28,000 garden pea plants.

DNA Data

At one point, Mendel tried to reproduce his pea plants experiment using another plant called the hawkweed. These experiments didn't work out as well as the pea plants research.

From Peas to People

Much has happened in the world of science since the dedicated monk Mendel observed pea plants in his garden. We now call the obvious, visible display of a trait the *phenotype* (FEEN-a-type), from the Greek word which means "to show" or "to display." We call the actual genetic makeup the *genotype* (JEEN-a-type), regardless of whether the trait is visible or not.

Genetic Jargon

A **phenotype** (FEEN-a-type), from the Greek word that means to show or to display, is the name given to a living thing's observable display of a trait. Its **genotype** (JEEN-a-type) is the trait's genetic constitution. An **allele** (a-LEEL) is a form of a gene. A living thing with two identical alleles is called **homozygous** (ho-mo-ZY-gus), and if it has two different alleles, it is **heterozygous** (HEH-ta-ro-ZY-gus).

Cell Mate

Some human family trees show that a few traits, such as eye color or the ability to roll one's tongue, seem to follow the genetic ratios that Mendel worked out when he experimented on pea plants.

What Mendel called the different forms of a trait are now known as *alleles*, which are basically different versions of a gene. For instance, smooth peas and wrinkled peas are two alleles because they are two different versions of the same trait. Flowers growing up and down the stem of a plant and flowers only growing on the top of the stem are two alleles of a different trait: flower distribution. In humans, straight and curly are two alleles for hair.

If a person or other organism has two alleles that are the same (for instance, if a pea plant has two tall alleles, or gene forms for height), then you can say the person or pea plant is *homozygous* for that gene. If the person or pea plant has two different alleles for the trait (for instance, if a pea plant has one tall and one short allele for height), it is said to be *heterozygous*.

Life After Mendel

Science has come so far that we've made the astounding discovery that humans are not pea plants. We differ from peas in many ways besides the fact that our skin isn't green and we're not rooted to the ground. Most of our human traits are determined by more than just one gene. Several genes come into play. Mendel would have had a hard time coming up with the mathematical ratios that explain all the variations in generations of human families.

Despite Mendel's lack of recognition in his own time, he nevertheless hit upon an innovative approach for finding out about heredity. Every scientist and student of genetics today should be thankful for the dedicated monk whose brilliant curiosity, love of nature, and unfailing perseverance led him to quietly and carefully tend hundreds of plants in a quiet monastery garden in Austria.

The Least You Need to Know

➤ Gregor Mendel was an Austrian monk who lived from 1822 to 1884.

➤ Mendel experimented with pea plants to observe how traits are passed from one generation to the next.

➤ Mendel noted that there are certain "factors" that are passed from one generation to the next. Today we call these factors *genes*.

➤ Mendel determined that traits are inherited in pairs and that each generation gets one form of the pair from one parent and the other from the other parent.

➤ Mendel noted that one form of a trait tends to be dominant and the other tends to be recessive, disappearing in some generations.

➤ Although unrecognized in his time, today Mendel is considered "The Father of Genetics."

Chapter 3

Mendel's Successors

In This Chapter

➤ How Mendel's Laws of Heredity were rediscovered

➤ What fruit flies and pus taught scientists about heredity

➤ How a blender proved that DNA is the code of life

A long, winding road led from Mendel's "factors" to the discovery that something called DNA was responsible for passing traits from one generation to the next. It took nearly a century before scientists understood the whole picture. Little by little, they fit together small pieces of the puzzle. They learned about genes, chromosomes, and other biological structures with Greek-sounding names. They learned that Mendel was no fool, after all. They learned how tiny things inside your cells that you can't see or talk to will determine what your kids will look like.

Thanks to this book, it will take you considerably less than a century to get up-to-date on what these scientists discovered. If you're a quick reader, you could be a biologist by the end of the week (or at least pretend to be one in front of your friends). You'll be able to impress acquaintances by using hard-to-pronounce words with ease and rattling off the names of pioneering scientists such as Sutton, Morgan, and Avery.

In this chapter, you'll learn all about the hunt for heredity that finally led to James Watson and Francis Crick unlocking the secret of the structure of DNA. So get ready to go back in time and see how scientists jumped over biological hurdle after hurdle until they reached their scientific goals.

Are You Cell Conscious?

Three hundred years ago nobody had video games, virtual reality, or access to the Internet, so scientists had a lot of time on their hands. They contented themselves by peering into the miniature worlds they found under the microscope. In 1665, an English scientist named Robert Hooke (1635–1703) took a peek at a section of cork under his microscope. Under the microscope, Hooke saw that the cork was made up of tiny rectangular divisions that reminded him of the small rooms in a monastery. He named these structures *cells*, after the monastic chambers that they resembled, and the name stuck.

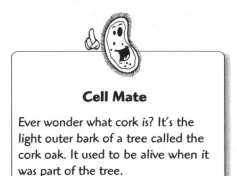

Cell Mate

Ever wonder what cork is? It's the light outer bark of a tree called the cork oak. It used to be alive when it was part of the tree.

After that, scientists examined all kinds of cells under the microscope. They noticed that there was usually a small, dark structure located inside each cell. In 1831, Robert Brown (1773–1858), a botanist from Scotland, came to the conclusion that all cells had this structure. He named it the *nucleus*, from the Latin word for "little nut."

Cell Theory

In 1838, Matthias Schleiden (SHLY-den) (1904–1881), after eyeballing a lot of different kinds of leaves and flowers under a microscope, came to the conclusion that all plants are composed of cells. One year later, Theodor Schwann (say SHVON if you really want to impress your friends) (1810–1882), another German scientist, determined that all animals are composed of cells, too. The concept that the cell is the basic unit of all plants and animals is known as the cell theory of life.

DNA Data

By now, you've most likely picked up on the fact that a lot of scientists have a thing about using Greek words for biological structures. Instead of using simple phrases like "colored bodies" or "little nut," they tend to make up new English words based on a combination of borrowed Greek words. Unfortunately, unless you're a classics scholar, this makes things pretty difficult to pronounce, to remember, and to understand. Fortunately, if you know what the Greek root words mean, the scientific terms are much easier to remember. So in this book, we're discussing as much of the etymology as possible. By the way, *etymology* is just a fancy Greek-type word meaning the study of where words come from.

By the end of the 1800s, scientists were discovering more and more about cells. A German biologist named Walther Flemming (1843–1905) showed the world that cells are something to dye for. Because making out the tiny innards of cells under the microscope wasn't always easy, Flemming thought it might be easier if he used bright colors to dye the cells he was studying, and his colorful idea worked. This is because some parts of the cell absorb more dye than others and therefore show up in stronger contrast to their surroundings. When he noticed long, thin structures that looked like well-cooked noodles, he called them chromosomes, from the Greek words for "colored bodies," which is how they looked after they were dyed.

Beneden and Weismann

By 1887, Edouard von Beneden (ben-AY-den) (1846–1910), a biologist from Belgium, realized that each type of animal or plant always has the same number of chromosomes (those "colored bodies" in the nucleus) as other animals or plants of the same type. For instance, humans have 46 chromosomes, corn plants have 20 chromosomes, and alligators have 32 chromosomes. See the following table for more chromosome numbers for different species.

Counting Chromosomes

Organism	Number of Chromosomes
Chimpanzee	48
Cabbage	18
Camel	70
Chicken	78
Cat	34
Dog	78

In 1892, a German scientist named August Weismann (1834–1914) published a paper proposing that the reproductive cells of the parent plants or animals contain the information that tells the offspring plant or animal how to develop. This was a pretty revolutionary concept for the time.

Every Little Breeze Seems to Whisper de Vries

Around this time, in 1886, a Dutch botanist named Hugo de Vries (rhymes with sneeze) (1848–1935) saw a field in his native Netherlands filled with an American plant called the evening primrose. Apparently someone had put one plant there and it spread, or some seeds were accidentally brought to the area, and the plant grew wild. What de Vries found interesting was that some of the plants looked very different from the rest.

At the time, people referred to weird or deformed varieties of plants, animals, or people as sports or monsters. Being a good sport himself, de Vries uprooted some of the sports and planted them in his garden. He conducted experiments along the lines of Mendel's pea plant experiments, even though he had never heard of the Austrian monk or his research with heredity.

By now, you're probably convinced that one of the main things that scientists do is make up new words for biological processes and parts of organisms. Although strictly speaking that's not required, de Vries did come up with a word of his own. When he got strange variations of the plants he tended, he called them *mutations*, from the Latin word meaning "to change." Scientists have used this designation ever since to refer to odd varieties of living things that are radically different from their parents.

Three Scientists, One Conclusion

In 1900, something stranger than the oddest mutation happened. De Vries and two other scientists from different countries, completely unknown to one another, independently published papers about their theories on heredity, each of which was almost identical to Mendel's findings. The other scientists were Karl Erich Correns (1864–1933), a botanist from Germany, and an Austrian botanist whose name was a mouthful, Erich Tschermak von Seysenegg (CHAIR-mock fon ZY-zen-egg), who lived from 1871 to 1962.

Before the three of them independently published their work, they wanted to know if anything similar had been written on the topic before. All three of them must have been thunderstruck to find that Mendel had done nearly identical experiments in the mid-1800s. Each of the three gave credit to the Austrian monk who was never acknowledged or appreciated as a true innovator in his own lifetime. Today, we honor his genius by referring to the Mendelian Laws of Inheritance.

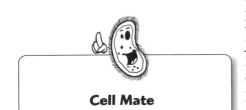

Cell Mate

Wondering why almost every scientific term comes from Greek and Latin? Traditionally, the classic languages have been the unofficial tongue of scientists and physicians. Until recently, even common prescriptions were written in Latin.

Now that the study of heredity was gaining more popularity, scientists needed—you guessed it—new words to describe things. In 1909, Wilhelm Ludvig Johannsen (yo-HAN-sen) (1857–1927), a botanist from Denmark, came up with a new word for the sections of chromosomes that were found to be responsible for producing one trait or another in the next generation. He decided to call what Mendel had dubbed "factors" by a different name: *genes*, from the Greek word meaning birth or generation.

This great Dane was also the one who came up with the words *phenotype* (FEEN-o-type) and *genotype* (JEEN-o-type). As we discussed in Chapter 2, "Mendel: The Father of Genetics," the word *phenotype* refers to the observable appearance of a plant or animal, and the word *genotype* refers to the genetic composition of the organism.

Following Mendel's Laws

After the rediscovery of Mendel's Laws of Inheritance, scientists made huge strides in the field of genetics. In 1902, an American biologist named Walter Stanborough Sutton (1877–1916) published a paper confirming that Mendel knew what he was talking about, even if he didn't understand all the ins and outs of genes or even that chromosomes existed. Working at Columbia University in New York City, Sutton showed that chromosomes, those spaghetti-like structures in the nucleus of a cell, come in pairs that are similar to one another.

DNA Data

Walter Sutton studied the behavior of chromosomes in grasshopper cells. He found that they behaved according to the laws of heredity that Mendel had postulated.

Many years previously, Mendel had come to the conclusion that each new plant inherited two "factors" for each trait from its parent plants, but often one of them showed in the appearance of the plant, and the other was "hidden." The new plant had received one "factor" for the trait from each parent. For instance, Mendel concluded that a plant's height was determined by the "factors" for tall or for short. Tall and short were two variations on a theme.

Sutton concluded from his own experiments that Mendel was correct. There were indeed pairs of chromosomes holding the information that would tell a new organism how to develop. Although Mendel came up with the theory, Sutton realized that Mendel's ideas had a physical reality in cells. Sutton came up with the chromosome theory, which states that genes, the parts of a cell that are responsible for hereditary information being passed to the offspring of an organism, are parts of the chromosomes. This theory proved to be true.

The Fruitful Fruit Fly

Another scientist who contributed greatly to the science of genetics in its early days was Thomas Hunt Morgan (1866–1945), an American biologist who worked at Columbia University. From the early 1900s to the mid-1930s, Morgan and some of his students, including Alfred Sturtevant (1891–1970), Calvin Bridges (1889–1938), and Hermann Muller (1890–1967), studied hereditary traits in fruit flies, the tiny insects that flit around your bowl of bananas. So the next time you shoo one away, keep in mind that it might be a descendant of a fly that was essential to our scientific progress.

DNA Data

Morgan worked in a small laboratory in the Zoology department at Columbia University in New York City. This lab was referred to as The Fly Room. It was filled with hundreds of milk bottles, which made cozy homes for the fruit flies.

Why a Fly?

Morgan decided to use fruit flies for his experiments because they have several qualities that make them well-suited to genetic research:

➤ They only have eight chromosomes, so they're easy to study.

➤ They reproduce very quickly, and they have lots and lots of offspring.

➤ They're teeny tiny, so they don't take up a lot of room in the lab.

➤ They don't need a whole lot of food to survive.

Imagine for a moment the immense problems Morgan would have encountered if he had his heart set on doing his genetic experiments on, say, elephants. First, they're much too big to fit in a modestly sized lab. (How many elephants can you fit in a laboratory?) Their pregnancies last 22 months, and they usually only have one baby elephant at a time. You also need lots and lots of food, not to mention special equipment like shovels to get rid of the elephant excrement. And if fruit flies escape from a lab, nobody cares or even notices, but with elephants, people living nearby would be up in arms, shouting, "Not in my neighborhood!"

So Morgan and his talented students conducted his experiments using the tiny insects with red eyes. They wanted to find out whether variations in new generations happen gradually or in short bursts of new traits. When Morgan's student, Calvin Bridges, noticed a single white-eyed male fruit fly among all the red-eyed normal flies, he decided to keep it.

When the white-eyed fly mated with red-eyed females, Morgan's group followed generation after generation of these flies to better understand the inheritance of new traits. This accidental finding of a white-eyed male, which was a mutation, eventually showed how important it was to use mutations when analyzing how traits get passed from one generation to another.

Morgan and his group also explained how some traits are sex-linked, that is, they pass only to one sex or the other. For instance, they found that the white-eyed mutation could only pass this trait to male offspring. Part of Morgan's research tried to answer why.

Morgan and his brilliant students also discovered that genes are arranged in a row inside the chromosomes in a specific order. (Science teachers today like to say that genes are like beads on a string of DNA.) As a result of all this ground-breaking research in genetics, Morgan was awarded the Nobel prize in 1933.

DNA Data

The scientific name for the fruit fly is *Drosophila melanogaster*. *Drosophila* comes from two Greek words meaning "loves dampness," and *melanogaster* means "black stomach." Although it might be easier to say "little critters with black stomachs that like dampness," it doesn't sound half as serious as the Greek name scientists use.

Geneticists' Greatest Hits

Date	Event
1865	Mendel discovers the Laws of Heredity
1902	Sutton's chromosome theory
1913	Sturtevant publishes first gene map (of the fruit fly)
1920s	Morgan shows that genes lie in a row on chromosomes
1941	Beadle and Tatum's one gene/one enzyme theory
1944	Avery discovers that genes are made of DNA
1952	Hershey and Chase confirm that DNA is the hereditary material
1953	Watson and Crick discover the structure of DNA

Of (Dead) Mice and (Live) Men

Working in the 1920s, a British scientist named Frederick Griffith did research on mice and the bacteria that cause pneumonia in mice, pneumococcus (NEW-ma-KAH-kus). Griffith found that the cells of these bacteria came in two varieties. Normally, these bacterial cells had a smooth coating surrounding them, but a mutant strain of the bacteria had a rough coat.

When Griffith injected the smooth-coated bacteria into the lab mice, they developed pneumonia and died. However, when he injected the rough-coated bacteria into other mice, the animals lived. So Griffith concluded that the smooth-coated strain was the cause of pneumonia, but that the rough-coated strain was harmless.

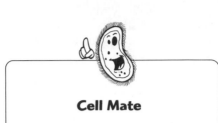

Cell Mate

As you can see by the history of scientific investigation and break-throughs in genetics, every researcher owes a tremendous debt to the researchers who came earlier. Mendel, Sutton, Morgan, and others all have paved the way for Watson and Crick's discovery of the structure of DNA.

Then Griffith killed the smooth-coated strain by heating it. When he injected the dead smooth-coated strain bacteria into some laboratory mice, they didn't get sick. So Griffith concluded that killing the bacteria rendered them harmless.

So far, so good. However, nature is full of surprises. When Griffith killed some smooth-coated strain and injected this into lab mice along with the live, harmless, rough-coated strain, he was amazed at what happened next. The mice got sick and died.

How could they develop pneumonia, Griffith wondered, when the rough-coated strain was harmless and the harmful strain was dead? In 1928, Griffith proposed that a transforming "principle" that came from the dead, harmful, smooth-coated strain had gotten into the live, harmless, rough-coated strain and had transformed it into lethal bacteria that killed the mice.

Griffith's mice.
(1) Mouse injected with rough strain lives.
(2) Mouse injected with smooth strain dies.
(3) Mouse injected with killed smooth strain lives.
(4) Mouse injected with killed smooth strain mixed with live rough strain dies.

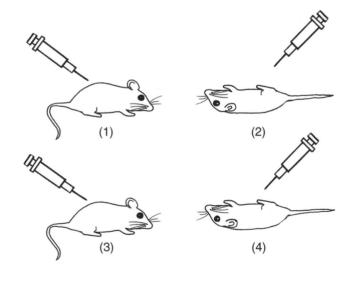

The Hunt for the Gene

Now that Griffith had performed his intriguing experiments, the stage was set for scientists to follow in his footsteps to solve the riddle of this strange, transforming "principle." From 1930 to 1944, a scientist named Oswald Avery and colleagues at the Rockefeller Institute in New York City built upon the groundwork of Griffith's experiments. They used the same type of bacteria that Griffith had chosen for his experiments. Avery's group also observed that the rough, harmless bacteria turned into the smooth, lethal kind. It must've been scarier than watching an old Lon Chaney Jr. movie and watching him turn into a werewolf.

In 1944, Avery and his team announced that they determined that the transforming principle was DNA. At the time, not everyone in the scientific community agreed with them, but later their theory proved to be true.

One Gene/One Enzyme

Around the same time, George Wells Beadle (1903–1989) and Edward Tatum (1909–1975), working at Stanford University in California, also contributed to the understanding of what genes do. Many years earlier, in 1908, a British doctor named Sir Archibald Garrod (1857–1936) came up with the idea that human diseases might be the result of a person lacking one *enzyme* or another. Garrod speculated that these enzymes were absent in some people due to improperly functioning genes inherited from their parents.

Beadle and Tatum proved Garrod's theory to be true. Working with bread mold, they produced mutations by exposing it to X-rays. After being irradiated, some new molds failed to grow properly, and apparently each mutant X-rayed mold lacked a specific enzyme that it needed to grow normally.

Genetic Jargon

An **enzyme** is a substance that helps the body carry out certain essential chemical reactions. We will discuss enzymes in more detail in Chapter 8, "Producing Proteins."

So Beadle and Tatum concluded that Garrod was right. In 1941, they published their findings, stating that each individual gene causes the production of a specific substance. They called this the one gene/one enzyme theory. (Today, we know that it would be more accurate to call it the one gene/one protein theory, but more on that in Chapter 8, "Producing Proteins.")

In One Ear and Out the Other

During the 1940s, a scientist named Barbara McClintock (1902–1993), working at the Cold Spring Harbor Laboratory in Long Island, New York, came up with some radical concepts in genetics that were not accepted until many years later. Working with Indian corn, she observed the patterns of different colors that varied from kernel to kernel in different ears of corn. Her research showed that genes, formerly thought to stay in just one place on the chromosome, could actually move from one spot to another on the same chromosome or even travel from one chromosome to another.

These jumping genes, or *transposons* as McClintock dubbed them, could jump into another gene for color and turn it off. If the jumping gene moved again, the color would be turned back on. McClintock realized that this theory could explain the different patterns of spots on the corn that she studied.

Much like Mendel's work, which wasn't appreciated at all in his time, Barbara McClintock's work on jumping genes was not accepted at first. However, her life as a scientist had a happier outcome than Mendel's. In 1983, when McClintock was 81 years old, she was awarded the Nobel prize for her discovery of jumping genes.

Hershey and Chase and the Bio-Blender

Although Oswald Avery proposed in the 1940s that DNA was responsible for hereditary traits, it wasn't until 1952 that scientists were convinced this was so. Two scientists working at the Cold Spring Harbor Laboratory in Long Island, New York—Alfred Hershey and his assistant, Martha Chase—not only advanced scientific knowledge, but also immortalized a kitchen appliance.

Hershey and Chase studied phages under a microscope. These tiny viruses look something like futuristic space-ships on spider's legs. Phages commandeer the innards of some poor unsuspecting bacteria to make more phages. Eventually, the newly produced phages burst out of the bacteria, killing them in the process. (Think of this as a real-science version of that horrible dinner-table scene in the movie *Alien*.)

Using a Waring Blender, Hershey and Chase mixed up a cocktail of bacteria and phages. The blender was set to just the right speed to separate the bacteria from the phages that were busy making mincemeat out of them.

While whirring around in the Waring, the phages broke off the bacteria and Hershey and Chase could find out whether it was the DNA or the protein of the phage which was important. The blender did the trick, and Hershey and Chase found out that it was only the phage DNA that got into the bacteria. Then they announced that DNA is responsible for the hereditary instructions that are passed down from one generation to future generations.

Genetic Jargon

A **transposon** (tranz-POH-zon) is the name that Barbara McClintock gave to genes that jump around from spot to spot on a chromosome or even jump from one chromosome to another.

Genetic Jargon

Hershey and Chase studied **phages** (FAY-jez). The word phage comes from the Greek word "to eat." These viruses are also called bacteriophages because they attack and infiltrate the bodies of living bacteria. They genetically instruct the captured bacteria to make more phages out of the bacteria's own material. Numerous phages are produced inside the bacteria, and then they burst out, killing the host bacteria in the process.

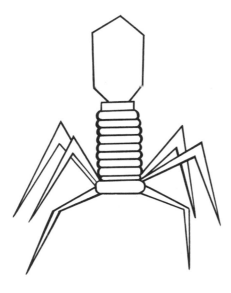

Just a Phage You're Going Through. Lucky for you, a phage is microscopic and only attacks bacteria.

"Gross" Science

Hershey and Chase's discovery didn't settle the question of what DNA looks like. A Swiss doctor named Johann Friedrich Miescher (MEESH-er) (1844–1895) had actually discovered the presence of DNA in cells way back in 1869, just four years after Mendel published his paper on the laws of inheritance. The way Miescher went about getting the subject of his experiment is a story straight out of the annals of "gross" science. He got the DNA from the nuclei (NEW-clee-eye) of white blood cells. As a doctor, he had an almost unlimited supply of the stuff because he got it from pus on old surgical bandages.

DNA Data

In the late 1800s, Friedrich Miescher went from dirty bandages to another source of DNA. He found out that salmon sperm cells had large nuclei, and it was easier to extract what he called nuclein from them. When his student, Richard Altman, extracted DNA from this substance, he named it nucleic acid.

But Miescher didn't call the substance he isolated DNA. He dubbed the stuff *nuclein* (NEW-klee-in) instead. Despite his discovery of DNA, Miescher didn't have a clue as to what it was made of, or what it was supposed to do, or what it really was.

The experiments of Sutton, Morgan, Avery and his colleagues, Beadle and Tatum, and Hershey and Chase all contributed to the great discoveries about heredity and DNA in the twentieth century. But it wouldn't be until 1953 that James Watson, an American scientist, and Francis Crick, a British scientist, would work out the details of the structure of DNA, the code of life. (Lots more on this topic will come in Chapter 5, "Nature's Blueprint.")

The Least You Need to Know

➤ Mendel's Laws of Heredity were simultaneously rediscovered by three scientists working independently in 1900, and this discovery led to a renewed interest and appreciation of his work.

➤ We now call Mendel's factors *genes*, the units of heredity that are composed of DNA.

➤ Sutton came up with the chromosome theory, which agreed with Mendel's findings.

➤ Morgan and his students discovered that genes are arranged like beads on a string and that some genes are sex-linked, meaning they can only be inherited by one sex.

➤ McClintock discovered that some genes can jump from location to location on a chromosome or from one chromosome to another.

➤ Avery showed, and Hershey and Chase confirmed, that genes are made of DNA.

Chapter 4

Me, My Cell, and I

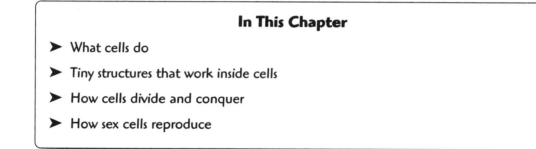

In This Chapter

➤ What cells do

➤ Tiny structures that work inside cells

➤ How cells divide and conquer

➤ How sex cells reproduce

To understand how DNA and genes work, first you need to learn about cells, where DNA is found. Just as you might understand more about your friends by going to their house or office and seeing how they live and work, genes and DNA make more sense if you visit their home: the cell.

Every type of cell has DNA, and right now your DNA is sitting within every cell of your body. In this chapter, you'll find out about the trillions of little pieces that make up you, your pets, the zillions of bacteria that have taken up residence in your body along with your cells, and more. Once you understand how cells work and reproduce and what's inside them, you'll be ready to graduate to a full comprehension of DNA's place in your cells and the part it plays in your life.

Let's Celebrate Cells

All living things are made of cells, even the ones that seem barely alive, like some of your co-workers. Plants, animals, and humans all have these structures. This includes the living person who is reading this book right now. Your body is composed of trillions of cells, the basic units of life. There's a whole world going on inside you right now, with different kinds of cells performing different functions, yet working together as a whole to make up the entity that's called you.

Cells often get together in groups to work in concert. You can think of these instances as big cell-ebrations. Groups of cells are called tissues, and groups of tissues are called organs. Your heart cells work together to pump blood cells throughout your body, giving you the oxygen you need to survive. The cells in your stomach are doing all kinds of rhythmic, inner belly dances to ensure that your food is well-digested, and your brain cells work together as you read books on decoding your genes or engage in other worthwhile intellectual pursuits.

But as small as some of these cells are—most can only be seen with a microscope—they have even smaller parts inside them. These tiny structures act as little factories that reproduce, do general repairs, take out the garbage, and make sure that everything inside the cell stays inside and things that belong outside stay outside.

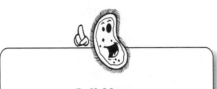

Cell Mate

Although most cells are too small to be seen without the aid of a microscope, there are some whopping exceptions. One cell is even big enough for breakfast. Yes, the yolk of that sunny-side-up or over-easy egg you ate this morning is actually one huge cell. But it's small stuff compared to the largest single-celled eggs of any living species. That honor belongs to ostrich eggs, which weigh about one pound apiece.

It's easy to forget that cells are dynamic, living things. The drawings you see in biology books and even the photos you've seen taken under powerful microscopes can give you an idea of what a cell looks like, but this is only a small part of what it means to be and act like a cell.

If you saw a photograph or a drawing of Arnold Schwarzenegger, it would show you his appearance at one particular moment, but you could never understand what his personality and performing styles and accent were all about unless you went to see one of his movies. In a similar way, you have to remember that a real cell, as opposed to a photo or drawing of one, is filled with action.

Wild, non-stop parties are going on inside your body at this very moment. Tiny structures are zipping around and dancing inside the cell membranes as we speak. Cells are growing and stretching and dividing in two, and all kinds of important events are happening within the nucleus.

Bacteria: Keeping It Simple

The simplest cells are one-celled life forms called *bacteria*. They are also called by a fancy name, *prokaryotes* (pro-CARRY-oats). This word is derived from Greek words that mean "before nucleus." The reason is simple: Prokaryotes don't have a nucleus, which is the small, dark body inside the cell that contains the genetic stuff.

So how do these nucleus-free bacteria pass any hereditary traits to future generations? The answer is that each bacterium usually has a circle of DNA floating freely inside it. It's like spaghetti that's all curled up into a ball, loose and flopping around but all in one piece.

One thing that bacteria are really good at is multiplying by dividing. In other words, bacterial sex is a lonely process that consists of growing very large and then splitting in two. In the absence of a sexual partner, bacteria literally go to pieces to "give birth." The two bacteria that result from this process are identical. They're fast, too. A bacterium can reproduce an exact copy of itself in only 20 minutes. That's a lot quicker than the nine months it takes a human embryo to develop into a baby.

Sex and the Single Cell

Sometimes bacteria can get a little raunchy and do something that's sort of like real sex. It's called *conjugation*, a term that comes from Latin words that mean "joining together," which is exactly what bacteria do in this process.

Here's what happens: Two bacteria get up close and personal. They're as close as two cars parked next to each other when the drivers are conversing. A tube forms from one and goes to the other one. It's as if the driver of one of the cars inserts a hose through the window of the other car and turns it on full blast. In the case of a hose from one car to the next, water would get the other car all wet. In the case of the two bacteria, the tube sends DNA, the hereditary stuff, from one to the other.

You've probably heard a lot lately about some types of bacteria being resistant to antibiotics, the medicine that's supposed to get rid of them. When a lot of bacteria are causing a disease in a human, some of them may be naturally resistant to antibiotics. Eventually, the ones that aren't resistant will die, and the resistant ones will survive and reproduce. Sometimes these survivors give their genes for antibiotic resistance to other bacteria via conjugation. This explains why it is very hard to eradicate some diseases that are caused by bacteria.

Genetic Jargon

Bacteria are the simplest life form. Because they don't have nuclei, they are also known as **prokaryotes** (pro-CARRY-oats), which in Greek means "before nucleus" because they don't have nuclei. They carry their DNA loose inside themselves. **Conjugation** (con-joo-GAY-shun) is one process by which bacteria reproduce.

Bacterial Good Guys

Not all bacteria are bad guys. It's true that there is a type of E. coli (EE KO-lie) bacteria called 0157 that produces a poison that makes people sick or, in really bad cases, dead. Young children and the elderly are especially susceptible to this bacteria, which some people are unfortunate enough to pick up in contaminated food.

But certain types of bacteria do important, even essential, jobs for us. For instance, another variety or strain of E. coli has a warm, cozy home in the digestive systems of normal, healthy people. This friendly strain of E. coli helps us by breaking down our food to aid digestion. Other useful bacteria help dead animals and plants decompose. We'd have a heck of a garbage disposal problem if it weren't for these microscopic demolition workers that reduce useless, dead matter into its basic components, which can then be used again by living things.

DNA Data

The friendly strain of E. coli bacteria that lives in the human digestive tract can reproduce in only 20 minutes. So if you start with just one at noon, you'll have two in 20 minutes. You'll have four at 12:20, eight at 12:40, and 16 at 1:00 p.m. If they continued reproducing every 20 minutes, there would be over 8 billion by 11 p.m. that night all generated from just one to start with!

You Carry Oats

Aside from prokaryotic bacteria cells that don't have nuclei, there is another category of cell, which is found in plants and animals. It's called a *eukaryote* or a eukaryotic (you-carry-AH-tick) cell. This type of cell has a nucleus that contains the cell's DNA (which is in the chromosomes, which we'll get into soon). All human cells are eukaryotes except for mature red blood cells, which lose their DNA as they get older.

Most eukaryotic cells have some basic structures in common. Take a look at the following illustration to see where all of these parts fit in. The three main parts are the nucleus, which holds all the hereditary material inside it, the *cell membrane*, which surrounds the cell, and the *cytoplasm*, which is the jelly-like part of the cell between the membrane and the nucleus. Inside the cytoplasm, happily swimming around in the jelly, are smaller structures called *organelles*.

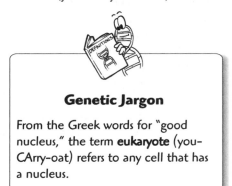

Genetic Jargon

From the Greek words for "good nucleus," the term **eukaryote** (you-CArry-oat) refers to any cell that has a nucleus.

A typical animal cell.

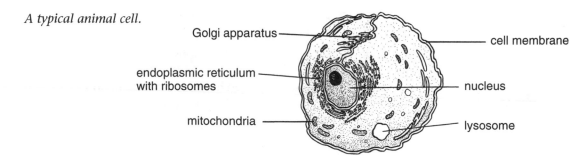

Golgi apparatus

endoplasmic reticulum
with ribosomes

mitochondria

cell membrane

nucleus

lysosome

The Cell Membrane

Cells are surrounded by the cell membrane. Think of it as a picket fence or guard rail that keeps things out, or a shrink wrapping that keeps everything inside where it belongs. You might even envision it as the steel bars on prison cells. The cell membrane is basically a barrier or boundary that makes sure that its organelles don't escape and take a wild ride in the bloodstream, and that the rest of the world knows that the cell's insides are off-limits. The cell membrane allows substances the cell wants to pass inside, and bars others that the cell doesn't want. In this way, the cell membrane is something like a security guard for the cell.

The Little Dark Nut

Every cell in the human body (except for older red blood cells) has a nucleus. As we mentioned earlier, the word *nucleus* comes from the Greek meaning "little nut." This dark spot in cells is the main organelle of the cell and is also the easiest one to see. It's responsible for issuing all of the instructions for the cell to grow and reproduce. If the nucleus is removed, a cell has no command center and no chance for survival.

Genetic Jargon

The **cell membrane** is the "skin" of the cell, holding its contents and keeping harmful things out. The **cytoplasm** (SY-toe-plaz-em) is the jelly-like inside of a cell contained within the barrier of the cell membrane. **Organelles** are the tiny structures inside the cell that perform functions that the cell needs to survive.

DNA Data

The nucleus has extra protection. Instead of relying on just the cell membrane to act as a security guard, the nucleus has its own membrane as well. This is something like having bouncers at the front door of a casino and extra security at the inner door where the serious players sit.

One of the things that makes this little nut the big cheese is the fact that it contains the hereditary material of the cell. The structures that look like strands of spaghetti inside the nucleus are the chromosomes. Inside the chromosomes, you'll find DNA. You can't see it, but it's definitely there. As you may remember from Chapter 1, stretches or regions of DNA are called genes (Mendel called them factors).

Cell Mate

About one-third of the cytoplasm is composed of organelles, floating around like islands in a "sea" of salty water.

Genetic Jargon

The job of the **endoplasmic reticulum** (en-do-PLAZ-mick reh-TICK-you-lum), or E.R. as it is known among friends, is to act as a kind of highway through the cell, going from the nucleus all the way to the cell membrane. It carries things back and forth to where they're needed. E.R. comes in two flavors: smooth and rough.

Ladies and Gentlemen, the Organelles

The organelles aren't the latest Motown group; they're structures in the cytoplasm that play essential roles in the life of the cell. They're like little organs for the cell, which is why they're called *organelles*. These organelles include mitochondria, lysosomes, and the futuristic-sounding endoplasmic reticulum.

Networking in the E.R.

E.R. usually stands for emergency room, but in the case of cells, it means *endoplasmic reticulum*. Not a whole lot of emergencies go on here, and it isn't even a room. It looks something like a big, thin pancake that's folded over and over again or the loose, hanging skin of some exotic breeds of dogs.

The endoplasmic reticulum is an organelle that takes up a lot of space inside the cell. The many ins and outs of this folded pancake provide lots of space for traffic to traipse around inside the cell. The E.R. is a series of networks of tubes that starts at the cell's surface and zigzags its way to the nucleus of the cell.

Along these networks, the surface of the E.R. is dotted with what look like chocolate chips. These are the ribosomes (RYE-ba-sohms). These tiny little dots help out in the production of protein, so they're like little protein factories. We'll discuss these organelles in Chapter 8, "Producing Proteins."

Cell Storage

The Golgi (GOAL-jee) apparatus sounds like hockey equipment, but it's really just another organelle. Unlike most Greek-named organelles, the Golgi apparatus got its name from Camillo Golgi, an Italian scientist (1844–1926). Under the microscope, the Golgi apparatus looks something like a stack of matching plates.

In the Golgi apparatus, or Golgi body as it is sometimes called, proteins get wrapped up in membranes. In this way, the Golgi apparatus helps to either store the proteins or to send them to other parts of the cell. These newly manufactured proteins might be needed outside of the cell, so they might travel through the E.R. and be let out of the cell by the cell membrane. The Golgi apparatus can also help to customize some of the proteins that are manufactured in another part of the cell.

Hanging Loose

The lysosomes (LIE-so-sohms) are another type of organelle. The name comes from the Greek meaning "loosening body." Lysosomes are like little sacks that store dangerous materials that can break down or "loosen up" unwanted material in the cell. The contents are released to make short work of bacteria and other foreign intruders. Sometimes lysosomes are deliberately used to destroy specific cells. For instance, when a tadpole turns into a frog, it doesn't need its tail any more. The lysosomes therefore release their destructive contents, and the tail cells are history.

The Cell's "Dill Pickles"

Other organelles that play an important part in the workings of a cell are the mitochondria (my-toe-KAHN-dree-a). They look a lot like small, fat dill pickles, and their name comes from the Greek words meaning "thread" and "small grain." Maybe the Greeks didn't have a word for dill pickles.

Cell Mate

The electron microscope is a powerful tool that can magnify many thousands or even a million times. After the invention of the first practical electron microscope in the late 1930s, cell studies skyrocketed. Organelles, in particular, could now be seen with more clarity than ever before.

DNA Data

Because mitochondria sometimes seem to act independently, some scientists believe that at one time they may have been cells on their own. A scientist named Lynn Margulis (born in 1938) came up with something called the Endosymbionic (EN-doe-SIM-bye-ON-ic) Theory. Margulis proposes that mitochondria were originally free-living bacteria that formed a close relationship with larger cells. The mitochondria gave the larger cells most of their own genetic material, and this wound up in the cell's nucleus. As a result, Margulis' theory says that what once was a bacterial cell on its own can no longer live without a larger cell as its home, because it gave most of its genetic material to that larger cell.

Scattered throughout the cytoplasm, the mitochondria are the little powerhouses of the cell. They produce the energy that fuels the activities of the cell. Therefore, the more energy a cell requires, the more mitochondria it has floating around in its cytoplasm. For instance, liver and pancreas cells in humans perform a lot of heavy work helping the body with digestion, so they have more mitochondria than most other cells. When more energy is needed in a cell, its mitochondria grow and divide until there are enough powerhouses to meet the demand for more energy. So, some biologists have noticed that mitochondria seem to act like independent cells, even though they're part of another cell.

Down at the Plant

We've already discussed single-celled bacteria and many-celled organisms such as animals and humans. Now it's time to talk about plant cells. We'd have a heck of a time without plants. Even if you're not a vegetarian, you ultimately eat things that eat plants. Even if you only eat meat—let's say you devour five hamburgers a day—there wouldn't be any cows to make those burgers from unless the cows had lots and lots of grass and other green plants to eat. So plants are essential to the well-being of every living animal, including us humans.

Genetic Jargon

Plants have organelles called **chloroplasts** (KLOAR-oh-plasts), which do not exist in animal cells. They contain a green substance called chlorophyll, which absorbs light energy from the sun. This substance is later turned into food for the plant in a process called **photosynthesis** (photo-SINTH-a-sis). Some chloroplasts in plant cells are used to store starch.

If you examine plant cells under a microscope, you'll see that they're similar to animal cells. They contain all the organelles that we've discussed so far. However, the two types of cells differ in some basic ways.

To begin with, one main difference between plant and animal cells is that whereas an animal cell is surrounded and protected by a cell membrane, a plant cell has both a cell membrane and a cell wall. That's kind of like the difference between having a flimsy picket fence and having a picket fence plus a tall, thick, concrete wall protecting your property.

In addition, plant cells have organelles called *chloroplasts*, which take energy from sunlight so the plant can make its own food. You can think of chloroplasts as the solar panels of the cell world. The process by which sunlight is transformed into usable energy to make food in the plant is called *photosynthesis*.

That's My Toe, Sis

At some point in a cell's life, it has a choice to make. It can either die, or it can grow to the point where it can divide into two new cells. Given these options, most cells (and people) would say the answer is a no-brainer. The cell's best choice is to choose to fatten up until it has enough material to turn into two brand-new, identical cells. In fact, this choice is so much more appealing that millions and millions of cells in your body divide and reproduce every second.

Just think: If you were a cell, you would get older and very, very fat, and then one fine day, there would be enough of you to go around and split in two, and then you'd be two young, healthy, and slender twins of yourself. You'd also be your own mother and father. Good thing this doesn't happen. It might lead to an identity crisis.

Divide and Conquer

When a cell divides, it's like a complicated square dance that starts with the process called *mitosis*. (If your female sibling accidentally steps on you, you might say, "that's my toe, sis," and that's how this unwieldy Greek word is pronounced.) You can follow the process of mitosis by looking at the following illustration.

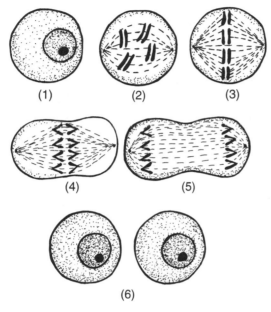

Mitosis.

What happens in mitosis is essential to the well-being of the new cells. Because cells can't survive without a nucleus and the hereditary material inside it, the existing nucleus and its contents must be copied.

Cell Parts and Their Jobs

Cell Part	Job Description
Cell membrane	Security guard; checks what goes in and out of the cell
Endoplasmic reticulum	Highway from the membrane to the nucleus
Ribosomes	Protein factories
Golgi apparatus	Storage and packaging center; stores or sends new protein traveling; "customizes" new proteins, too
Lysosomes	"Demolition workers"; break down unwanted materials in the cell
Mitochondria	Powerhouses; generate energy for the cell
Nucleus	The "brains" or control center of the cell; houses hereditary material in the form of chromosomes, which contain genes, which are stretches of DNA, which code for proteins

The first step in mitosis is that the chromosomes, which are usually thin and almost invisible when the cell is at rest, get blobby and fat and start to look like little X-shaped figures. What's happening here is that each half of the X-shape is identical. The chromosomes have duplicated themselves, but the two copies are still joined together in the middle.

Then, two little dot-like structures called the *centrioles*, (SEHN-tree-oles), which are joined together by long filaments called the *spindle*, start to pull apart. As they pull apart, they move to opposite ends of the cell. At the same time, the membrane around the nucleus breaks down and disappears.

At this point, the chromosomes get attached to the spindle and are lined up in pairs in the middle of the cell. If this were a square dance, we'd be at the stage where the dancers line up in pairs, ready and waiting to continue their dance routine's next step. But in this case, the next step is for the partners in the square dance to perform a tug of war. Attached to the spindle, the chromosome pairs get pulled apart to separate ends of the cell. In this way, each opposite side of the cell gets a full set of chromosomes (the "dancers" who were originally lined up in the middle).

Now the dance nears its conclusion. Each set of chromosomes at opposite ends of the cell is identical to the other. New nuclear membranes form around these two sets of chromosomes. The whole thing looks like two sunny-side up eggs that got connected in the frying pan.

Finally, the rest of the cell begins to pinch together in the middle so that it starts to resemble a figure eight. Imagine what would happen if you were to take a piece of dental floss and tie it around a round piece of bread dough. As you tightened the knot, the dough would get thinner and thinner and eventually break off into two pieces. This is similar to what happens to the cell. It gets thinner and thinner in the middle until it breaks into two brand new cells, each with the same hereditary material inside its new nucleus.

Genetic Jargon

Mitosis refers to the process of duplicating all of a cell's chromosomes during cell division. When the cell finally divides, the two daughter cells, as they are called, each have a copy of all the chromosomes that were in the original cell.

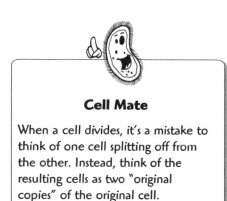

Cell Mate

When a cell divides, it's a mistake to think of one cell splitting off from the other. Instead, think of the resulting cells as two "original copies" of the original cell.

My, Oh, Meiosis

You just learned about how most cells reproduce, or divide into two new cells. Cell reproduction is generally done for growth or repair. However, not all cells divide in this way. Sex cells—the sperm and eggs, not detention rooms for perverts—divide in a different way.

Too Much of a Good Thing

Just imagine what would happen if sperm cells and egg cells split into two and kept the same number of chromosomes as other cells do. When the sperm and egg united for fertilization, the new, developing organism would have twice as much genetic material as was contained in the parent cells.

For instance, humans have 46 chromosomes. If a sperm cell gave 46 chromosomes to its offspring, and the egg it fertilized also gave that number, the child would have 92 chromosomes in all. It would get mighty crowded in the child's cells, and the child's children would have it even worse.

So instead of dividing by mitosis, sex cells have their own more complicated type of square dance that includes an extra division. The way sex cells divide is called reduction division, or—here's another Greek word to impress your friends and biologist friends—*meiosis*. You can follow the process of meiosis by looking at the following illustration.

Genetic Jargon

Meiosis (my-OH-sis) refers to the process of first duplicating all the chromosomes in sex cells, and then reducing their number by half so that there are four new cells, and each one of them has half the number of chromosomes in the original cell. Meiosis starts out like mitosis, but then it has one more division.

Meiosis.

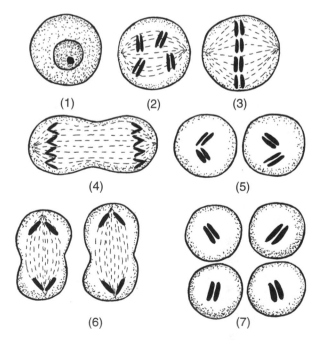

(1)　　(2)　　(3)

(4)　　　(5)

(6)　　　(7)

The Dance of Love

Here's how meiosis works. The chromosomes go through the same dance that they would in regular cell division, mitosis. As in mitosis, the chromosomes stand at the center and wait opposite their identical chromosome partners. Then they're pulled apart into two separate cells.

At this point, it's possible for the chromosomes to swap genetic material. This would be something like the female square dancer taking off her red bandanna and swapping it for her partner's blue checkered neckerchief or the male square dancer trading his wide, brown leather belt for her black, thin leather belt. Each "dancer" represents a complete set of genes on a chromosome. One set is the maternal version and the other set is the paternal version, but they're both just versions of the same kinds of genes. They can make these swaps because they have similar, yet varied, pieces of clothing to exchange. In much the same way, the chromosome partners (one from the father and one from the mother) are similar enough that they can exchange genes, which can take each other's place on the corresponding chromosome. This process is called *crossing-over*, and it causes more and more genes to get mixed.

In this constant mixing and matching of stretches of DNA, the more combinations that occur, the better. When nature produces organisms that are basically the same yet varied in some ways, it helps the species as a whole to be more adaptable to changes in the environment. The more varied the individuals in a species, the greater the chance that at least some of them will survive in the event that conditions in their outer world should change.

Genetic Jargon

Crossing-over occurs when chromosomes from the mother and the father cells line up in pairs during meiosis, when egg and sperm cells make copies of themselves. The two chromosomes can swap some genes so that some of the genes originally on the father's chromosome are now on the mother's chromosome and vice versa. This process helps to create a lot of variation in a species, and variety is good for the species as a whole to deal with environmental changes that may require new traits to cope with them.

Crossing-over.

Then something different from mitosis happens. There's a second division in which the chromosomes go to the center of each of the two new cells. In this division, the two strands of "spaghetti" that were joined are pulled apart as each of the two cells divides into two more cells.

This process gives you four new cells, and each one of them has half of the number of chromosomes that were in the original cell. In humans, you start out with 46 chromosomes, which are duplicated, but in sex cell division, these go to four cells, each of which ends up with 23 chromosomes. Then when the sperm fertilizes the egg, the developing child will have 46 chromosomes again—half from the sperm and half from the egg—which is the number that humans are supposed to have. Thus, the child that develops will have a new combination of traits, some from the father and some from the mother.

The Least You Need to Know

➤ All living things—animals, plants, and complete idiots—are made up of cells.

➤ Bacteria are one-celled organisms that lack a nucleus and are called prokaryotes.

➤ Everything living other than bacteria has a nucleus and falls into the category of eukaryotes.

➤ The cell is equipped with tiny inner structures called organelles that act like its organs and carry out various functions so the cell can live and thrive.

➤ When cells divide to reproduce, so does the heredity stuff in the chromosomes. This process is called mitosis.

➤ When sex cells reproduce, they divide twice and end up with half the normal number of chromosomes in each sex cell. (This way, the offspring will get one half the chromosomes from the mother and one half from the father.)

Part 2
The ABCs of DNA

Most jeans are made from denim, but your genes are made from DNA. What is this DNA stuff all about? And how do genes work? And what happens when they split at the seams? In this section, you'll learn how Watson and Crick started a molecular revolution in 1953 when they found out how DNA is put together.

Chapter 5

Nature's Blueprint

In This Chapter

➤ How atoms make up molecules

➤ How scientists like Kossel, Levene, and Chargaff studied DNA

➤ How Rosalind Franklin's research led the way to DNA

➤ Watson and Crick fool around with models and discover the structure of the DNA molecule

In Part 1, "Welcome to the Age of Genetics," we discussed how Mendel and his followers wondered about heredity, and the advances they made in this field. We also went over what's going on inside cells. That brings you up to speed with the scientists who were studying heredity in the early 1950s. They had all kinds of clues that might lead to figuring out how heredity worked, but the big picture was still incomplete.

One of the greatest puzzles that needed solving had to do with DNA. Knowing that DNA was the mysterious factor that somehow caused traits to pass from one generation to the next, scientists still had a lot of questions about it. What exactly did this DNA stuff look like? What was it made of? And how did all of its moving parts work together?

To understand the quest for DNA, you need to know about how everything on this earth is divvied up into atoms and molecules, which eventually get built up into cells, tissues, organs, rocks, hills, mountains, and other stuff. You'll learn that and much more in this chapter, including how the DNA puzzle was solved.

In the Beginning, There Was Atom

The Greeks had a word for it. The word is *atom*. (Remember that almost everything you read about in science seems a lot simpler if you just find out what it means in Latin or ancient Greek.) So an atom is the smallest particle of an *element*. Atoms are much too small to be seen. You will never see two atoms walking hand in hand down your street, but you've seen immense groups of them which make up people and places and things.

The wild thing about an atom is that there's more of it that *isn't* there than is there. That's because atoms have a whole lot of empty space in between the tinier little pieces that make them up.

The Atom as Solar System

If you were paying a little attention in high school chemistry, you may remember that atoms look kind of like miniature solar systems. Our solar system has the sun in the middle and nine plants circling around it. An atom has a nucleus in the center instead of the sun, and its "planets" are particles called electrons.

Different atoms have different numbers of electrons (the planets) whizzing around different-sized nuclei (the suns). So we could say that atoms come in different flavors, such as copper, iron, and lead. Granted, these are not the most appetizing flavors. Unfortunately, there's no atom called Rocky Road.

Atomic Alliances

When two or more atoms get together, they form what is called a *molecule*. This process of getting together is called bonding. It's the same stuff that guys do in men's groups, only it's a lot more intimate and it lasts a lot longer than some of the friendships that form in those groups.

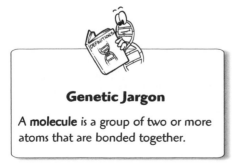

Sometimes a molecule can just consist of an atom of one element getting together with another atom just like itself. For instance, two oxygen atoms that form a close, meaningful relationship constitute an oxygen molecule. Teachers and scientists call this O_2.

Other times, two different kinds of atoms decide to get together for a long-lasting party. For instance, if one oxygen atom bonds with two hydrogen atoms, you have H_2O, which even the most scientifically challenged person will recognize as water.

A water molecule.

Oxygen and water are pretty small as far as molecules go, but there are also larger molecules. For instance, when you start to talk about living things, the molecules get larger and more complicated. Ninety-nine percent of what's in people or pea plants or peacocks consists of just six elements, but they're put together in some pretty big molecules.

Now, no matter how big or small they are, molecules are not just like little groups of toothpicks stuck into olives that can pretty much lie flat on a table. They're three-dimensional and take up space. Different atoms fit together in many different ways, something like so many pieces of plumbing. Think of how the water pipes under your sink fit together or how gas pipes in your house have different extensions, and you'll start to get the idea of how molecules are put together.

Proteins are molecules that perform important functions in living things, and the way they interact with one another has a lot to do with what different traits a person, plant, or porcupine will have. In many cases, these protein molecules are thousands of times larger than oxygen or water molecules.

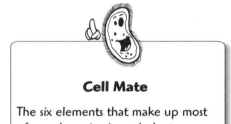

Cell Mate

The six elements that make up most of people, animals, and plants are hydrogen, carbon, nitrogen, phosphorus, sulfur, and oxygen.

What Shape Are Your Molecules In?

All molecules have different shapes. Keratin (KEH-ra-tin), the protein that's in hair and nails, has a spindly shape, for example, and collagen (KAH-la-jin), which is in ligaments and bones, has a shape that's something like three corkscrews winding around each other. Each shape gives us a clue as to how the molecule functions and what it's for.

Scientists are particularly interested in the shape of *protein* molecules, because many proteins function as *enzymes*, speeding up certain actions and reactions that occur in the body. The human body needs lots of enzymes to help digest food and do other important tasks.

To understand how enzymes work a little better, imagine that you're standing in front of a locked door with a note on it saying that you're entitled to everything that's inside. You want to get inside because through a small window in the door you can see a room filled with designer clothes, bottles of champagne, and a new Rolls Royce all waiting for you. You bang on the door, and eventually, one of two things will happen. Either you'll wait for five hours until the owner, a generous millionaire, comes home and opens the door to give you your gifts. Or instead of waiting, you could break down the door, but this will take a relatively long time, too.

But suppose you have a friend who's a locksmith, and he has a key that fits the door. He could come by, use his special key to unlock the door, and you could rush in, drink the champagne, eat the chocolates, put on the designer clothes, and take a spin in your new Rolls. The locksmith has nothing to do with these gifts, and you would've gotten in the room sooner or later without him, but you would've waited a much longer time.

The locksmith is like an enzyme that enables you to do something that you were capable of doing without his help, but he's certainly made things more convenient. In the same way, proteins sometimes function as enzymes, helping chemical reactions within your body, or in plants or animals, to happen in a good way and in a timely fashion.

Genetic Jargon

A **protein** is a type of large molecule that fulfills an important role in a cell. Many proteins are enzymes. An **enzyme** is a protein that can speed up specific chemical reactions in cells.

DNA Data

Without enzymes, some of our bodily processes would take so long to happen that we might not survive. For example, some enzymes help our bodies to speed up the digestion of food. Enzymes in the saliva break down the starch in food, enzymes in the liver help to break down fat, and some enzymes in the stomach break down proteins.

Protein molecules are often the shape of "keys" to doors that you want to get into. If a molecule doesn't have the right shape, it's not going to get you into the door you need to open, and it might get you into another room instead. Even molecules that aren't proteins or enzymes have unique shapes. And as with enzymes, each shape gives us a clue as to how it functions and what it's all about.

The Great Race

Scientists are still trying to figure out the molecular structure of certain substances. This isn't the kind of job that can be done overnight. Max Perutz (born in 1914), for example, spent 25 years figuring out the molecular structure of hemoglobin, an important protein in red blood cells. His long-term dedication led to a Nobel Prize. So if scientists want to figure out the structure of a molecule, they can be sure that this is going to be a long-term commitment.

By the 1950s, several scientists were poised on the brink of discovering the structure of DNA. There had been so many contributions to finding out how the Code of Life is composed and structured, but all of the major clues left out small pieces of the puzzle.

You read in Chapter 3, "Mendel's Successors," that in 1869, Friedrich Miescher (MEESH-er), who investigated what was inside the pus he took from discarded bandages (Ewww!), discovered that there was a substance called DNA (only he didn't call it that). Years later, in 1895, Miescher mentioned in a letter to his uncle that he suspected there might be large molecules that were responsible for carrying traits from one generation to the next. He even guessed—correctly—that these molecules, whatever they were, got their message across in the same way that the 26 letters of our alphabet can be combined to carry messages in many different languages.

Miescher had no idea just how close he was to working with a molecular "superstar." Like most scientists of his day, Miescher didn't think that DNA was anything special. It seemed to him and most everyone else that proteins, which they already knew could do all kinds of cool things as enzymes, were more likely to be the magic stuff of heredity. It wasn't until much later that the secret of the DNA molecule would be deciphered.

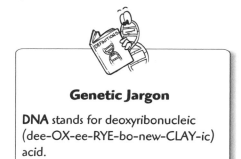

Genetic Jargon

DNA stands for deoxyribonucleic (dee-OX-ee-RYE-bo-new-CLAY-ic) acid.

Kossel Covers the Bases

From 1885 to 1901, Albrecht Kossel (1853–1927), another German scientist, got together with some of his students to analyze the chemical structure of DNA. He built upon some of the research that Miescher had already done. Kossel mistakenly thought—just like Miescher and most other scientists of the time—that proteins were the most likely suspects to be the carriers of hereditary characteristics, and that DNA wasn't likely to have anything to do with the passing of traits from one generation to the next.

While trying to figure out how DNA is put together, Kossel discovered that some of its components were four substances called bases. These have nothing to do with baseball, but they have a lot to do with heredity, even if Kossel didn't recognize that at the time.

These bases may not be the same as home plate in baseball, but they do have one thing in common: They're just as flat. One thing that these bases don't have in common with the sport is the fact that they have long, unwieldy names. They're adenine (A-den-een), cytosine (SIGH-toe-seen), guanine (GWAH-neen), and thymine (THY-meen). Lucky for you, even scientists can't be bothered with these long names, so the bases are usually abbreviated as A, C, G, and T. Think of these abbreviations as the letters of a very, very short alphabet. Even though the idea behind this can get very complicated, this "alphabet" is still easier to memorize than the 26 letters you finally learned after your parents sat through your countless, agonizing repetitions of the ABC song.

DNA Data

The names of the four bases, or nucleotide (NEW–klee–oh–tide) bases as they are more formally called, have a weird history. The name adenine comes from the Latin word for gland, because that's where it was first found. Thymine was named after the thymus gland, where it was first found. Cytosine is found inside the nucleus of a cell, but originally, some scientists thought that it came from the cytoplasm in the cell, so they gave it a name that sounded like the word *cytoplasm*. Guanine has the most bizarre history of all. You can find lots of guanine in guano, or bird droppings. It would've been more appropriate to call the stuff bird dropine, but somehow that doesn't sound all that scientifically impressive.

Although the discovery of the A, C, G, and T bases of DNA netted Kossel the Nobel prize in 1910, neither he nor anyone else in the scientific community at the time realized the great importance his discovery would have many years later as a vital piece of the DNA puzzle.

Levene's Sweet Contribution

The next scientist who contributed a piece of the puzzle was Phoebus Aaron Levene (1869–1940), a Russian scientist who had moved to New York City to escape anti-Semitism in his native land. He studied chemistry at Columbia University and then from 1905 through 1939, he worked at the Rockefeller Institute, also in New York City.

Levene also studied DNA, and his contribution was the discovery that some of the components of DNA are sugars. Although not the sweet white stuff that you put on your corn flakes, deoxyribose (dee-OX-ee-RYE-bose) is related to table sugar and makes up part of DNA.

Levene mistakenly thought that DNA was a small molecule and that each one of these mini-structures had only one A, one C, one G, and one T. Years later, Watson and Crick would prove that he was way off and that DNA is a very large molecule with lots of As, Cs, Gs, and Ts.

So now scientists knew what DNA was composed of. They knew there was a sugar and some As, Cs, Ts, and Gs, and they also knew that it contained phosphate (which is commonly found in bones, for example). The only problem was that they had no idea how all these pieces fit together. Remember, the way a molecule is put together says an awful lot about the way it functions and what it's capable of doing.

Cell Mate

Scientifically speaking, there are lots of different kinds of sugar. Many are not appropriate for sweetening food. The names of sugars end with the suffix -*ose*, such as fructose and sucrose.

Imagine that you order a kit for a prefabricated house, and you have everything delivered to the site of your future dream home. When the boxes arrive, you open them and find that you have hundreds of wooden beams, all the siding, a great tangle of electrical wiring, and a number of windows and doors. You have thousands of nails, screws, and floor tiles. But there's one small problem: nobody gave you a diagram or instructions on how to fit the pieces together.

It was even worse for the scientists trying to figure out how to piece together DNA. You at least have the advantage of knowing what a house is supposed to look like. If you didn't, you might wind up putting the floor tiles on the chimney or standing the ceiling beams against the walls. If you mistakenly thought you had to put the door up on the roof, the house wouldn't even serve its intended purpose for you.

This is pretty much the situation that scientists faced after Miescher, Kossel, Levene, and others had figured out the components of DNA. They knew what all the different pieces were, but they had no clue as to how they were supposed to fit together as a molecule. More clues were needed to lead to the discovery of DNA's structure. Eventually, these missing pieces of the puzzle were provided.

Chargaff's Ratios

Around this time, a scientist named Erwin Chargaff (born in 1905), who was born in the Soviet Union, was doing important work. He came to the United States and worked at Columbia University in New York City. Chargaff was very interested in Oswald Avery's findings. In Chapter 3, we discussed how Avery discovered that the "transforming factor" that caused harmless bacteria to turn into microscopic serial killers was actually DNA.

So Chargaff took up the study of DNA. Chargaff's research showed that Levene was wrong when he concluded that there was only one A, one C, one G, and one T to a

DNA molecule. Instead, Chargaff discovered that the number of As was always the same as the number of Ts and that the number of Cs was always the same as the number of Gs. However, the number of As and Ts could be different from the number of Cs and Gs.

DNA Data

Erwin Chargaff once went out to eat with Watson and Crick, the scientists who later discovered the structure of the DNA molecule. He was completely unimpressed with them. He thought they had a poor knowledge of chemistry, and this suspicion was reinforced when Crick admitted that he didn't know the chemical differences between A, C, T, and G. Crick then brashly told Chargaff that he could always look them up.

X–Rated Crystallography?

Another clue to DNA, maybe the most important one of all, came from a very special imaging tool. It had been around for a while, but it had yet to be used to help solve the DNA puzzle.

In the early 1900s, a British scientist named Sir William Bragg (1862–1942) and his son Lawrence (1890–1971) developed something called X-ray diffraction (diff-RACK-shun), or X-ray crystallography (kris-ta-LOG-ra-fee). This tool made it possible for scientists to obtain a visual image of things that were too small to be seen by the naked eye or even through a microscope. Bragg realized that when X-rays pass through substances and then land on a piece of photographic paper, they create a pattern unique to that substance, which gives a clue as to the molecular structure of that substance.

Before someone could take one of these X-ray pictures, they had to form crystals from the substance that they were studying. In the case of DNA, a scientist would mix up a bunch of DNA with a type of salt, and then let it sit. Following this procedure, DNA crystals would form, in much the same way that snow crystals form on a cold window-pane in the winter.

So a scientist hoping to "see" what DNA looks like at the molecular level takes some DNA crystals and puts them in a holder, and then directs an X-ray beam at them. The X-ray then interacts with the atoms inside the DNA crystal, and the beam is scattered around in different directions. Some of the scattered rays register darker, and others register lighter on the X-ray film. A trained X-ray crystallographer should be able to get some information about the position of atoms in the original molecule by looking at the pattern and also by using some fancy mathematical calculations.

DNA Data

In the late 1940s, the Medical Research Council of Great Britain decided that X-ray crystallography was an important tool in figuring out the structure of different molecules. They opened a special unit for these studies in the Cavendish Physics Laboratory in Cambridge, England. Lawrence Bragg was its director. He and his father won the Nobel Prize for Physics in 1915, when Lawrence was only 25 years old.

Franklin's Contribution

One of the most talented scientists in this field at the time was a British chemist named Rosalind Franklin (1920–1958). Without her great contribution to the imaging of DNA, the puzzle of its structure might have taken many more years to be solved. Rosalind Franklin learned X-ray diffraction technology at the State Chemical Laboratory in Paris, where she worked from 1947 to 1950. Then in 1951, she was hired to work in the Biophysical Laboratory in King's College in London with Maurice Wilkins. Here, Franklin took her thorough knowledge of the technology and used it for research on DNA.

There is still some controversy as to the role she played in the discovery of the DNA molecule's structure. If you read one book, you'll come away thinking that Franklin was on the wrong track and therefore missed coming up with the correct structure herself. If you read another book, you'll come away with the impression that Franklin was discriminated against because she was a woman in science and that Watson and Crick couldn't have come up with their final, correct conclusions about the structure of DNA without her data. Although no one can say which view of Franklin's work is closer to the truth, the one fact that no one disputes is that her X-ray diffraction images of DNA and other substances were of superb quality.

Getting Credit

Some people argue that Franklin should have won the Nobel prize along with Watson, Crick, and Wilkins for the great contribution she made to the discovery of DNA's structure. Unfortunately, Franklin died before the prize was given out, and it cannot be awarded posthumously. Would she have been awarded the prize had she lived? We'll never know.

When James Watson (born in 1928), an American geneticist and biophysicist, and Francis Crick (born in 1916), a British biophysicist, worked on the puzzle of DNA in the Cavendish Laboratories in Cambridge, England, Franklin's work played a key role. After seeing Franklin's best X-ray images of DNA crystals, which were shown to them by Maurice Wilkins, and then reading a report that Franklin had written, Watson and Crick learned what turned out to be the keys to solving the DNA puzzle.

Watson and Crick: Playing with Beautiful Models

Using all of the incomplete clues at their disposal, such as knowing about the presence of an equal number of As and Ts and an equal number of Cs and Gs, and then concluding from Franklin's images that DNA might be a double helix (a double spiral, like a twisted rope ladder), Watson and Crick got to work by "playing." They made models out of metal and cardboard cut-outs and tried to work out different solutions to how all the different parts of DNA might look and where they would have to go in order for them to work together.

Remember how we talked about getting all the materials for a house, but not knowing how they were supposed to be put together? Watson and Crick did the equivalent of picking up the pieces of wood and the floor tiles and windows and cross-beams and trying to figure out which pieces might logically go together with all of the other pieces. It was as if they had looked at a very fuzzy, vague photograph that Rosalind Franklin had taken of a "house" and read her ideas on what this particular "house" might look like. They also learned from other scientists that there were certain numbers of "supplies" that needed to be in equal amounts.

Triple Helix or Double Helix?

Working with this puzzle, Watson and Crick had a couple of false starts. At one point, they came up with the wrong solution. In addition, they heard that Linus Pauling (1901–1994), a brilliant American chemist, was also working on the problem. Pauling came up with his own idea of how the DNA molecule was constructed. He said that it was a triple helix—that is, three spirals all wound around each other in one continuous chain. This turned out to be incorrect.

Genetic Jargon

The word **helix** (HE-liks) comes from a Latin word that comes from a Greek word that simply means "spiral." So a double helix is a double spiral.

Watson and Crick's idea was a simple yet elegant solution. They published their findings in an article in the prestigious British journal *Nature* on April 2, 1953. They said that the DNA molecule was a double *helix*. As we mentioned earlier, this structure looks something like a twisted rope ladder or a spiral staircase. The rope in the ladder, or the rails on the staircase, are made of a sugar and a phosphate that alternate in a repeat pattern. The rungs of the ladder, or the steps on the staircase, are made up of base pairs. (This does not mean low-lifes who hang around in groups of two).

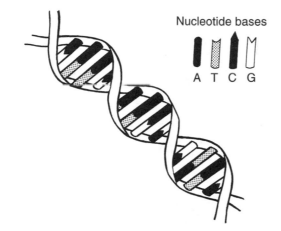

Nucleotide bases

A T C G

The DNA molecule.

A base pair is a rung or step on the "ladder" of DNA. It's made up of either A combined with T or C combined with G. As Chargaff had noted, As and Ts always come in the same amounts in a DNA molecule and the Cs and Gs also come in the same amounts. So Watson and Crick figured out that A and T were somehow able to loosely bond together so that each one made up a half-rung or half-step. T couldn't pair with C or G because their chemical structures didn't allow them to bond, and A couldn't pair with C or G, either. A was meant for T. The fact that there's always T&A remains a delight for male geneticists everywhere.

The same holds true for C and G; they are chemically attracted to one another. Therefore, other rungs on the ladder consist of one half C and one half G. This arrangement was in perfect agreement with Chargaff's findings that C and G always come in equal numbers in a DNA molecule.

The Language of Genetics

It turns out that all the As, Cs, Ts, and Gs are like a mini-alphabet that spells out messages for the cell. Different combinations of these letters spell out codes, and these codes tell the cell which specific proteins it should make. Proteins are important for all the workings of a cell and, in essence, control a great deal of what happens within a living organism. Proteins are also responsible for many of the traits that living things have. So the DNA molecule is responsible for passing heredity information by instructing the cell to make certain proteins that will influence the growth, development, and appearance of the living thing it is part of.

Because of this landmark discovery of the structure of the double helix, James Watson, Francis Crick, and Maurice Wilkins received the Nobel prize in 1953. This tremendous discovery, many years in the coming, ushered in the Age of Genetics. It made it possible for the scientists who followed Watson and Crick to do research that would enable them to mix and match genes, to try to find out the inner workings of disease and health alike, and to search for the secrets of heredity that had escaped scientists and laypersons alike for so many centuries.

Dr. James Watson.
Photo courtesy of Bill
Geddes, 1998.

The Least You Need to Know

➤ Atoms are the smallest particles of an element.

➤ Groups of atoms are called molecules.

➤ DNA is the molecule responsible for the hereditary instructions in cells.

➤ Scientists including Miescher, Kossel, Levene, and Chargaff contributed important clues about the composition of DNA.

➤ Rosalind Franklin, a British scientist, took X-ray crystallography images of DNA that led Watson and Crick to determine the molecule's elusive structure.

➤ James Watson, an American, and Francis Crick, an Englishman, discovered that DNA is a molecule shaped like a double helix (it looks like a twisted rope ladder with "rungs" in between).

Chapter 6

Assembling Your Genes

In This Chapter

➤ How DNA molecules replicate themselves

➤ How DNA's parts act like a four-letter alphabet

➤ How DNA deals with errors during its replication

➤ How DNA sequences sometimes get changed by copying errors and outside influences

---

In Chapter 4, "Me, My Cell, and I," we discussed how cells reproduce. In this chapter, you'll learn how DNA, the molecule that genes are made of, makes copies of itself to go into all of those new cells. Finding out how DNA replicates was a direct result of Watson and Crick's landmark discovery of the structure of DNA, the Molecule of Life. Once you understand how this incredible molecule passes on the genetic code to succeeding generations, you'll be able to understand how scientists have used this knowledge to forge ahead in the genetic revolution.

Splitting at the Seams

Watson and Crick hoped that when they figured out the structure of DNA, it would give them some clue as to how the molecule replicates, or makes copies of itself. If you or I took a peek at the DNA molecule and asked ourselves how it copies itself, there would be a resounding "I dunno," heard for miles around. (It helps to be a scientific genius when thinking about questions like this.) Watson and Crick, however, knew almost immediately what was going on. Even in their first article in *Nature* magazine, which announced the discovery of the double helix structure, they hinted that they could guess how DNA replicates itself and that this clue came from the way the As paired up with Ts and the Cs paired up with Gs.

DNA Data

Watson and Crick didn't discuss replication in the first draft of their article in *Nature* magazine. According to Crick, this omission was because Watson was afraid that they might have gotten the structure of DNA wrong again. They had suggested a different structure earlier on, and it turned out to be wrong. Shortly afterward, Watson's fears disappeared, and the two scientists wrote an article for *Nature* magazine on May 30, 1953, explaining DNA replication.

As we discussed in Chapter 5, "Nature's Blueprint," DNA is made up of two spiral strands that wind around each other like a twisted rope ladder. The rungs of the ladder consist of *nucleotides*: As, Ts, Cs, and Gs. Because of their chemical structures, As can only fit together with Ts, and Cs can only fit together with Gs. Because of their unique structures, A can't pair with C or G, and T can't pair with C or G, either. Each rung is either half A and half T or half C and half G.

Genetic Jargon

A **nucleotide** (NEW-klee-oh-tide) is one of the building blocks of the DNA molecule. The four nucleotides are A, C, T, and G. Because of their chemical structures, A and T can pair only with each other, and C and G can pair only with each other. These pairs are known as base pairs.

Cell Division and Multiplication

As you learned in Chapter 4, "Me, My Cell, and I," when cells duplicate themselves, the first thing that happens is that the chromosomes (the rod-like structures in the nucleus) make copies of themselves so that the two cells that result from replication will have the exact same instructions on how to develop and act as the parent cell did. Most cells replicate by the process called mitosis, meaning that the chromosomes duplicate themselves first. Sex cells undergo meiosis, meaning that they duplicate themselves into four cells, each with half the number of chromosomes as the original parent cell.

Either way, copies of the chromosomes are carried over into the new cells, and these chromosomes contain all the genes of the parent cell. Every cell in your body contains all of the estimated 80,000 or so genes that you inherited from your mother and father, and these genes are stretches of DNA, all linked up together and sitting around in the chromosomes. (Not all genes are activated in each cell. In Chapter 16, "Send in the Clones," we'll discuss how some scientists are trying to activate genes that have been "shut off" in order to create clones of adult organisms.)

Breaking Up: Not So Hard to Do

Watson and Crick developed a concept of what happens when cells divide along with their chromosomes, and DNA gets to reproduce itself. To begin with, each strand of the DNA molecule has a head and a tail. The head of one strand is always lined up with the tail of the other, and vice versa. They wind in different directions, but if you turned them upside down, you'd see that they are perfectly symmetrical. In other words, if you drew a line down the middle, each side would look like a mirror image of the other.

When DNA is about to duplicate, it's literally as if it were splitting at the seams. The whole strand begins to unravel. The twisted ladder of the molecule untwists, and the two sides then pull apart like a molecular zipper. The rungs of the ladder—the As paired with Ts and the Cs paired with Gs—split in half, and what you have left is a long row of As, Cs, Ts, and Gs that are missing a partner to make complete rungs on the ladder.

Cell Mate

If all the DNA in just one human cell were unraveled and stretched out in a straight line, it would measure about six feet, but it would be so incredibly thin that you wouldn't be able to see it. If the total DNA in all the trillions of cells in one human being were laid out in a straight line, it would stretch to the sun and back more than a thousand times.

DNA Data

Francis Crick tells the story of how he went to dinner with James Watson at a club for scientists in England, and when people asked about the structure of DNA, all Watson could do—inspired by a little too much wine—was repeat that it was beautiful.

DNA replication.

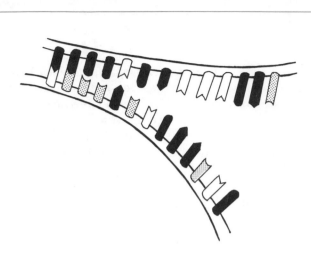

Stealing Bases

Now that the DNA is split up, will its whole existence unravel? No. Nature will come to the rescue. There are always some free-floating As, Cs, Ts, and Gs in the nucleus of the cell, and they are attracted to their chemically compatible partners with the help of an enzyme that directs the base pairing like a molecular matchmaker. As you recall, A pairs only with T, and C pairs only with G. If an A floats towards a C, it just bounces off, because, like divorced molecular couples, they're just not compatible. The A keeps floating around until it finds a T, and the two hook up loosely as a rung on the DNA ladder.

DNA Data

Our description of DNA replication is somewhat simplified, but it should give you an idea of what goes on. One of the details we've left out is that there are enzymes that cause the DNA to "unzip" in the first place. Other enzymes work on small sections of DNA to help it unravel at different locations along the strand. It's like building a highway from New York to California by having small crews start in different locations all over the United States so that eventually they can all hook up. Still other enzymes make sure that the unzipped strands don't get twisted into knots.

In this way, both separated strands get a complete set of matching As, Cs, Gs, and Ts. Now there are two completely identical molecules of DNA where before there was only one. Each one can twist back into its ladder configuration, and now you have two DNA molecules ready to give hereditary instructions to new cells.

For example, if you begin with a part of a strand of DNA that reads AACTGCTATTG, then its other half-rungs would have to read TTGACGATAAC. Because the first letters on the first strand are As, they can pair up only with Ts, and so on down the line. When these strands split apart, the missing half-rungs are replaced with the free-floating As, Ts, Cs, and Gs that they can pair up with, and you end up with two identical molecules.

Assembly Line Jeans

To understand DNA replication a little better, let's imagine a factory where they make blue jeans instead of genes. There are two styles that they make at this factory: dark blue with cuffs and wide pockets, and khaki with no cuffs and small pockets.

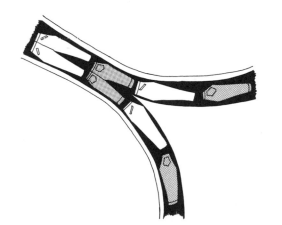

Replication of jeans.

As you walk into this factory, you see pairs of jeans on a long conveyor belt. The two sides of each pair of jeans are sewn together very loosely. Suddenly, something very strange happens. The entire conveyor belt begins to split down the middle, and the two sides of each pair of jeans—which are attached at the hip to the conveyor belt—are pulled apart. The seams were loose to begin with, and that's why they come apart so easily. Now you have two conveyor belt halves with half of each pair of jeans attached to them.

Obviously, the khaki, uncuffed right leg (think of this as a G nucleotide) isn't going to fit together with anything but the other khaki, uncuffed left leg (which you can think of as a C). If you put one blue, cuffed right leg (an A) together with the right or left-legged uncuffed khaki (a C or a G), they'll never fit together. It has to go with its match: the blue, uncuffed left leg (a T).

Luckily, lots of extra half-pairs of jeans are sitting around on the factory floor, so the factory workers immediately start running around to find the correct halves that will fit together with the ones that were pulled apart on the split conveyor belt. When all the dark blue, cuffed halves have been paired together, and all the khaki, uncuffed pairs have met their matches, the workers sew them together loosely, and the conveyor belts magically come together again. The factory now has twice as many pairs of jeans as it had before.

The Four-Letter Alphabet

You now understand how DNA replicates itself by splitting apart and replacing its partners on the rungs of the twisted ladder or double helix. But what do the arrangements of all these As, Gs, Cs, and Ts mean? Why does the DNA molecule go through all the trouble of duplicating itself in exactly the same order? The answer is that DNA's rungs, the nucleotide bases, have a special language, a sort of code that spells out instructions governing how a cell—and consequently the organism that consists of that cell and others like it—will grow, including when and how it will develop in specific ways.

DNA Data

Genes do a lot more than determine things like how tall a person or plant or animal will be, although this is certainly one of their roles. They also make it possible for a human being to have only one head and that it will grow in proportion to the body. Genes also determine that a person should have a right and a left arm and that there should be only one heart that pumps blood in the right way. In other words, genes are responsible for the design of different cells—and therefore for the parts of the body these cells comprise—as well as how they develop.

The DNA Alphabet

In Chapter 5, we said that the bases A, C, T, and G are like a four-letter alphabet, much like our own 26-letter alphabet. You might be asking yourself what kind of information can possibly be transmitted with just four letters. (This is a natural time to bat around words like cat, tag, and tact.) Rest assured that this four-letter alphabet can be put together in a huge number of combinations that mean a great deal to the cell.

If you suspect that four letters couldn't amount to much, just consider that each gene can be hundreds or thousands of letters long. Consider that Morse code uses only two symbols—dots and dashes—but can convey any complex message you can think of. All computer data is nothing more than a series of ones and zeroes, even if it's a digital version of a Rembrandt. And just think: Morse code and computers work with an alphabet only half the size of the DNA "alphabet."

If you think of the As, Ts, Cs, and Gs as letters, then combinations of these letters spell out words and sentences. These words and sentences are like the genes, which are sections of DNA that are all connected together on the DNA strand. If you were to put lots and lots of these sentences together until they formed a fascinating book like a Complete Idiot's Guide, then you would have a chromosome. Think of that book (the chromosome) as one of many volumes in a set of encyclopedias, which represents an entire *genome*. (As you remember from Chapter 1, the word *genome* refers to the totality of DNA contained in the cells of a living organism.)

Reading the Encyclopedia Genetica

Usually, one strand of the DNA molecule is the *sense* strand, meaning that it carries the letters that give the instructions, and the other strand is the *antisense* strand. However, stretches of DNA sometimes carry sense on one side of the helix, and sometimes on the other. Still other DNA molecules have been found that seem to make no sense at all on either strand. For most of the length of DNA, neither side codes for anything.

When you put together a bunch of nucleotides, you form genes. A gene is a long series of these four letters that carries an instruction to the cell. So what exactly does the gene tell the cell to do? As pioneering scientists like Beadle and Tatum found out, the gene not only tells the cell to manufacture a particular type of protein, but also when to produce it and how to produce it.

Proteins are used extensively in living things. Proteins are large molecules that are made up of smaller units called *amino acids*, much in the same way that DNA is made up of long stretches of smaller units called nucleotides. Different proteins are found in all parts of cells, and these proteins are needed for many substances in the body, such as hair and nails.

> **Genetic Jargon**
>
> **Amino** (a–MEAN-o) **acids** are the chemicals that are the building blocks of proteins. Chains of different amino acids linked together make up different proteins.

Mutant Madness

Consider this: for even one of your cells to make a copy of all of its DNA, it has to make perfect copies of an estimated six billion nucleotide bases (all those pairs of As, Cs, Gs, and Ts). There's a lot of room for error here. The mechanisms behind all this copying are almost perfect, but nonetheless the odds are that mistakes can, and sometimes do, happen.

Imagine if you had to type out every single word in an entire encyclopedia. Don't you think you might make one or two or maybe hundreds or even a few thousand mistakes? Then put yourself in the position of the poor DNA molecule, which takes about seven hours to copy all of the genes in just one cell. Doing this year after year, you might just write an A instead of a C or a T instead of a G.

Now imagine how these mistakes—small as they may seem in the grander scheme of things—would affect the proper reading of the encyclopedia if you wrote "cat" instead of "act." What if you deleted whole sentences? It might not even affect the reading, but then again, it could really mess up the sense of the encyclopedia.

To counteract these occasional errors, DNA has its own repair system. Its job is to straighten out those messy misprints that can randomly occur. In the same way that publishing companies hire proofreaders to look for mistakes in manuscripts and books, the DNA copying system has built-in proofreaders to make sure that any accidental errors will be spotted and corrected. These proofreaders are enzymes.

Several DNA repair enzymes work together by performing different tasks within the strand of DNA. One type of enzyme searches and destroys. It looks for damaged or mismatched As, Cs, Ts, and Gs. When it finds one, it cuts it out of the DNA strand. A second type of enzyme then recruits whichever letter should be in the place of the mistake and replaces the old one with a new, appropriate letter. To finish off the process, yet another enzyme makes sure that everything that has been replaced is connected the way it should be.

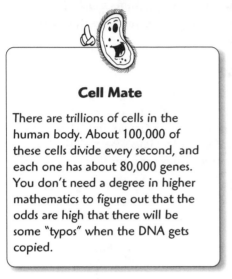

Cell Mate

There are trillions of cells in the human body. About 100,000 of these cells divide every second, and each one has about 80,000 genes. You don't need a degree in higher mathematics to figure out that the odds are high that there will be some "typos" when the DNA gets copied.

Outside Forces

In addition to the errors that can occur during replication, there are also outside threats to copying DNA correctly. For instance, if you stay out in the sun too long, especially without sunscreen, the ultraviolet light from the sun can damage your DNA. Certain chemicals can also bring about changes in your DNA.

DNA Data

Some scientists think the reason why more of the sun's ultraviolet rays can get to your skin and cause DNA damage than ever before is due to a depletion of the ozone layer. Ozone is a gaseous element that absorbs and screens out ultraviolet radiation from the sun before it reaches our skin. Earth has an ozone layer high up in its atmosphere. However, in the 1970s, scientists discovered a hole in the ozone layer over the south pole. Australia is the populated area closest to the south pole, and there has been a higher incidence of skin cancer there than there was prior to ozone depletion.

X-rays can also damage your DNA. In Chapter 3, we described how a scientist named Thomas Hunt Morgan and his students studied heredity in fruit flies. One of his students, Hermann Muller, realized that X-raying these tiny creatures caused mutations that the scientists could study in future generations. (Mutations are changes in genes that can be inherited by future generations. They often cause unusual varieties of plants or animals. You'll learn more about mutations and how they affect DNA later in Chapter 9, "Irregular Genes.")

Tailored Genes

In addition to being copied wrong or being damaged by outside influences such as X-rays or the ultraviolet rays of the sun, DNA can change in other ways thanks to the fact that some genes can change their positions on the chromosome. Barbara McClintock discovered this phenomenon. As you may remember from Chapter 3, Barbara McClintock was an American geneticist who studied the hereditary studies of Indian corn in the 1940s.

She observed an odd behavior in some genes that led to the rearrangement of DNA sequences in some of the corn that she studied. Indian corn, also known as maize, has many differently colored kernels. Some are dark, and others are light. McClintock came to the conclusion that this variation occurred because some of the corn's genes moved around on the chromosome. Before her investigations, it was believed that genes stayed put wherever they happened to lie on the chromosome.

These movable stretches of DNA that McClintock discovered are called *transposons* and are sometimes referred to as "jumping genes." McClintock's conclusion was that all of the kernels of Indian corn had cells with genes for a dark color to be present. When a "jumping gene" moves next to one of these genes for dark-colored kernels, the dark-colored kernel is disabled, so the corn has a light-colored kernel instead. If the jumping gene jumps back again, the dark-colored gene will be turned back on again. McClintock also believed that these jumping genes could be replicated before they jump, so there could be numerous copies of them in the chromosomes.

Genetic Jargon

Transposons (trans-POE-sons), or jumping genes, are stretches of DNA that travel from one position on the chromosome to another. Barbara McClintock first discovered this phenomenon in the 1940s, but few scientists accepted her theory at the time.

Although McClintock's work was not widely accepted at the time, McClintock's research was rewarded with a Nobel prize nearly 40 years after her work with Indian corn. Since then, scientists have discovered that jumping genes can occur in other organisms as well.

Viruses: Molecular Terrorists

In addition to faulty copying, environmentally or chemically induced mutations, and "jumping genes," DNA sequences can be changed by something that crosses the fine line between the living and the nonliving. These strange creatures are called *viruses*. Their name comes from the Latin word meaning poison or slime.

Viruses can't comfortably be classified as living because they're unable to reproduce on their own. Like very tiny vampires, they need to attach themselves to living cells in order to make more of their own kind, so they can spread to other cells and do it again. Viruses therefore cause different types of diseases, ranging from the inconvenient but easy to overcome common cold and measles to very serious diseases such as hepatitis and AIDS.

Viruses are much smaller than bacteria, and bacteria aren't exactly huge to begin with. A picture of viruses attacking a bacterium would look like a bunch of marshmallows sticking out of a roast pig. Most viruses are so small that they can't even be seen through a regular microscope. You need an electron microscope, which can magnify around a million times, to get a good look at them. But you might not want to see what they look like. Some of them look like they're straight out of science fiction scenarios.

A virus is basically a bunch of genes wearing a protein coat. You might say viruses are all dressed up with someplace to go. Where they want to go is usually a nice living cell that they can invade like molecular terrorists carrying out a plot to take over a country.

The takeover works like this: Something in the virus's coat tricks an unsuspecting cell into letting it come inside. Normally, a cell would have some kind of a "security guard" that would recognize a virus or other foreign body as a terrorist and ban it from entering, but the virus's coat is constructed in such a way that it confuses the cell.

Genetic Jargon

A **virus** is a microscopic bag of genes that has a protein coat. It tricks a cell's "security guard," or membrane, into opening its "doors." Then these molecular terrorists make the cell sick or even dead when the cell is duped into making more of the virus with its own cellular resources. Scientists don't consider viruses living things because they cannot reproduce on their own and rely on their hosts to carry out functions for them.

Once inside the cell, the tiny terrorist is like a miniature striptease dancer. It takes off its outer coat and inserts its own naked DNA (or in some cases, RNA, which we'll learn about in Chapter 8, "Producing Proteins") into the cell it's just invaded.

Now the virus takes over the cell. The virus's DNA tells the bacterial cell to make more of the tiny terrorists. In a short time, the cell uses up much of its own energy in the service of its invader. In a matter of hours, thousands of these new terrorists-from-within can be generated.

Eventually, the new terrorists burst out of the unlucky cell, killing their gracious host in the process. (Talk about ungrateful guests!) Now these hordes of new invaders go off to hunt down more cells in the area, repeating their underhanded takeover process again and again.

During this takeover process, viruses sometimes inadvertently cause changes in the DNA sequences of cells they invade. The virus starts out by inserting its own DNA into the DNA of the host cell. Eventually, when the virus makes its getaway, it might take along some of the host cell's genes to which its own DNA was attached. When it injects its DNA into its next victim, some of the genes from the first host cell will become part of the new victim's genome.

Nature has worked out ways to promote diversity in sequences of DNA. This diversity winds up being good for organisms because it means they can adapt to changes in the environment and it might allow an organism to better deal with factors in the outside world that its ancestors couldn't handle. So mutations, "jumping genes," and viruses mixing genes around help to ensure that new living things change instead of remaining old-fashioned, stick-in-the-mud models that only worked well in yesterday's environment. All of this leads to a rich diversity of life that hopefully keeps getting better and better equipped to deal with a new tomorrow.

Mutant Misconceptions

Don't confuse viruses with bacteria. Bacteria are single-celled organisms without nuclei. Viruses aren't even cells—in fact, scientists don't even consider them to be alive. That's why you can't treat a bacterial illness the same way as a viral illness.

The Least You Need to Know

➤ DNA makes copies of itself by first splitting its two strands apart like a zipper and then adding the appropriate As, Cs, Ts, and Gs to each of the single strands until there are two identical DNA molecules.

➤ The nucleotide bases—the rungs on the DNA ladder—are called A, C, T, and G. They are like a four-letter alphabet that combine to instruct cells how to make specific proteins.

➤ Sometimes errors in copying DNA occur, and this leads to mutations, or permanent changes in the DNA.

➤ Sometimes outside forces such as X-rays, excessive ultraviolet light from the sun, or harmful chemicals cause mutations in DNA.

➤ Barbara McClintock discovered that genes, which are sequences of DNA, can rearrange themselves on chromosomes.

➤ Viruses are microscopic particles that invade cells and trick them into making more viruses. Sometimes viruses "steal" genes from cells and carry them to other cells that they subsequently invade.

Chapter 7

S, M, L, and XL Genes

> **In This Chapter**
>
> ➤ How genes come in different lengths
>
> ➤ Swapping DNA
>
> ➤ Coding and noncoding parts of DNA
>
> ➤ Repetitions in DNA sequences
>
> ➤ Comparing human chromosomes to those of animals and plants

You learned in Chapter 5 that genes are made of DNA. You also know how DNA makes more copies of itself, by "unzipping" and getting spare As, Ts, Cs, and Gs to get together with the two sides of the "zipper." Now you're going to learn more about the types of genes and their sizes.

Just as blue jeans come in different sizes and styles, the genes we're talking about can come in different sizes and styles, too. In this chapter, we'll examine some of the different small, medium, large, and extra-large genes that you won't find in your local clothing store. You'll read about some animal and plant genes and how they relate to human genes, and you'll learn about the not-so-scary ghost genes.

One Size Does Not Fit All

If there were only one model and size of blue jeans available, you'd have one heck of a time finding something that you would like and that would fit you. If a clothing store only carried navy blue jeans in a size small, then the store wouldn't meet the needs of everyone, and eventually it would be put out of business. Instead, blue jeans come in

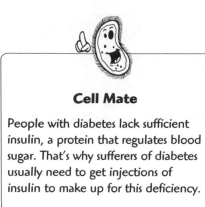

Cell Mate

People with diabetes lack sufficient insulin, a protein that regulates blood sugar. That's why sufferers of diabetes usually need to get injections of insulin to make up for this deficiency.

Genetic Jargon

A **nucleotide** (NEW-klee-oh-tide) is one of the building blocks of the DNA molecule. The four nucleotides are A, C, T and G. Because of their chemical structures, A can only pair with T, and C can only pair with G. These pairs are known as base pairs.

Cell Mate

Factor VIII and Factor IX help in the coagulation (clotting) of blood. A deficiency in either of these factors leads to different types of hemophilia, a disease in which the sufferer is in danger of bleeding profusely, even from a small cut.

many different sizes and styles to meet the demands of tall people who need lengthy jeans, as well as short people who need shorter jeans. The same is true for genes made of DNA; they don't come in just one size.

Remember, as we discussed earlier, each gene codes for a specific protein, and only that specific protein. Humans, for instance, have thousands of different proteins in their bodies that are needed to perform tasks necessary for survival, such as digesting food and clotting blood. If there were just one length of DNA, it probably wouldn't be able to code for all the different-sized proteins that an organism needs to function properly.

Genes come in an enormous range of sizes. For example, the gene that codes for insulin, a protein that regulates blood sugar levels, is 1,700 base pairs long, which means there are 1,700 rungs on this gene's twisted ladder of DNA. You'll remember from Chapter 6, "Assembling Your Genes," that base pairs are the "rungs" on the "twisted ladder," or double helix, that makes up the shape of the DNA molecule. Each rung is made up of two halves which are constructed in one of two specific ways. A rung can be either one half A with one half T, or it can be one half C with one half G. The reason for this is that the chemical structure of these As, Cs, Ts, and Gs will allow A to pair only with T, and C to pair only with G, and vice versa.

So the 1,700 base pairs of the insulin gene means 1,700 rungs on the twisted rope ladder that is the shape of DNA. Since each base pair is made up of two halves, you could say that the insulin gene is made up of 3,400 nucleotides, or As, Cs, Ts, and Cs.

At 1,700 base pairs long, the insulin gene is considered a smaller gene. There are also other longer and much longer genes. For example, the gene that codes for Factor IX measures 34,000 base pairs and is considered to be medium-sized. The gene that codes for Factor VIII is much longer at about 186,000 base pairs. The largest gene that's been identified so far is the one that is involved in Duchenne muscular dystrophy, a degenerative disease that occurs mostly in males. That gene stretches to an enormous two million base pairs in length.

Is Bigger Better?

If you think about the fact that there are so many different lengths of genes, you might begin to wonder whether all these different lengths mean anything. Is a longer gene somehow more important or better than one that is half that length? Or are shorter genes more efficient because they're more compact? Does size matter at all?

For the answer, let's go back to our comparison to blue jeans. It would be like asking whether long jeans for tall people are better than shorter jeans for smaller people. The obvious answer is that the longer jeans use a lot more material, but that doesn't necessarily mean that more is better or that longer jeans are better than shorter jeans.

To evaluate blue jeans properly, you'd have to look at other factors besides the length. You might look at the workmanship of the pants. Are the seams stitched properly? Are the button holes the proper size for the buttons? Are the pockets sturdy and located where they're supposed to be?

In the same way, when you're talking about genes made of DNA, you need to go beyond the question of length to see whether one gene is somehow more important than any other. In the case of molecular genes, part of the answer to the "Is bigger better?" question lies in the fact that some DNA carries genetic information, and some DNA does not.

Junk DNA

As you've learned, each gene codes for a specific protein that's needed for specific purposes in the organism that houses the genes, and the DNA that makes up these genes provides the code. However, not all of the DNA within a gene codes for proteins. The coding parts of genes—*exons*—are the most obviously useful type of DNA and are therefore the type of DNA that most researchers are busy studying these days.

The other, more mystifying types of DNA have been given the highly undignified title of *junk DNA*, because they don't seem to code for anything. Junk DNA comes in two flavors: introns and spacer DNA.

We Interrupt This Program...

Introns are stretches of DNA within the gene that sit in between the protein-coding sections, the exons. Introns are something like the intermissions between acts of a play or the commercial breaks that interrupt your favorite TV shows. Because most

Genetic Jargon

Although it may sound like a brand of gasoline, an **exon** is a part of a gene that codes for proteins.

Cell Mate

Occasionally, one of these noncoding intron sequences might contain one or more parts of other genes. In some cases, an intron can even contain another gene in its entirety. In such a case, you would have a gene within an intron within a gene.

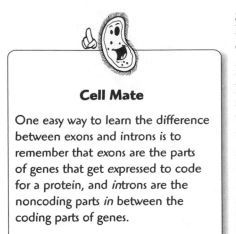

Cell Mate

One easy way to learn the difference between exons and introns is to remember that *exons* are the parts of genes that get *expressed* to code for a protein, and *introns* are the noncoding parts *in* between the coding parts of genes.

genes have more introns than exons, a typical gene is like a half-hour sitcom that's interrupted by many hours of commercials.

Even though there's been a pause at the theater, the play still remains one complete drama. Even though the sitcom that you're watching on TV is stopped a few times so you can hear about the virtues of a brand new toothpaste or a fantastic hair dye, you still manage to get the whole story. For example, you might find a gene on a strand of DNA, and then inside it might be some molecular gibberish such as AATGTCGCGTA, and then connected to that would be more of the gene that codes for a protein. There might be another "commercial break" (intron) after this and then some more coding parts. There could be just a few introns or many of them separating the coding parts of the gene.

Exons and introns.

exon	intron	exon	intron	exon	intron	exon

Spacing Things Out

In addition to introns, which act like commercial breaks in a TV sitcom—because they are found right in the middle of coding sections of DNA—there are also noncoding DNA sequences in between separate genes on a strand of DNA. These sequences are known as *spacer DNA*, because they occupy spaces in between separate genes.

Just Junk?

The surprising fact is that the vast majority of an organism's DNA sequences is composed of junk DNA. Probably only about five percent of a human's total DNA is made up of genes, the coding part of the double helix. Scientists learned this fact in the late 1970s, and they were shocked to find that DNA was not all neat and tidy and composed of stretches that were efficiently constructed in a compact, meaningful way. This would be something like picking up a classic book such as Tolstoy's *War and Peace* and finding that only a few pages scattered throughout the book contained Tolstoy's brilliant story of old Russia but the rest of the book had hundreds of pages of gibberish in between the real writing.

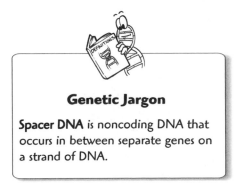

Genetic Jargon

Spacer DNA is noncoding DNA that occurs in between separate genes on a strand of DNA.

Some scientists object to the belittling term *junk DNA* for noncoding DNA. It's very possible, they argue, that as scientists learn more and more about the workings of DNA, they will discover that these noncoding sections have a subtle, less obvious function, but an important function nevertheless. Who knows? Even though introns can't code for a protein, they might be responsible for helping exons that do, and scientists just haven't figured this out yet.

DNA Data

In 1977, two scientists independently made the same discovery that genes, the segments of DNA that code for proteins, don't necessarily consist of unbroken stretches of DNA. The two scientists were Richard Roberts and Phillip A. Sharp. Both were awarded the 1993 Nobel prize for Physiology or Medicine for their simultaneous discovery of the fact that genes can be interrupted by lengthy stretches of noncoding DNA.

Who Needs All This Junk?

Because so little is known about introns and spacer DNA, scientists argue about their significance. One thing that scientists know for sure is that some organisms, like most bacteria, don't have any noncoding DNA, and they seem to do just fine without it.

Most bacteria have their DNA in one continuous stretch. For instance, E. coli (which look a lot like cocktail frankfurters, but are much smaller and presumably taste different) have their DNA in a circle containing almost all of their 4,400 genes with very little spacing in between. So apparently, introns and spacer DNA aren't necessary for life as we know it.

So what the heck is all this extra, seemingly meaningless DNA doing if it doesn't code for proteins? One theory is that genes are made up of sub-units, and that by mixing and matching these sub-units, you can get variations. Many scientists feel that there might be a few hundred types of exons that get shuffled around like a pack of playing cards. So in the same way that you can get thousands of different hands when you're playing poker because the dealer shuffles the cards, this theory holds that sub-units of genes get shuffled around, resulting in a huge amount of variation.

Suppose you hold an ace of spades in your hand. This ace can function in different roles in a poker hand. For example, an ace of spades held with two other aces is part of a three-of-a-kind. But if that same ace of spades is held with four other spade cards, you're holding a flush. It all depends on what other cards are in your hand. The same may hold true for sub-units of genes, which can produce different combinations with various other sub-units. This would be a great way to generate more and more genetic diversity, which as we've discussed is very important for the continuation of a successful species on earth.

Repeats and Other Genetic Mysteries

Much of junk DNA is repetitive, which means that sometimes the As, Ts, Cs, and Gs form a pattern that repeats itself. For instance, if you look at a long stretch of DNA, you might keep seeing the pattern AATTGGCGCGCGTAGAATCGGG in several locations on the noncoding DNA.

Not much is known about this phenomenon or why it occurs. What scientists have learned, however, is that there are long repeats, composed of about 50 to 100 bases, and also short ones, composed of about two or three bases. The number of these repeats varies considerably from person to person. For example, one person might have the noncoding, seemingly meaningless sequence ATTCCGAAT repeated in seven different spots throughout the length of one particular chromosome or another. Another person might have 12 repeats of a different sequence on a different chromosome. These repeats seem to occur in unique ways in different individuals.

Genetic Jargon

One way that scientists are able to genetically check a person's identity (for example, in forensics or paternity) is by checking their **tandem repeats**, which are DNA sequences unique to each person that appear again and again.

In Tandem

Another type of repeat is called a *tandem repeat*. Think of a tandem bicycle, which is a bicycle built for two. One rider sits straight behind the other. In the same way, tandem repeats of base pairs are repeated immediately after one another. It's like a genetic stutter.

Scientists can now take advantage of these unique repeat patterns in different people to use DNA for identity and paternity testing. It's helped solve crimes when there were no witnesses, and it can positively identify the father or other relatives of a child when paternity is in question. You'll read more about this in Chapter 17, "Stories Genes Can Tell," when we discuss ways to identify people through their DNA.

Ghost Genes

There's another mysterious type of DNA sequence that's known as a *ghost gene*. Despite the name, these are not things that go bump in the night. Nonetheless, it's fair to say that they are shadows of their former selves.

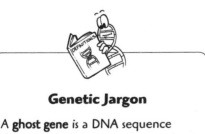

Genetic Jargon

A **ghost gene** is a DNA sequence that used to function as a gene that could code for a protein, but it has changed and can no longer function as a coding gene.

Sometimes scientists analyze a particular sequence of DNA and realize that it must have at one time been a gene, even though it can't function as one any more. The way they know this is that there's enough order in some DNA sequences to suggest that they once had a function. But over time, and as a result of slow and steady changes, the gene reaches a point where it just isn't itself anymore and can no longer code for a protein.

Genetic Evolution

Everything seems to evolve, and genes are apparently no exception. Because of this gene evolution, scientists can make comparisons between different organisms, such as humans and chimpanzees, to see where we all fit into the evolutionary scheme of living things.

When scientists make these comparisons, some interesting patterns emerge. If you take a look at the number of chromosomes that different organisms contain, you might be in for a surprise. You might assume that humans would have the most chromosomes, and that mammals that seem to be lower on the evolutionary totem pole would have a smaller number of chromosomes, and so on down the line.

But this isn't always the case. If you look at the following chart, you'll find a number of scientific surprises. Chimpanzees, for instance, have 48 chromosomes, but humans have only 46. Does this mean that chimps are more evolved than we are? Part of the answer lies in the fact that humans and chimps share a common ancestor. At one point in the course of evolution, about five million years ago, our branch split off from that common ancestor, and two of those 48 chromosomes fused together, so we were left with two less than our relatives, the chimps.

Cell Mate

Scientists have determined that we're not so far removed from some of our animal relatives. Humans and chimpanzees, for instance, have about 99 percent of the same DNA, giving new meaning to the phrase, "I'll be a monkey's uncle."

A Comparison of Chromosome Numbers

Organism	Number of Chromosomes
Human	46
Chimpanzee	48
Cherry	170–180
Opossum	22
Minnow	36
Mouse	40
Orangutan	48
Guinea pig	62
Horse	64
Mulberry	224–308
Baboon	42
Fern	512

Just looking at the number of chromosomes an organism has doesn't tell you how evolved it is. There's no real rule here. The numbers seem to jump around so much. So there must be another way of comparing the DNA of these different organisms that would make more sense in terms of evolution.

Another way of comparing organisms is to ask just how much DNA each contains in its cells. Looking at things this way, mammals have more DNA than reptiles, and reptiles have more DNA than fish. But even this approach has its exceptions. Some amphibians, for example, have more DNA than humans, and if you compare two closely related species of reptiles, one might have twice or even three times as much DNA as the other.

DNA Data

In case you're wondering, reptiles are cold-blooded, egg-laying critters like snakes, lizards, and crocodiles. Amphibians, on the other hand, are animals that spend some time in the water and other time on land. Frogs are amphibians. They live on dry land, but their eggs hatch into tadpoles that swim around in the water until they grow up and move on to the land. One of the strangest amphibians is the axolotl (AX-a-LOT-ull), which has been called the Peter Pan of amphibians because it never grows up. This type of salamander starts out like a tadpole, but it usually doesn't change into a land-living adult in the way that frogs and most other amphibians do.

One really bizarre fact is that some plants have more DNA than people. There is, however, an explanation. You learned that in meiosis, sex cells (sperm and egg cells) divide so that the copies have only half the number of chromosomes that were in the original. When a human sperm and egg unite to form what will eventually become a human child, that baby receives one set of 23 chromosomes from the father and one set of 23 chromosomes from the mother, for a grand total of 46 chromosomes for that individual.

If there are extra chromosomes, there can be dire consequences for a developing human. An extra chromosome will almost always result in abnormal development. For instance, a child that develops from a fertilized egg cell with an extra copy of chromosome 21 will be born with Down Syndrome.

But for reasons that are not properly understood, plants can tolerate extra copies of chromosomes. Some plants can contain three, four, or even seven sets of the chromosomes that they received from their parent plants without any apparent ill-effects.

Many of our food crop plants, such as potatoes, have several sets of chromosomes. This is one explanation of why some plants have a lot more chromosomes than humans.

Another factor to consider when comparing the number of chromosomes found in different species is the fact that sometimes things aren't always what they seem. For instance, some salamanders have more DNA than humans, but it turns out that they actually have a lot more noncoding DNA than we do. So if we want to compare humans with other animals or plants, it might make sense to compare the amount of coding DNA, actual genes. We have more genes than any other species. In Chapter 8, "Producing Proteins," you'll learn more about these coding regions of DNA and see how they translate into proteins that plants, animals, and people use to function on a daily basis.

Cell Mate

Scientists have given each chromosome found in humans a different number. They go according to size, with chromosome 1 being the largest, and chromosome 2 being the next largest, and so on.

The Least You Need to Know

➤ Genes come in small, medium, large, and extra-large lengths.

➤ During meiosis (the reproduction of sex cells), chromosomes swap genetic information.

➤ Stretches of DNA that code for proteins are called exons, and stretches of DNA within genes that don't code for anything are called introns.

➤ Some noncoding sequences of DNA get repeated, sometimes at different locations on the chromosome and sometimes over and over again at the same location.

➤ Humans have more chromosomes than some plants and animals and less than others, but humans have the most coding DNA of any organism.

Chapter 8

Producing Proteins

In This Chapter

➤ What proteins are, and why you need them

➤ How proteins are made up of amino acids

➤ How messenger RNA carries instructions from genes

➤ How proteins are made in the cells' factories

➤ How proteins fold up into specific shapes that affect their function

Now you know the structure of the DNA molecule and how it makes copies of itself. We've also discussed that stretches of DNA that code for proteins are called genes and that the order of the rungs on the twisted ladder that is the DNA structure—the As, Ts, Cs, and Gs—are like a four-letter alphabet that spells out a set of instructions to the cell.

Each gene codes for a protein. That means that the genes tell the cell which protein to make and how to make it. That's why DNA and genes have often been called the blueprints of life. Some scientists have also compared genes to a recipe in nature's cookbook of life.

Genes tell the cell what ingredients to go out and buy, exactly how much of each ingredient to use, and how and when to combine these ingredients. This doesn't necessarily mean that the angel food cake or cheddar cheese soufflé or chocolate mousse is going to turn out right every time or at all. But if the recipe is followed correctly without any mistakes, all should go well, and the desired result will be as planned. In this chapter, you'll learn just how these recipes or blueprints in the DNA in each cell get translated into the final product, proteins.

Genetic Jargon

Keratin (KEH-ra-tin) is a protein found in your hair, fingernails, and toenails. **Collagen** (KAH-la-jin) is another protein that's found in skin. Collagen doesn't get replaced, so as it gets older, it loses its elasticity. This is why we get wrinkles as we get older.

Genetic Jargon

Antibodies are substances in the blood that defend the body against invading foreign substances such as harmful bacteria and toxins. **Hormones** are substances formed in the tissues or glands of the body that usually circulate through the bloodstream. These chemical messengers influence the different functions and growth of cells and organs. Antibodies and hormones are made of proteins.

Types of Proteins

As you learned in previous chapters, each gene codes for a specific protein. So what exactly are proteins, and why are they so important to the plant or animal or person or bacteria that needs them? Virtually every part of your body is made entirely or at least in part of proteins. Just about everything you can see on your body is made of these substances.

For example, *keratin* is a protein that makes up most of your hair and nails. *Collagen*, another protein, is a component of skin. Muscles are also composed of proteins. There are other proteins in your eyes that are sensitive to light and still other proteins that make you sensitive to certain smells. Some proteins are responsible for specific chemical reactions, such as the hundreds of proteins that are used in your digestive system to break down food.

All of your organs are made of proteins, too. For instance, the heart, which is essentially a strong, pumping muscle, is composed of protein. And *antibodies*, the microscopic soldiers that defend the body, are also made of proteins. In addition, enzymes, which you learned speed up chemical reactions in the body, are composed of proteins. Some *hormones*, which are formed in organs and then transported to other parts of the body to perform specific tasks, such as getting the body to grow, are made of proteins as well.

Humans aren't the only creatures made up of, and in need of, proteins. Plants and animals and bacteria use pretty much the same or similar proteins, with some slight differences, that you do for survival.

DNA Data

In 1840, a Dutch chemist, G. J. Mulder (1802–1880), coined the name *protein* from the Greek word *protos*, which means "first," because proteins were thought to be of primary importance. As we discussed in Chapter 5, "Nature's Blueprint," a number of scientists incorrectly thought that protein was the substance responsible for the passing of hereditary traits, because it was so complex and intriguing.

Building Blocks: Amino Acids

The way proteins are built is similar to how DNA is put together. They are both made of smaller units that connect to form a long chain. While DNA is made up of sub-units called nucleotides (the As, Ts, Cs, and Gs), proteins are made up of other sub-units—those building blocks of proteins called *amino acids*. There are 20 amino acids in all, and different combinations of these linked in a chain in different ways make it possible to form tens of thousands of different proteins.

Some proteins are made up of a chain of just a few amino acids; others are a combination of over a thousand amino acids. The amino acids for a specific protein are always in a specific order on the chain. The different combinations of amino acids yield an enormous variety of proteins for the thousands of jobs that living things need to do for survival.

Genetic Jargon

Amino acids are the building blocks of proteins. There are 20 amino acids, and they are linked together in different combinations to form different proteins.

Molecular Origami

When you read about DNA's structure in earlier chapters, you learned that its structure is important because it helps define how the DNA molecule can replicate and function in different ways. The same is true of proteins. Each one has a specific structure that arises from the way the amino acid chain is folded up, and this structure defines how the proteins function.

In contrast to the winding double helix structure of DNA, proteins have been likened to molecular origami because each different amino acid chain can be folded up in different twists and turns, like the Japanese art of paper folding that can take a simple square sheet of paper and transform it into a swan, boat, or Christmas tree, depending on what gets folded and where. These different shapes help to minimize the proteins' energy and maximize efficiency.

For instance, proteins often find themselves in environments where they are either surrounded by fat or by water. Some amino acids in a protein chain like water, and others prefer fat. This preference has to do with the way the protein is folded. When a protein chain folds up, it does so in a way that will make it compatible with its environment.

If a protein is near a cell membrane, which is made of fat, its fat-compatible amino acids will twist toward the outside of the molecule so that they can be near the fatty membrane. If a protein is in water, its water-compatible amino acids will generally fold to the forefront and hide the fat-compatible amino acids in the middle of the molecule.

If a protein is an enzyme, its shape acts like a lock that can be "unlocked" by another molecule that fits into it. Antibodies use their shape to defend the body from foreign invaders. In this case, the shape of the antibodies help them act as a key that fits the lock of a dangerous substance, such as harmful bacteria, that doesn't belong in the body. This "lock and key" interaction renders the invading substance harmless.

Because there are tens of thousands of different proteins, and each one has a unique, complex shape, scientists still have not discovered the structure of most of these molecules. But scientists can use X-ray diffraction to analyze the shape and take guesses at what the structure of the molecule could be. You'll remember that this is the same process that Rosalind Franklin used with DNA crystals to help determine their structure.

Genetic Jargon

When a cell manufactures protein, this process is called protein **synthesis**. The word synthesis comes from two Greek words meaning "to place together." Appropriately, proteins are made by taking a few, or many, amino acids and placing them together in a specific order in a chain.

Some Assembly Required: Making Proteins

Proteins are essential to life, and most of them need to be made by the cell frequently. Here's where genes play their part by starting up the process of synthesizing whichever proteins the body needs. Although genes themselves don't physically make the proteins, they carry the instructions for their *synthesis*. Because each gene codes for a specific protein, and proteins are so essential to survival, any mistakes in genes are likely to cause trouble for the organism. (This is the subject of the next chapter.)

So how do genes pass on instructions to the cell to manufacture proteins? And what exactly does the cell do to make these essential substances? And where does the cell manufacture these important substances?

To begin with, the genes themselves do not manufacture proteins. Remember that genes are stretches of DNA and that each cell contains a complete copy of all the genes required to make the entire organism that it resides in.

However, not all of our genes work all of the time. In an early stage of development, shortly after the sperm fertilizes the egg cell that will grow into an adult organism, most of the genes in each cell are shut off. There's an important reason for this. What if the genes that code for the cells that make up your hand got turned on in your back? Then you'd have an extra hand that could scratch your back. This would be convenient, but it would look kind of weird. Or how about if the genes that code for your nose to grow in the middle of your face remained turned on in the cells in your feet? Then your feet might literally smell. So, most genes do not function in every cell all the time, even though they're all there.

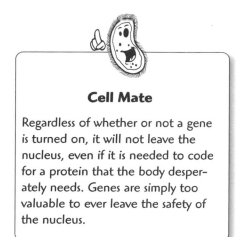

Cell Mate

Regardless of whether or not a gene is turned on, it will not leave the nucleus, even if it is needed to code for a protein that the body desperately needs. Genes are simply too valuable to ever leave the safety of the nucleus.

A Priceless Possession

Suppose you had an original copy of the Declaration of Independence. It's worth a fortune, so you wouldn't want to risk losing it by carrying it around wherever you went. You'd keep it locked up in a safe in your wall.

Now imagine that someone you know wants to read what's written in the original document. You might go to your safe, unlock it, and carefully take out your copy of the Declaration of Independence. Rather than let the original document out of your possession, you could slowly and carefully transcribe the required sections onto a new piece of paper. After you finish writing out the sections your friend wants to read, you put the valuable document back in the safe.

Now you have a handwritten copy, and you can send a messenger to deliver this to your friend. You have just made a disposable copy, but you still have the priceless original intact in your safe. This is similar to the way genes work.

The genes, more valuable than an original copy of the Declaration of Independence, carry the instructions for making proteins, so you wouldn't want to risk having them floating around in the cytoplasm. They're offered a lot more protection hidden away in the nucleus. Your genes will not leave the nucleus, just as you would not take your original Declaration of Independence outside of its safe and risk losing it.

However, proteins are manufactured in ribosomes, which are organelles located in the cytoplasm of the cell. (You can review all of this in Chapter 4, "Me, My Cell, and I.") Ribosomes are outside of the nucleus. To keep the genes in the nucleus, yet allow them to send code to the ribosomes in the cytoplasm, a messenger makes a copy of the needed gene and then carries the copy to the ribosomes. The chemical that is responsible for this is called RNA, or ribonucleic (RYE-bo-new-CLAY-ic) acid.

Genetic Jargon

Uracil, abbreviated as U, is a nucleotide, a sub–unit, of RNA. RNA uses U instead of T. In the same way that T pairs with A in DNA, U pairs with A in RNA.

If RNA sounds like DNA, deoxyribonucleic acid, this is because the two are closely related. There are some differences, however. RNA has a single strand, whereas DNA is double-stranded. Also, RNA is a much shorter molecule than DNA. It might consist of from fifty to several thousand nucleotides (As, Cs, Gs, and Ts); DNA in a gene can have more than a million nucleotides.

Another important difference is that RNA's nucleotides are slightly different from DNA's. RNA has three of the same nucleotides: As, Cs, and Gs. But instead of T, RNA has a nucleotide called U, which stands for *uracil* (YOUR-a-sill). The U nucleotide pairs with A in the same way that T does.

Don't Shoot the Messenger

Genetic Jargon

Messenger RNA (mRNA) is a special type of RNA that is made by the cell as a disposable copy of the valuable set of instructions in a gene inside the nucleus. The process by which a messenger RNA strand is made is called **transcription**. Messenger RNA takes its message, in this case a code for a gene that's needed for the production of a protein, from the DNA in the nucleus to the ribosomes in the cytoplasm.

The first step in transferring a gene's information to the ribosomes is that enzymes go to the strand of DNA in the cell's nucleus and "unzip" the one gene that's going to be copied. The rest of the strand stays put. This is something like the way DNA replicates itself, but in this case, only one gene is unzipped instead of the whole strand.

Now a strand of *RNA* is built using the "unzipped" DNA as a template. The As, Cs, and Gs are matched up, but because RNA is different from DNA, Us are used instead of Ts in its nucleotide sequence. The result is a strand of RNA with letters that are the opposites of the ones copied from the DNA. For example, if a gene in the DNA strand reads AATCCGCG, then the RNA, which will use the letters that match up with these, will read UUAGGCGC. This process of copying the letters that pair up with the DNA strand is called *transcription*.

When the transcribing RNA is finished copying the selected gene, it disengages from the DNA and is almost ready to travel to the protein factories in the cytoplasm of the cell, outside of the nucleus. At this point, before leaving the nucleus so the manufacture of the protein can begin, the RNA needs to have its message "edited." You'll remember that genes are made up of some sequences called exons, which code for proteins, and introns, which don't code for anything. In the same way that an editor at a publishing company crosses out sections of text that make no sense, a molecular "editor" takes out the introns, the noncoding parts of genes, before the protein can be synthesized. When the noncoding introns are taken out, the coding parts, or exons, are stitched together. The new, edited message is called messenger RNA.

Following Instructions: The Genetic Code

The messenger RNA carries instructions copied from the gene that's safe inside the nucleus out into the cytoplasm, where it searches for a ribosome protein factory. As you learned in Chapter 4, ribosomes look something like little chocolate chips and are found on the endoplasmic reticulum, which is the organelle highway from the nucleus to the cell membrane. If you see a ribosome up close, it looks more like a snowman with a small head and a larger, round body.

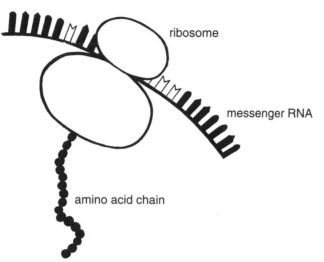

ribosome

messenger RNA

amino acid chain

Ribosomes are the cell's protein factories. They look like miniature snowmen.

The messenger RNA attaches to a ribosome and directs it to manufacture the specific protein that the copied gene codes for. The next step is *translation*, which is the process by which a ribosome makes a protein using the information provided by the messenger RNA. To understand translation, you'll need to know something about what's called the genetic code.

Cracking the Code

You learned that there are 20 amino acids that combine in different ways to make proteins. During translation, the As, Cs, Gs, and Us (remember, RNA doesn't use T like DNA) are read as three-letter "words" called *codons*. In a way, RNA and DNA speak the same language of a few letters. These three-letter codons each represent different amino acids. This is known as the genetic code.

Genetic Jargon

A **codon** (COE-don) is a three-letter "word" that represents a specific amino acid. The way these codons signify different amino acids is known as the genetic code.

For instance, the three-letter "word" or codon CAA represents the amino acid gluta-mine. The codon UAC represents tyrosine. Each of the 20 amino acids has one or more codons that tells the ribosome to use it in assembling a specific protein.

There are 64 codons, so you might have figured out that sometimes two or more of them represent the same amino acid. These codons are like synonyms, in that they code for the same amino acid. For instance, AAU and AAC both represent the amino acid asparagine, and both CAU and CAC represent the amino acid histidine.

DNA Data

We know about these codon "synonyms" because of two scientists named Marshall Nirenberg and Johann Matthaei. In 1966, they started to crack the genetic code and found out that codons represent amino acids. By 1965, the three-letter codes for each of the 20 amino acids were determined. This led to Nobel prizes for both scientists.

The 20 amino acids are (in alphabetical order): alanine, arginine, asparagine, aspartic acid, cysteine, glutamic acid, glutamine, glycine, histidine, isoleucine, leucine, lysine, methion-ine, phenylalanine, proline, serine, threonine, tryptophan, tyrosine, and valine.

Cell Mate

After the messenger RNA has delivered its instructions for making a specific protein, sometimes it will continue working by going off to another ribosome and repeating its job. Other times, it will just break down into the As, Cs, Gs, and Us that it's made from. These nucle-otides can then be recycled for other tasks in the cell.

The Final Product

By following the genetic code, the ribosome can manu-facture the protein. As the ribosome starts to assemble the protein, it reads the first codon and obtains that amino acid. Then it continues down the line, and as each new codon is read, a new amino acid is attached to the chain that is forming. (Keep in mind that this is a basic explanation for a very complex process.) When the ribosome reads the codon that tells it to stop, its job is finished. The resulting chain of amino acids constitutes a protein.

After the ribosomes link all the amino acids that form a protein, the protein folds into its characteristic shape. As we mentioned earlier in this chapter, the shape of the protein is very important, because it dictates the way the protein will be able to interact with other substances.

Finishing Touches

Although some proteins are complete and fold properly after they've been manufactured in the ribosome, others require more work. They're like unfinished furniture that you need to apply paint and varnish to after you bring it home.

Just Add Sugar

One example of a protein that needs a little extra finishing work is insulin, which helps regulate blood sugar. After insulin is manufactured in the cell's protein factory, it needs to have a tiny piece snipped out of it. After this is done, it has two pieces left, and these two pieces of the chain get spliced together. Then the insulin is ready to go.

Some proteins need other types of finishing touches before they're able to do their jobs. *Interferon* is a protein that helps improve the body's natural defenses against disease. It can fight dangerous invaders such as viruses and can also stop the growth of tumors. Interferon and some other proteins won't work properly unless they have certain sugars added to their amino acid chains.

Genetic Jargon

Interferon (pronounced in-ter-FEAR-on) is a protein that helps improve the body's natural resistance to disease.

Ready for Export

Yet another type of finishing touch needs to be applied to some other proteins. Although most proteins are manufactured to work within the cell itself, some need to leave the cell to help out in other parts of the body. Such proteins need to be modified before they depart.

Imagine that you have a long chain that you made in your basement workshop. You made it for one of your friends in the next town, who needs to use it to fence in an area. You wouldn't just put her address on the chain and put it in the mailbox. You'd wrap it up in a sturdy package before sending it out. The same holds true for protein chains that are going to be sent away from the cell where they were made. They need to be packaged. Getting the protein ready for travel is the job of the Golgi apparatus, which you read about in Chapter 4, "Me, My Cell, and I".

These get the wandering proteins wrapped up and ready for export. The Golgi apparatus is alerted when it sees a protein with a special group of amino acids on its end. These amino acids, or *signal sequences*, are something like a zip code. It's a specific group of about 20 to 24 amino acids that tells the cell that this particular protein needs to get packaged for export.

If everything goes as planned in the protein manufacturing process, the protein is up and running. However, if there are any mistakes in the gene or the messenger RNA copy, this will affect the organism, sometimes in a harmful way. You'll read in the next chapter about what can happen when there are copying errors that affect the manufacture of the proteins that are needed for specific jobs in the body or even for the survival of the organism.

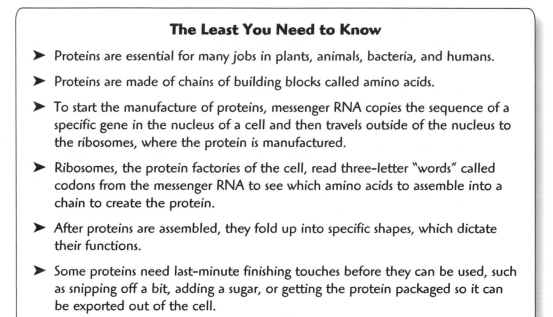

The Least You Need to Know

➤ Proteins are essential for many jobs in plants, animals, bacteria, and humans.

➤ Proteins are made of chains of building blocks called amino acids.

➤ To start the manufacture of proteins, messenger RNA copies the sequence of a specific gene in the nucleus of a cell and then travels outside of the nucleus to the ribosomes, where the protein is manufactured.

➤ Ribosomes, the protein factories of the cell, read three-letter "words" called codons from the messenger RNA to see which amino acids to assemble into a chain to create the protein.

➤ After proteins are assembled, they fold up into specific shapes, which dictate their functions.

➤ Some proteins need last-minute finishing touches before they can be used, such as snipping off a bit, adding a sugar, or getting the protein packaged so it can be exported out of the cell.

Part 3
Faulty Blueprints: Mutations and Genetic Diseases

When your cells multiply, all your genes get copied along with them. Every so often, your inner copy machine makes a mistake. Usually, it's no big deal, but sometimes these copy errors can result in diseases.

This section tells you how and why mistakes can be made in your DNA, and exactly what these mistakes mean to you. You'll learn about some genetic diseases, but you'll also see that sometimes little mistakes can be good for the future of the human race.

Chapter 9

Irregular Genes

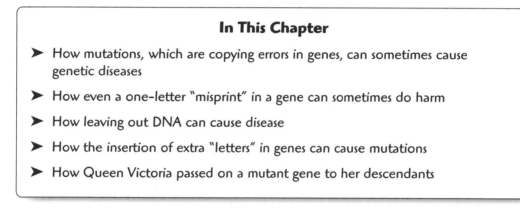

In This Chapter

➤ How mutations, which are copying errors in genes, can sometimes cause genetic diseases

➤ How even a one-letter "misprint" in a gene can sometimes do harm

➤ How leaving out DNA can cause disease

➤ How the insertion of extra "letters" in genes can cause mutations

➤ How Queen Victoria passed on a mutant gene to her descendants

Now you know how the DNA molecule is structured and how it makes copies of itself. You also learned that stretches of DNA are called genes and that each gene codes for a different protein. We discussed that there are thousands of proteins in the human body (and in animals and plants) and that each protein is needed for a different job that ensures that the organism functions correctly. So now you can see why genes have to work properly for you to be in good health.

You learned in Chapter 6, "Assembling Your Genes," that copying errors can occasionally occur in DNA. In this chapter, you'll read about the different kinds of errors that can occur. If a gene is copied incorrectly, it may or may not code for the correct protein. In some cases, the absence of the correct protein will hardly be noticed, but in others, the consequences can be serious or even lethal.

What Are Mutations?

You learned in Chapter 6 that mutations are basically bad copies of DNA. If a mutation occurs in a person's body cells, they will affect only that person and not his or her offspring. However, if a mutation occurs in sperm or eggs, then the mutation will be passed on to the next generation.

When DNA replicates itself, it literally splits at the seams. The twisted "rope ladder" structure untwists and splits down the middle like a molecular zipper. When this happens, the rungs of the ladder (the nucleotide pairs of A and T or C and G) open up, and each nucleotide gets a new partner. As before, A always pairs with T, and C always pairs with G.

But as we've seen, sometimes the copying mechanisms don't always get things straight. That's because our cells (and those of other higher organisms) have a large number of base pairs that must be copied. Also, there are a large number of cells in higher organisms, and all those cells have DNA that needs to be duplicated every so often. So there's a high chance of making a mistake.

The DNA machinery has some built-in repair mechanisms that edit out the mistakes, but occasionally, an error slips through anyway. How many times have you picked up a book and noticed several misprints? The same thing happens when the genetic alphabet of As, Ts, Cs, and Gs spells out "words" and "sentences."

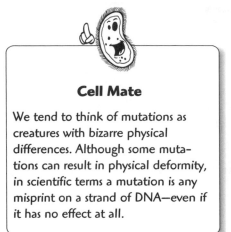

Cell Mate

We tend to think of mutations as creatures with bizarre physical differences. Although some mutations can result in physical deformity, in scientific terms a mutation is any misprint on a strand of DNA—even if it has no effect at all.

Ruined Recipes

You read in Chapter 8, "Producing Proteins," how the genetic code tells the ribosomes (the cell's protein factories) the recipe for making specific proteins. These recipes are spelled out in three-letter "words," or codons, which consist of combinations of As, Ts, Cs, and Us. You know that there are only 20 amino acids, but sometimes more than one codon can tell the ribosomes to use the same amino acid in the recipe. These codons are like synonyms, because they're different, yet they have the same meaning.

Now that you understand more about how proteins are made, just imagine how getting the recipe wrong might affect the making of the final protein molecules. Suppose a certain gene somehow has miscopied the message for the ribosomes, which have to put together a combination of amino acids. One of the codons for a particular amino acid in this protein should have been CUU, which is the codon for the amino acid leucine (LOU-seen). (If they ever make a sitcom about amino acids, "I Love Leucine" would make a great title, don't you think?)

Due to a copying error, however, the ribosome got the message CUC instead of CUU. As it turns out, this error will not affect the production of the intended protein, because CUU, CUC, and even CUA all tell the ribosomes to assemble the amino acid leucine in the amino acid chain. CUU, CUC, and CUA are "synonyms" that carry the same genetic code.

Other amino acids also have several codons that represent them. For instance, serine (SEHR-een) can be produced when the ribosome reads UCU, UCC, or UCA. Even the "stop" sequences that tell the ribosomes when to quit assembling amino acids into proteins can have different codons to represent them. UAA, UAG, and UGA are all molecular stop signs.

So very often, a "misspelling" in the third letter of a codon will not produce harmful consequences. Mutations like these are called *silent mutations*. But in other cases, misspellings can cause problems. If the first and second letters of the codons are scrambled or misspelled, often the ribosomes substitute a completely different amino acid for the one that was intended. This substitution can result in good, bad, or indifferent consequences.

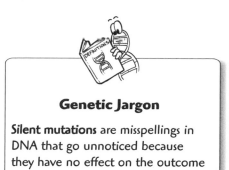

Genetic Jargon

Silent mutations are misspellings in DNA that go unnoticed because they have no effect on the outcome of a protein.

The Copied Cookbook

Imagine that you have a cookbook with the recipes for the world's best chocolate cakes. You read the first recipe, which is for chocolate mousse cake. It tells you to use two cups of flour, one cup of sugar, two-thirds of a cup of cocoa powder, two eggs, and other ingredients. You put these together according to the instructions in the recipe, and after 40 minutes in the oven, you have a delicious chocolate mousse cake that tastes the way the chef who wrote the book intended.

Now imagine that a friend of yours wants to make the same cake. You photocopy the recipe and send it to her. Then another friend of hers requests the same recipe. Your friend photocopies the photocopy and sends it off to her friend. Friend after friend requests subsequent copies.

Finally, a friend decides to write out the recipe for her sister. The photocopying has gone on for quite a few generations of copies, so the copy this woman holds is difficult to read. Therefore, she makes a few mistakes when copying the recipe. Instead of two cups of flour, she writes four cups of flour. Instead of two-thirds of a cup of cocoa powder, she incorrectly lists 23 cups of cocoa powder. Worse yet, instead of a cup of sugar, she writes a cup of salt. Finally, she writes that the cake needs to bake in the oven for 20 minutes instead of 40. The resulting cake would be a far cry from the moist, delectable one that the original cook had envisioned. The same can be true of badly copied recipes in genes for the manufacture of proteins.

Cell Mate

There are over 4,000 known diseases of genetic origin. Despite this large number, however, the occurrence of these diseases is rare, partly because many sufferers do not survive long enough to bear children and pass the defective genes along to the next generation.

There can be small genetic errors or large ones. There can be mistakes that occur from a misspelling of just one of the nucleotides—a mistaken A, T, C, or G—or there can be extra DNA stuck in the middle of a gene. There can also be repeats of meaningless letters that cause the gene to malfunction. Other mutations have to do with problems on the chromosomes that house the genes.

Some genes are more likely to mutate than others. For instance, large genes that have a lot of noncoding DNA inside them are more likely to have copying errors than smaller genes. And the meaningless repeats that are part of the noncoding DNA mutate a lot. Scientists have also learned that most diseases of genetic origin are caused by the mutation of more than one gene.

Genetic Jargon

Point mutations are disease–causing mutations that occur when just one nucleotide (an A, T, C, or G) is copied incorrectly in a gene.

Genetic Jargon

Anemia refers to a reduction in the number of red blood cells.

Sickle–cell anemia occurs when some of the red blood cells take on a long, thin, sickle or crescent shape instead of their normal round shape. These cells clog the blood vessels and damage the internal organs.

One Letter Off

The simplest form of mutation that can cause genetic diseases is called a *point mutation*, which is a one-letter "misspelling" in a gene. One disease that results from a point mutation is sickle-cell anemia. This disease affects the shape of the red blood cells, which are responsible for bringing oxygen to the cells in the body, among other tasks. Normally, red blood cells contain hemoglobin (HE-mo-globe-in), which is a mixture of a protein and iron.

Normally, red blood cells look something like bright red English muffins or red jelly donuts. They're round and slightly indented in the middle. When someone suffers from *sickle cell anemia*, however, some of their normal, round red blood cells take on a long, thin crescent or sickle-shaped form.

These mutant cells are easily broken, and this breakage leads to *anemia*, or a reduction in the number of red blood cells. Because of the odd shape of the cells, they can clog and damage blood vessels and can even cause harm to organs such as the lungs. This clogging can be very painful to people who suffer from the disease, who are also prone to infections. Brain damage and heart failure can also result.

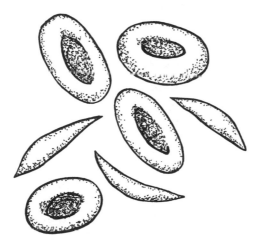

*Healthy red blood cells
and sickled cells.*

Dominant and Recessive

To get sickle-cell anemia, a child must receive two
copies of the mutant gene: one from the mother
and one from the father. Sometimes parents each
have one copy of the mutant gene, but they are not
aware of this problem because the disease only
manifests when two copies of the mutation are
present. In other words, this disease is a recessive
trait, so it occurs only when two copies of a mutation for it get together, like the shortness trait in
Mendel's pea plants (see Chapter 2).

Another type of anemia is called *thalassemia* or
Cooley's anemia. In this genetic disease, the red
blood cells are smaller than usual. Like sickle-cell
anemia, Cooley's anemia is recessively inherited,
meaning that persons who inherit this disease must
receive one mutated copy of the gene from the
mother and one from the father. If they carry just
one mutated gene, they will not develop the
disease. However, they may later pass the disease on
to their children, if they marry someone who also
gives the child a copy of the mutated gene. Other
recessively inherited diseases include Tay-Sachs
disease and cystic fibrosis.

Other genetic diseases are dominantly inherited,
which means that only one copy of the mutant
gene is necessary for the disease to occur in a child.
An example of a dominantly inherited genetic
disease is Huntington's disease.

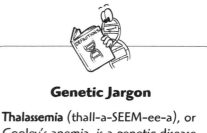

Genetic Jargon

Thalassemia (thall-a-SEEM-ee-a), or
Cooley's anemia, is a genetic disease
characterized by abnormally small
red blood cells.

Cell Mate

When a disease is recessively inherited, a child needs to inherit two
copies of the gene, one from each
parent, to develop the disease. If a
disease is dominantly inherited, only
one copy of the mutant gene is
needed to cause the disease.

105

"Bad Air"

Although both sickle-cell anemia and thalassemia are severe and debilitating diseases, the mutations that cause them contain a blessing in disguise. The deformed blood cells make it harder for a person with one or two copies of the gene to succumb to malaria.

Genetic Jargon

A **parasite** is any organism that lives off another organism, but does not contribute anything good to the host. In some cases, it makes the host sick or eventually destroys it. The word itself comes from the Greek words meaning "one who eats at the table of another."

A **carrier** is a person who carries only one copy of a mutant gene for a recessively inherited genetic disease. These diseases develop only if a person has two copies of the mutant gene. A carrier does not develop the disease, but may pass the gene for the disease to the next generation.

The name malaria comes from the Italian words meaning "bad air," because it was once thought that people contracted the disease from bad air emanated by swamps. Although this disease is common in tropical areas, especially near rivers and swamps, it's not the swamps themselves that cause malaria. Malaria is a disease caused by a microscopic *parasite*, which is an organism that lives off its host, something like the friend-of-a-friend or distant cousin who lives on your sofa for a month, eats your food, costs you money, and just won't leave.

The parasite lives inside a certain type of mosquito. When these mosquitoes bite a human, the parasite enters that person's bloodstream. The parasite eventually gets into the person's red blood cells and takes their iron for themselves. The tiny freeloaders multiply quickly while living off their unfortunate host. Fever and chills accompany this ailment, and if the parasites get into a person's brain, the disease will be fatal.

When the malaria parasite finds its way into the bloodstream of a person with sickle-cell anemia or thalassemia, however, it meets with some resistance because the abnormal shape of the cells makes it difficult for the parasite to grow. The same is true if the parasite gets into the bloodstream of a *carrier* of either of these two genetic diseases.

Dangerous Deletions

Some mutations occur when certain parts of a gene are deleted. As you read in Chapter 7, "S, M, L, and XL Genes," the mutated gene that causes Duchenne muscular dystrophy is the longest one that's been identified. It's two million base pairs long. That is, it has two million pairs of As with Ts and Cs with Gs. There are also 50 introns in this gene, which as you learned are something like a gene's version of commercial breaks on TV. They're interruptions of noncoding DNA within the coding portions of the gene. It turns out that of the two million base pairs, only 15,000 code for something, and the rest are noncoding. This is quite rare. Most genes don't contain that many separate coding parts.

Unfinished Proteins

If one or more of the coding parts of a gene is deleted, as sometimes happens with a bad copy of a good gene, then the amino acids that the missing parts would have coded for will be missing as well. Thus, the message to ribosomes in the cells is incorrect in a big way. If the amino acids are missing, then the finished protein won't work properly or possibly won't work at all. This would be like making the chocolate mousse cake we mentioned earlier without the flour, cocoa powder, and eggs.

The parts that are deleted from the mutated gene that causes Duchenne muscular dystrophy can vary from person to person. In some, the first coding area of the gene is missing, but in others, the third or fourth area is missing. Regardless of which coding part is gone, any of these deletions will lead to a serious problem.

DNA Data

Muscular dystrophy is not one disease. It's a group of several related diseases characterized by a gradual deterioration of the muscles. Duchenne muscular dystrophy is the most severe of these illnesses; it is also the most common of this group. Advanced cases of Duchenne muscular dystrophy can include problems with breathing as well as problems with the heart.

Cystic Fibrosis

Another genetic disease that arises from a deletion in a gene is *cystic fibrosis*. This mutation results in large amounts of mucus accumulating in the lungs of its victims. In 1989, a large group of scientists led by Francis Collins and Lap-chee Tsui discovered the gene responsible for this disorder.

It turns out that hundreds of different types of mutations can cause cystic fibrosis, but 70 percent of the people who suffer from it have the same mistake in their DNA: a small deletion in the gene. Just three letters are missing.

Genetic Jargon

Cystic fibrosis (SIS-tick fy-BROE-sis) is a genetic disease characterized by the buildup of large amounts of mucus in the lungs of the sufferer. Patients with this disease rarely live beyond their 20s or 30s.

Stuck in the Middle

Another way that mutations cause genetic diseases is through the addition of extra As, Cs, Ts, and Gs that were not meant to be in the gene. This addition can occur in a number of ways. Some viruses carry letters from one gene to another. Jumping genes also can travel from one part of a chromosome to another or even to an entirely different chromosome.

The Elephant Man

David Lynch's movie, *The Elephant Man,* was based on the true story of Joseph Merrick, whose skull was so deformed and whose nose was so distorted that people said it resembled an elephant's trunk. He may have suffered from a disease called *neurofibromatosis* (NEW-ro-FY-broe-ma-TOE-sis). In some cases, this disease is caused by the insertion of some noncoding repetitions of DNA sequences. All these meaningless letters disrupt the function of a gene.

Woody Guthrie's Tragedy

The disease that caused the death of folk singer Woody Guthrie in 1967 is called Huntington's disease. It causes a degeneration of the nervous system. Sufferers usually do not develop the disease until they are in their 30s or 40s. In 1993, the mutation that causes this disease was found. This mutation stems from inserted DNA, which disrupts the function of the normal gene.

There are copies of the nucleotide sequence CAG repeated over and over again. Although this sequence occurs in a normal person's genes 10, 20, or 30 times, Huntington's patients have the CAG sequence repeated from 40 to 100 times, and the number of repeats increases with each generation. This increase means that the son or daughter who inherits this disease will have more CAG repeats than their father or mother who carried the gene. As we mentioned earlier, Huntington's disease is dominantly inherited, so it will occur in a child even if only one parent carried the mutated gene.

Stuttering Genes

Genetic stuttering, or repeated sequences, is also the cause of a disease called *fragile-X syndrome.* This is the second leading cause of mental retardation after Down syndrome. Fragile-X syndrome is the result of thousands of genetic stuttering repeats, causing an area near the tip of the X chromosome to break off at a weak point.

DNA Data

Fragile-X syndrome is the second most common genetic disease that results in mental retardation. It is characterized by a specific facial appearance that includes a jutting jaw and large ears. The most commonly occurring form of genetic disease that leads to mental retardation is Down syndrome.

This disease occurs more often in boys than in girls because boys have one X chromosome and one Y chromosome and girls have two X chromosomes. If one of these X chromosomes is defective in a girl, she can still rely on her backup copy, if that one is normal. But boys have only one X chromosome, and if it is defective, they will develop a severe case of the disease.

Trouble on the Chromosome

As you remember, genes are stretches of DNA, and DNA is contained in chromosomes. In some cases, mutations occur because of problems with an entire chromosome instead of just one or several genes. Usually, problems in an entire chromosome are extremely serious. Often an embryo with this kind of mutation will stop developing and spontaneously abort.

The most common mutation of this kind is the one that causes Down syndrome. This condition is caused by the presence by an extra copy of chromosome 21. Chromosomes are numbered by size. The largest is chromosome 1, and the rest are numbered in descending size order. Although chromosome 21 is relatively small, it contains thousands of genes.

Instead of receiving two copies of chromosome 21 (one from the father and one from the mother), people with Down syndrome accidentally receive three copies. This means that people with Down syndrome have 47 chromosomes in all. Normally, people inherit two sets of 23 chromosomes (one from the father and the mother) for a total of 46 chromosomes. The most obvious result of this mutation is mental retardation, which can occur in many varieties, ranging from mild to severe. In addition to slowed mental development, heart problems, digestive disorders, and other problems can occur.

DNA Data

People with Down syndrome have a characteristic appearance. They usually have short necks and small noses, but large lips and tongues. They often have malformed kidneys or hearts. They also have distinctive patterns on the palms of their hands and the soles of their feet.

Queen Victoria and the Mutant Gene

When you hear the word *Victorian*, you probably think of quaint British furniture and prim and proper ladies sipping tea while eating ever so dainty mini-sandwiches and cookies. The word comes from Queen Victoria, queen of Great Britain and Ireland, who reigned from 1837 until her death in 1901.

A Royal Mutation

In 1840, Queen Victoria married her cousin Prince Albert. Eventually, the couple became the proud parents of four sons and five daughters. Unfortunately, Queen Victoria was the first in the royal family to pass on to some of her children a mutation for a disease called hemophilia. Apparently the mutation started spontaneously within her egg cells, or she received a mutant copy of a gene when she was conceived.

There are several forms of this genetic disease, but the symptoms always include a failure of the blood to clot. This lack of clotting can cause a person to bleed profusely from the slightest cut and even to suffer internal bleeding from a minor accident. A person with this disease could even bleed to death.

Hemophilia is caused by a mutation on a gene on the X chromosome, which is the female sex chromosome, and the hemophilia gene is recessive. As you know, females have two X chromosomes, but males have one X chromosome and one Y chromosome. When a male child inherits an X chromosome with this defective gene, he will definitely get the disease. If a female child inherits one X chromosome with the mutation and one normal X chromosome, she will not develop the disorder, but she will be a carrier of the disease.

Each of Queen Victoria's daughters had a 50/50 chance of being a carrier of hemophilia. The queen's sons each had a 50/50 chance of getting the disease. Queen Victoria's son Leopold had the disorder, and it is assumed that her daughters Alice and Beatrice were carriers.

The Russian Connection

Queen Victoria's granddaughter Alexandra was married to the czar of Russia, Nicholas II, in 1894. In 1904, their son, czarevitch Alexis, was born with the cruel disease. The ruthless monk, Rasputin, exerted a strong influence over Alexandra by taking advantage of his alleged ability to heal the young Alexis of his incurable disease.

Through the marriages of some of Queen Victoria's many children and grandchildren, hemophilia was passed to other royal families in Europe, including those of Spain and Prussia. The members of today's British royal family are all descendants of King Edward VII, Queen Victoria's first son, who was not a hemophiliac. Because of this, the disease no longer plagues Britain's monarchs, but the mutation still exists among some of Queen Victoria's many distant descendants.

The Least You Need to Know

➤ Sometimes copying errors in genes can cause genetic diseases.

➤ Some genetic diseases are caused by one-letter "misspellings" of nucleotide sequences called point mutations.

➤ Some genetic diseases, such as muscular dystrophy, are the result of deletions of whole sections of genes.

➤ Other genetic diseases, such as Huntington's disease, are caused by extra DNA in the genes.

➤ Some genetic diseases involve an entire chromosome, such as Down syndrome.

➤ Queen Victoria of Great Britain passed on to many of her descendants a mutation for hemophilia, a genetic disease that is characterized by excessive bleeding.

Chapter 10

The X in Sex

In This Chapter

➤ What little girls and little boys are really made of

➤ Mutations on the X and Y chromosomes that lead to unusual variations in gender

➤ Genetic diseases that differ according to the sex of the parent that passed on the mutant gene

---

Is there a gene that compels men to drive around in circles for hours without asking for directions? Which chromosome is responsible for the tragedy of women who shop too much? Does one stretch of DNA code for little boys to jump up and down on couches, and another for men to sit on the same furniture, watching countless reruns of lurid kung fu movies? Is there a gene that enables women to talk on the phone for 90 minutes at a stretch while simultaneously applying deep purple polish to their nails? What mysteries of the two genders lie hidden in the genes? And are genes or the way that people are raised responsible for the degree of masculinity or femininity that we see in children and adults?

As you learned in the last chapter, incorrect copies of DNA sometimes cause genes to code for the wrong proteins. These mistakes can have mild consequences, serious consequences, or no consequences at all. So what happens when there are misprints in the genes that code for sex?

In this chapter, you'll learn why you're a male or a female, the genetic differences between men and women, and also what happens in the rare cases that there are gender misprints in the DNA. You'll understand how the X and the Y chromosomes play a large part in determining whether a fertilized human egg will turn into a male or a female.

What Is Sex?

We're all familiar with the nursery rhyme that gives us a somewhat unscientific view of how nature differentiates between males and females:

> *What are little girls made of?*
> *Sugar and spice and everything nice*
> *That's what little girls are made of.*
> *What are little boys made of?*
> *Frogs and snails and puppy dogs' tails*
> *That's what little boys are made of.*

Well, at least the nursery rhyme got one ingredient right. Girls *are* made of sugar. One of the components of DNA is deoxyribose (de-OX-ee-RYE-bose), a type of sugar. But boys are made of it, too. So if the nursery rhyme doesn't have all the answers, then what makes humans come in two different gender varieties?

Sex as a Matter of Convenience

Some life forms don't even have sexes. Just check out some microscopic organisms, such as the lowly amoeba (pronounced a-MEE-ba). It basically does nothing but eat and grow. Then it gets bigger and bigger until one day, it splits into two. So it's both mother and father to itself and another, new amoeba. The tiny organism doesn't have to frequent singles bars or flirt with co-workers so it can eventually become a parent.

Even some life forms that have two sexes are a little bit more flexible about when and why they should be males or females. Take the snail, for instance. If there's a dearth of females, a male snail can turn into a female and mate with the love-starved males that abound. If the ratio of males to females changes, it can always turn back into a male.

DNA Data

Temperature determines sex after fertilization in most turtles and all crocodiles. Most turtle eggs that are in a cold environment will produce males, but the same type of egg will produce females when the temperature sizzles. In contrast, with Mississippi alligator eggs, cold produces females and heat produces males.

Imagine if humans had this gender-switching capability. A woman would go into a singles bar and, noticing too many other women around, would suddenly grow taller and sprout a mustache and other male attachments. She would turn into a man to take advantage of the excess number of ladies in the bar that night.

In humans, however, gender is usually a very basic affair. There are males, and there are females. However, if you examine human DNA, you'll realize that sex is a pretty complex business, even though it all starts out in a comparatively simple and straightforward way.

Xs and Ys

As you learned, humans have 46 chromosomes: two sets of 23 chromosomes each. One set was inherited from the mother and the other set was inherited from the father. Most of these chromosomes are alike, except for those called X and Y. The Y chromosome is much smaller than the X. If a child-to-be inherits one X chromosome from the mother and another X chromosome from the father, it will be a girl.

On the other hand, if a child inherits an X chromosome from the mother and a Y chromosome from the father, it will develop into a boy. Obviously, the mother can only contribute an X because this is the only type of sex chromosome that she has, being a woman. But about half of the father's sperm carries an X chromosome, and the other half contains a Y chromosome, so the father can contribute either. There's about a 50/50 chance for either a boy or a girl to develop.

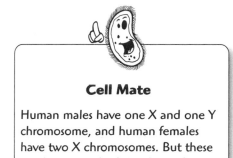

Cell Mate

Human males have one X and one Y chromosome, and human females have two X chromosomes. But these combinations don't hold true for all species. In butterflies and birds, for example, the combinations are the exact opposite of those in humans. Male birds and butterflies have two Xs, and the females have an X and a Y.

Unisex Cells

As it turns out, the sex chromosomes are not the direct cause of the gender of the emerging child. But the chromosomes do carry genes that will set in motion a cascade of events that involves switching other genes on or off, and hormones, which are chemical messengers, are sent to affect the growth of sex organs and help spark other developments.

Genes and Gender

In one sense, every human was at one time destined to become a female. But some of us carried genetic instructions to switch over to another type of development and became males instead. How does this happen?

DNA Data

In 1906, two researchers independently discovered that males had an X and a Y chromosome, and women had two Xs. Interestingly enough, one of these researchers, Nettie Marie Stevens (1861–1912), was a woman, and the other, Edmund Beecher Wilson (1856–1939), was a man. But even though the difference in chromosome distribution was recognized, it wasn't until 1959 that scientists realized that the Y chromosome is absolutely necessary for men to develop.

Even though some fertilized human eggs receive a Y chromosome instead of a second X, development will still go on for about six weeks as if the developing embryo were a unisex individual. There are cells in the embryo that can become either male or female internal sex organs, and other cells have the potential to develop into male or female external organs. Everything at this stage remains genderless.

The process of gender development is something like a car driving down the highway. If the car stays on its current path, it will reach a destination at the end of that road. Genetically speaking, this destination is female gender. But if the car takes the exit, the final genetic destination will be male gender.

Somewhere around the seventh week of development, one gene on the male Y chromosome, the *SRY gene*, changes the embryo's destiny forever. This gene acts like a miniature switch that turns on a series of chemical and physical events to change the course of the embryo's development from potentially female to all male. When this gene functions, it leads to the development of *testicles* in the embryo, and this development in turn leads to more and more chemical changes that eventually bring about the development of the external male reproductive organs.

Genetic Jargon

The **SRY gene** plays a crucial role in the sex determination of human males. *SRY* stands for "sex-determining region Y."

Testicles are the male internal reproductive glands that are the source of sperm. The word *testicles* comes from the Latin word meaning witness. In a sense, they are a witness to a man's virility.

Macho Chemicals?

If a developing embryo has the Y chromosome and therefore the SRY gene, it is on its way to becoming a male. It must receive signals in the form of hormones at precise times, or problems can ensue. Parts of the embryo that would have turned into the ovaries of a female must be shut off and disappear at the proper time, so the embryo will form testicles instead.

You've probably heard of the hormone *testosterone* because it's been blamed for everything from causing a chemical predisposition for hogging the remote control on the TV to creating an overactive libido in males. While all of that might be debated, one thing is certain. Testosterone directs the development of male parts such as the external genitals and the prostate gland.

In the absence of the SRY gene, the testicles will not develop. As a result, the hormones that go along with this development will not be present, and the male external reproductive organs will not develop. In this case, the embryo goes to its default setting. In the same way that your computer will revert to using a typeface like Geneva and a one-inch margin all around the page unless you change its settings to something else, an embryo will develop into a female, which is its default setting, unless given instructions to do otherwise.

> **Genetic Jargon**
>
> **Testosterone** (tes-TOS-ter-own) is the male sex hormone, or chemical messenger, responsible for many sexual characteristics in men.

Testing for Gender

Sometimes, on very rare occasions, there are departures from the obvious choices of either XX or XY. Take the case of a capable athlete from Spain named Maria Jose Martinez Patino. In 1985, she was about to run the 60-meter hurdles in the World University Games in Kobe, Japan. The 24-year-old woman was the Spanish national record holder in the athletic event that year, and she looked forward to participating in the upcoming competition. But first, she had to pass a routine test.

Failing the Sex Test

In the early 1960s, sex tests were instituted for participants in the Olympics and some other athletic competitions. This testing was in response to rumors that some of the women athletes were actually men in female clothing, hoping to take advantage of their physical prowess to beat the female competitors. At first, women were told to undress and parade in front of a group of gynecologists. As you can imagine, many people found this procedure demeaning and sexist, so it was dropped.

Instead, a series of biological tests were used. The first that was instituted involved scraping some

> **Cell Mate**
>
> Ethical considerations often arise from seemingly simple situations. Testing for gender for sporting events probably seemed very straightforward by the people who decided to do it, but it led to very intense scientific debate about what is male and what is female.

117

cheek cells from the athletes and then staining these cells with a special type of dye. When the cells were viewed under the microscope, a dark spot would indicate that there were two X chromosomes in the cells. If no such dark stain appeared, it would mean that the cells contained an XY combination.

This sex test didn't seem especially complex, but sex is not always as simple as you would think. In about 1 out of 1,000 people, there is a deviation from the normal pattern. Maria Patino is one such person.

How and Y

Maria Patino has a very rare condition that resulted in her carrying an X and a Y chromosome instead of two Xs like most women. Maria became a woman instead of a man because of a mutation. When she was a developing embryo, the gene on the Y chromosome caused testicles to form, and then male hormones were produced. But Maria's genetic mutation made it impossible for her cells to respond to these male hormones.

As a result, the embryo went to the default setting and continued to develop into a female. Maria has the body of a fully developed woman, but she cannot bear children because she never developed ovaries. The testicles that had formed in an early embryonic stage withered away because of the mutation.

Genetic Consequences

Eventually, Maria's genetic story was picked up by the press in Spain, and this publicity led to heartbreaking events in her life. Her boyfriend broke up with her upon hearing the news, and many friends abandoned her. She was banned from competing in future sporting events, and her coach stopped training her. When she walked down the streets of her native country, people stared at her. It seemed that her life in sports was over, and her social life was ruined.

But Maria refused to accept this genetic discrimination. She went to see a geneticist from Finland named Albert de la Chapelle, who argued that Maria should be allowed to compete. The geneticist explained that women like Maria who failed that specific sex test do not have an unfair advantage over XX females in terms of muscle strength. After all, the test was meant to determine whether an athlete had an unfair advantage over the others, and this was not Maria's case.

Some women with two X chromosomes are unable to conceive. Other women might have more testosterone than most females. Both men and women have male and female hormones. It's actually the ratio of one to the other that is meaningful when it comes to sexual development.

In 1988, nearly three years after she was first banned from competition, Maria Patino won back the right to participate. In 1989, she set a new record in Spain for indoor hurdles, and in 1992, she qualified for the summer games in Barcelona. Finally, she was recognized as the woman that she is.

One mutation does not warrant genetic sex discrimination, and Maria Patino's story shows that in matters of gender, things are not always a case of either/or. There are also some blurred boundaries in the complex process of sex determination, and some unusual combinations do occur.

Superfemales and Other Variations

There are other changes that sometimes appear involving whole X and Y chromosomes. For instance, a female may inherit three X chromosomes instead of just two. These XXX females are sometimes referred to as *superfemales*. They're rarely able to bear children, although some of them can.

DNA Data

In some insects, tremendous gender problems arise when there's a mutation on the sex chromosomes. There are actually flies that are female on one side and male on the other side; there are even flies with a female head and a male body.

In rare instances, occurring about once in every 2,000 live births, a woman inherits just one X chromosome and no other sex chromosome, neither another X nor a Y. As a result, a female will develop, but her reproductive system will not mature the way it is supposed to. Women with this type of mutation are called XO females.

This XO mutation is referred to as *Turner's syndrome*. Women with this disorder are often short of stature. In addition to having reproductive problems, they often experience other complications such as kidney, heart, or thyroid problems. Fortunately, not every woman with this syndrome will exhibit all of the possible related health risks.

Genetic Jargon

Turner's syndrome is a disorder that occurs from a chromosomal mutation. A woman with this disorder inherits only one X chromosome instead of two. Women with this mutation are short of stature, infertile, and can experience problems with their hearts, kidneys, or thyroids.

Male Variations

Men are not exempt from mutations that affect gender. You've learned that in the majority of cases, a male will result from the presence of one X chromosome and one Y chromosome. This is because the Y chromosome carries the SRY gene, which is like a master switch that sets in motion the first step in a long series of events that creates a male.

However, there are rare instances when a man has two X chromosomes and no Y chromosomes. At first glance, this would seem impossible because the Y chromosome contains the only gene that will result in a male. But in the 1980s, when researchers were better able to understand the order of the As, Ts, Cs, and Gs in DNA, they discovered that on rare occasions, it was possible for the SRY gene from the Y chromosome to get into the X chromosome, making the developing embryo male.

> ### Cell Mate
>
> Some living things are unisex when they're born. There's a type of marine worm that starts out as a sexual neuter. Those that eventually set up a home for themselves on the bottom of the sea turn into females. Others turn into tiny males that attach themselves to, and never leave, the female's body.

Some gender variations.

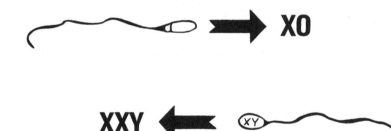

Two Xs and One Y

What would happen if a fertilized egg received two Xs and one Y? This XXY combination is called *Klinefelter's syndrome*, and it occurs in about 1 out of 1,000 live male births. The disorder gets its name from Dr. Harry Klinefelter (born 1912), who studied this condition in 1942.

There are variations of this disorder other than the XXY combination. These variations are even more rare than the XXY mutation, but they still fall under the category of Klinefelter's syndrome. Some men are born with two Xs and two Ys, for a total of 48 chromosomes. Others inherit three Xs and one Y, which also adds up to 48 chromosomes. Other mutations combine four Xs with one Y, giving each cell a total of 49 chromosomes instead of the normal 46.

The mutation that causes Klinefelter's syndrome results in the development of a male, but he is usually sterile. Most men who inherit this condition are tall, but they are not necessarily athletic. Sometimes they lack the usual amount of facial hair that occurs in men. It also seems that men with Klinefelter's syndrome are predisposed to speech problems. Men with the XXY mutation are often shy and lacking in confidence.

Another problem that can occur in cases of Klinefelter's syndrome is the possibility of overdeveloped breasts. Although this condition occasionally occurs in normal boys undergoing puberty, it usually disappears. In boys with Klinefelter's syndrome, however, the development may continue, and in some cases, patients have opted for surgical removal of this excess growth.

Genetic Jargon

Klinefelter's syndrome results from an inherited mutation that gives the offspring two Xs and one Y, or in rarer cases combinations of even greater numbers of Xs and Ys. Males who inherit this mutation suffer from sterility and sometimes enlarged breast tissue and speech problems.

Too Many Ys

There are also instances when a male has one X and two Y chromosomes. Some researchers have theorized that if one Y codes for being male, which often involves more aggressiveness than being female, then maybe two Ys could potentially cause a male to be ultra-aggressive.

In 1968, Dr. Stanley Walzer, a child psychologist, and Dr. Park Gerald, a pediatrician, started a genetic screening program for male infants born at the Boston Hospital for Women. If the infants had an extra Y chromosome, the researchers planned to study these male children as they grew up. As you might expect, though, telling the parents of a baby boy that he carries a chromosomal aberration that could lead him to a life of violence is problematic. There were obvious ethical overtones that did not sit well with some people.

In 1974, two geneticists, Dr. Jonathan Beckwith of Harvard University, and Dr. Jonathan King of M.I.T., wrote an article called, "The XYY Syndrome: A Dangerous Myth," for *New Scientist*, a British journal. In the article, they said that Walzer's XYY study—as well as other XYY studies in the past—was scientifically flawed. Beckwith and King contended that the study was simplistic because it tried to attribute complex human behaviors to one genetic cause, and it never took the potential influence of the environment into account.

After these protests, in 1975, the Boston Hospital study was stopped. Today, most scientists agree that the majority of men with an extra Y chromosome will lead normal lives and not exhibit overaggressive behavior.

Imprinting: A Lasting Impression

Gender can also play a part in how mutations affect the children of an individual who passes genetic misprints to them. Some diseases will develop if a certain mutation is

Genetic Jargon

Imprinting, or genomic imprinting, refers to how mutations in genes can result in one disease if inherited from the father but result in another disease if the same exact mutation is inherited from the mother.

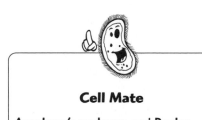

Cell Mate

Angelman's syndrome and **Prader-Willi syndrome** are two diseases caused by a mutation on chromosome 15. They both cause mental retardation. Angelman's is inherited from the victim's mother, and Prader-Willi is inherited from the victim's father.

inherited from the mother, and other diseases arise from the exact same mutation if it is inherited from the father. This phenomenon is called *imprinting*, or genomic imprinting. Imprinting is not well understood, but somehow, the gene bears the stamp or imprint of the gender of origin and causes a different expression of that genetic mutation in the child that inherits it.

For instance, it seems that children who inherit the gene for Huntington's disease, a severe disease that attacks the nervous system, will develop the disease on different time schedules depending on whether it came from the father or the mother. Children of men with the disease may begin to show the first symptoms of the genetic disorder earlier on than those who inherit the mutation from their mothers.

Another example of imprinting occurs with a mutation on chromosome 15. When a child inherits this defect from the mother, it will result in a genetic disease called Angelman's syndrome, which causes mental retardation and seizures and is characterized by a strange type of laughter.

However, if a child inherits this mutation from the father, it will result in another disease altogether, called Prader-Willi syndrome. This disease also causes mental retardation, but unlike a child with Angelman's syndrome, the child with Prader-Willi syndrome becomes extremely overweight and tends to be born with small hands and feet. These diseases are so unlike one another that for many years, scientists thought that they must be caused by different mutations, when in fact they are caused by the very same genetic problem on the chromosome.

The Greek philosopher Plato (427–347 B.C.) wrote in his Symposium that there were at one time three sexes: males, females, and androgynes (AN-dro-jines), which were supposed to be a mixture of the two. In a fit of anger, Zeus, the king of the gods, split these androgynes into two, and they now spend the rest of their lives looking for their other half. It's highly unlikely that there were ever three sexes, but as you've seen from this chapter, gender is not always as simple as you might have thought. In addition to the usual XX or XY, there are a number of genetic variations that blur the distinctions between male and female.

The Least You Need to Know

➤ Normally, women have two X chromosomes, and men have one X and one Y.

➤ Sometimes embryos have more than one X or more than one Y chromosome. These mutations affect gender and often lead to rare genetic disorders.

➤ A Spanish athlete named Maria Patino was banned from Olympic competitions in 1985 because she had one X and one Y chromosome, a combination which normally occurs only in men. She was later reinstated.

➤ Some women have Turner's syndrome, which results from a genetic mutation in which only one X chromosome is inherited.

➤ Some men have Klinefelter's syndrome, which is caused by a genetic mutation in which two X chromosomes and one Y chromosome are inherited.

➤ With some types of genetic mutations, the offspring develops one disease if the mutation is inherited from the father and a completely different disease if the mutation is inherited from the mother.

Chapter 11

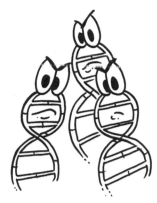

Gene Gangs

> ### In This Chapter
>
> ➤ Methods of mutation: a review
>
> ➤ What can happen when mutant genes get together
>
> ➤ Why your parents aren't to blame: genetic diseases that aren't inherited
>
> ➤ How cigarettes, fatty foods, and other environmental factors can lead to cancer
>
> ➤ What happens when cells divide incorrectly, but multiply anyway
>
> ➤ How yeast is a main ingredient in cancer research

As we've already discussed, diseases like cystic fibrosis, thalassemia, and sickle cell anemia result from the mutation of one gene. This causes a problem in the production of needed proteins. When an important protein is completely missing or is substituted with a defective protein, serious diseases can result. These one-gene disorders are called *monogenic* diseases.

But not all genetic problems work this way. In many cases, genetic diseases are the result from a combination of several different mutated genes. In order for such a disorder to occur, it often means that a mutation in one gene is complicated by a mutation in another one. These different mutated genes work together to bring about serious disorders.

These gene gangs give rise to what are called *polygenic* diseases, which are diseases caused by several mutated genes. In this chapter, you'll learn about these wayward gangs of genes and how they work together to cause a number of genetic illnesses.

Genetic Jargon

Monogenic (mahn-a-JEHN-ic) diseases are caused by the mutation of one gene. **Polygenic** (polly-JEHN-ic) diseases are caused by several mutated genes working together.

Cell Mate

Sometimes genetic disorders arise from different "misspellings" that cause the same physical problems. For instance, there are about 700 different mutations that can cause cystic fibrosis.

Genetic Jargon

Incomplete penetrance occurs when a mutated gene or genes does not affect a person, and the associated disease does not develop. Other genes or environmental factors may compensate for the mutated gene.

From Defects to Diseases

Scientists have learned that some genes seem to be more likely to mutate than others. As you read in Chapter 7, "S, M, L, and XL Genes," genes come in a wide range of sizes and styles. These differences might make some genes more susceptible to copying errors than others.

Different Causes, Same Disease

Mutations that cause diseases can vary. For instance, the mutation that causes sickle cell anemia is always the same one in the same gene, no matter where in the world the afflicted person comes from or where that person developed the disease. But other genetic disorders can arise from different mutations on different genes that cause the same symptoms and physical problems. For instance, there are 700 different mutations in the same cystic fibrosis gene that will cause the development of the disease to varying degrees. Different forms of hemophilia (the uncontrollable bleeding disease you read about in Chapter 9, "Irregular Genes") can be caused by mutations in different genes, and there are also mutations in genes that don't lead to problems.

To complicate matters, some genetic mutations can result in a serious disease in one person, but have no effect on another person. This phenomenon is known as *incomplete penetrance*—a complicated phrase which hides the fact that scientists still don't understand why this happens. Even when a genetic mutation causes the same disease in different people, it can manifest with different degrees of seriousness.

Mutations can result from a wide variety of completely unrelated factors. Causes can be external, such as too much exposure to X-rays or ultraviolet rays from the sun. Or mutations can be caused by certain chemicals. In some cases, mutations can even be caused by viruses that swipe genes from one organism and transplant them into another organism that they invade.

A great many mutations are also caused by miscopying of genes during cell division, and other mutations are the result of the failure of the cell's repair mechanisms to catch and correct these errors before they can cause damage. Mutations in the repair genes themselves are implicated in some types of cancer, such as ovarian and colon cancers.

A Relative Matter

As you've learned, some genetic disorders are dominantly or recessively inherited. Inheriting just one copy of a mutant gene will give a dominant genetic disorder to a child. Both parents must pass on the same defective gene for their child to inherit a recessive genetic disorder.

This may explain why marrying a close relative can cause genetic problems in the offspring. If a person carries a mutated gene for a recessive disorder, the chances are low that they will have children with someone who also carries the same rare mutated gene. If their child inherits only one copy of the recessive mutated gene, that child will not develop the disease.

But if that same person were to marry a close relative, such as a cousin, there is a higher chance that this mutated gene might run in the family. If this happens, their children can inherit two copies of the defective gene and develop the disease.

Born or Bred?

Are genetic diseases caused by genetic mutations alone, or by the environment, or by a combination of the two? Today, scientists think that in addition to the genetic components of disease, there are external factors. Mutant genes can be expressed in a variety of ways depending on environmental factors. Pollutants, for example, can increase a person's chances for developing diseases. Diet and lifestyle also seem to play a part in whether a genetic defect will develop into a full-blown disease—although the genetic and environmental effects can vary from gene to gene.

Genes alone do not always determine how, when, or if a disease will develop in a person. For instance, a predisposition for type l diabetes seems to run in some families, especially those with northern European roots. However, there are cases of identical twins—twins who come from the same egg and have exactly the same DNA—in which one twin developed the disease and the other did not.

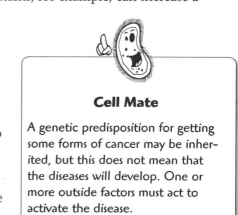

Cell Mate

A genetic predisposition for getting some forms of cancer may be inherited, but this does not mean that the diseases will develop. One or more outside factors must act to activate the disease.

Does Genetic Mean Inherited?

It's important to understand the difference between what is genetic and what is inherited. Any disease that involves the genes can be classified as genetic, but that doesn't necessarily mean that the disease was inherited from the victim's parents. Nor does it necessarily mean that the person will pass on that disease to the next generation.

Take the case of cancer. Cancer is not just one disease; dozens of diseases fall under this category. Cancer is a term that refers to many different types of tumors that can appear in different parts of the body and can grow in different ways. But all of the diseases that fall into this group, whether lung cancer or breast cancer or others, have one thing in common: they involve a problem with cell division.

Although normal cells only divide to reproduce themselves a finite number of times, cancer cells are clearly out of control. Something has gone wrong with the natural mechanisms that would normally regulate cell growth. This means that there are problems with certain genes switching on or off.

Genetic Jargon

Retinoblastoma (RET-in-o-blast-O-ma) is a rare childhood cancer of the eye.

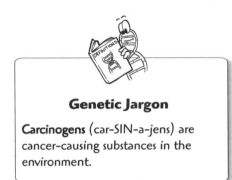

Genetic Jargon

Carcinogens (car-SIN-a-jens) are cancer-causing substances in the environment.

However, even though the functions of genes are involved in all these cases, very few cancers are inherited. For instance, only about five percent of all breast cancer incidences are considered to be inherited. In other words, most cancers are not passed from generation to generation. Those that are must be passed from parent to child through the sex cells, the eggs or the sperm.

Even when these mutations are inherited, external causes must prompt them to cause harm. An example is *retinoblastoma*, a rare childhood cancer of the eye. Even if a child inherits a mutated copy of the gene responsible for this disease, there is about a 10 percent chance that the disease will never develop.

It is generally agreed that more cancers are acquired than inherited, but some scientists feel that a genetic predisposition to certain cancers may be passed down from generation to generation. In other words, some people may inherit an increased risk to develop a disease. However, this risk does not necessarily mean that the person will get the illness. The person will develop the disease only in response to cancer-inducing occurrences, such as exposure to *carcinogens* in the environment that contribute to the abnormal growth of cells.

Genetically Dangerous Jobs

Many years ago, even before scientists had the knowledge they have today, some people couldn't help but note that certain types of cancer struck people under certain circumstances. In 1700, an Italian named Bernardino Ramazzini (1633–1714) noted that nuns often developed breast cancer. He concluded that the unmarried lifestyle of nuns coupled with the fact that they would never bear children might account for the high incidence of the disease.

In England, it was noted that certain professions seemed to carry a risk of health problems. Chimney sweeps, for instance, had a high incidence of scrotal cancer.

A British physician, Percivall Pott (1714–1788), concluded that the soot that got on the skin of chimney sweeps was the cause of their disease. It turns out that he was correct, because soot and tar are now recognized as containing carcinogenic chemicals. In Germany, miners were noted to be at high risk for developing lung cancer.

The Cell Cycle

As you've learned, many mutations are the result of genes being miscopied during cell division and the subsequent failure of the cell's repair mechanisms to fix these errors. By first understanding how cells divide under normal conditions, you'll be better equipped to understand what can go wrong in this process.

The life of a dividing cell is described by what scientists call the *cell cycle*. Although it may sound like a mode of transportation, the cell cycle is a kind of schedule for cell division. Each type of cell has its own timetable when it comes to dividing and reproducing itself. For example, a cheek cell in a person will divide about once every 24 hours, but most adult cells don't divide at all. Other cells, such as liver cells, divide only when they are needed to replace other liver cells that have been lost to disease or injury.

Fall into the Gap

The cell cycle consists of four phases, and special proteins regulate the passage of the cell from one stage of the cell cycle to the next. The first stage of the cell cycle is called G_1 (for Gap 1). In this phase, the cell begins to grow. During the S (synthesis) phase, the chromosomes double in preparation for the cell division that's about to take place. The next phase is the G_2 phase (for Gap 2). At this point, special proteins are made that are needed for the cell to divide.

Then the actual cell division, mitosis, occurs in the M phase. (You read about mitosis in detail in Chapter 4, "Me, My Cell, and I.") During this phase, the nucleus and the rest of the cell divide. The result is two new cells called *daughter cells*, which are exact copies of the original cell. There are special proteins that regulate the passage of the cell from one stage of division in the cell cycle to the next phase.

Genetic Jargon

The **cell cycle** refers to the four phases that a cell goes through when it divides to produce two new cells. These phases are the G_1 phase (for Gap 1), the S (for synthesis) phase, the G_2 phase (for Gap 2), and the M (for mitosis) phase.

Cell Mate

In the average adult, about 100,000 cells are dividing every second. An equal amount of cells are dying off and need to be replaced.

The cell cycle refers to the different phases a cell goes through as it divides into two.

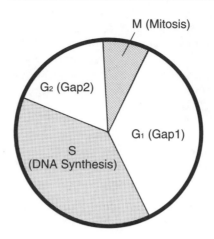

Out of Control

One reason why scientists became interested in cell division was to better understand the mechanisms behind cancer. Basically, cancer occurs when cell division gets out of control. Some cells in the body don't divide at all and these cannot form cancer cells. For instance, you don't usually hear about cancer of the muscle cells or the nerve cells, because these are among the cells that stay put and do not duplicate themselves.

Cells in the lining of the lungs and the milk ducts in women reproduce frequently. Cells that are in contact with the outside environment, such as skin cells and those that line the digestive tract, also continue to grow because in a sense they are "disposable."

If you cut yourself, for example, your skin cells divide until there are enough new ones to heal the cut. After these skin cells are formed, they need new blood vessels to supply them with nutrition. Genes produce proteins that induce these vessels to grow so the new skin cells can have a blood supply. Genes code for all of the proteins required to replace the missing cells and create a new blood supply, until the body gets the message that the mission is accomplished. At this point, the genes that are responsible for the new growth of skin cells and blood vessels turn off.

Scientists have determined that the more a type of cell divides, the more likely it is that a cancer will originate in these cells. One reason may be that there is a higher incidence of mutation when there are many cell divisions. Mathematically speaking, if a cell divides many times, the DNA has a greater chance to be copied incorrectly. Another reason is that during duplication, the DNA in cells is at its most vulnerable to cancer-causing agents.

Cell Mate

Epithelial (e-pi-THEE-lee-al) cells continue to grow and divide throughout a person's lifetime. Included in this class of cells are the skin cells that cover all of your body, and the linings of several internal organs such as the uterus, lungs, colon, and milk ducts in the breasts.

Studying Low Life

A certain type of yeast that brewers use to make beer has been useful to scientists as well, and not because they can always take a drink of beer if their experiments aren't working out right. Scientists study yeast cells to better understand how human cells divide correctly during a normal cell cycle and also to see what happens when these cells get out of control. Even though yeast is a single-celled organism like bacteria, it's a eukaryote—it has nuclei—and is much more similar to us.

What's Brewing in the Lab?

The reason scientists use yeast is because some of the proteins involved in its cell division are remarkably similar to those that play a part in human cell division. When nature finds a convenient, reliable mechanism, it tends to use it in many different species of organisms. If it works for one, then nature just lets the mechanism occur in other organisms.

Scientists prefer to use brewer's yeast because of its convenience and speed. Doing certain kinds of experiments on humans would, of course, be unethical, but there are no ethical problems involved in experimenting with yeast, or at least none of the yeast cells has complained in a way that scientists could understand.

Of Mice and Mutations

Scientists can learn a lot from yeast cells, even though yeast never gets cancer. Scientists experiment by causing mutations in the yeast cells, by exposing them to X-rays. (As you read in Chapter 3, "Mendel's Successors," X-rays can mutate genes.) By studying the mutations that can arise in these yeast cells, scientists can focus on the mutations that are associated with abnormalities in cell division.

For instance, some mutated yeast cells grow too fast or too slow. Scientists investigate which genes in yeast are responsible for these changes in the rate of cell division. After the scientists find the gene they're looking for in yeast, then they look for something similar in mice. And they look for mutations in the human equivalent of yeast genes, in tissues from human cancers.

Rather than experiment on humans, researchers often use laboratory mice to learn how genes can function or malfunction in the system of a mammal. It would be difficult to study the causes of, say, lung cancer in yeast, because so far, no one's been able to figure out a way to get yeast cells to smoke cigarettes.

Ganging Up

Several mutant genes can gang up on a person to cause a series of events that can lead to a serious illness. Such interaction among multiple defective genes results in disorders such as cleft palate, cleft lip, and spina bifida (SPINE-a BIFF-id-a), a disease of the spinal column.

These gene gangs can wreak havoc on a person in various ways. Some genes increase the normal effects of other genes, which sometimes leads to an abnormality. In other cases, a mutated gene might interfere with the normal function of another healthy gene or might completely suppress normal expression of a normal gene that the body requires for continued health. To better understand how groups of mutant genes interact to cause disease, let's look at how these genes cause cancer.

Cancer-Causing Genes

Scientists now feel that two types of genes are involved in the development of cancer: One type is called an *oncogene* and the other type is called a *tumor-suppressor gene*, also called an anti-oncogene. One way to explain how these genes function is to consider cells as being something like cars. If everything is working the way it should, the gas pedal responds to the driver, and the car runs in a controlled way at an appropriate speed. Now suppose the gas pedal has somehow gotten stuck to the floor, and the engine has responded by racing to its full capacity. An oncogene is like this faulty gas pedal. As a result, the car (the cell) goes completely out of control. It speeds crazily down the highway, and there is no stopping it while the gas pedal is stuck to the floor.

When the gene functions the way it should, it makes a protein that's needed for the cells to divide normally. But when an oncogene is activated by a mutation, this can lead the cell to divide uncontrollably.

The other types of genes that are implicated in the development of cancer are called tumor-suppressor genes. These can be compared to brakes on the car. When the brakes have been damaged severely (the tumor-suppressor genes have mutated), they no longer function to stop the car.

In this case, instead of losing control due to the broken gas pedal (the oncogene), the car races out of control because the mechanisms that would normally hold it back are malfunctioning. So even though the oncogenes and the tumor suppressor-genes work in different ways, when either of them malfunctions, the final effect is the same: The cells divide without stopping.

Genetic Jargon

An **oncogene** (AHN-ko-gene) is a gene that can cause cancer when it's mutated. The word *oncogene* comes from the Greek word *onkos*, which means mass or tumor. This is also where we get the word *oncology*, the study of tumors. A **tumor-suppressor gene**, or anti-oncogene, normally stops the growth of cells, but it can fail to do so when mutated, and can lead to the development of cancer.

The ras Race

In 1971, Nixon declared a war on cancer. Because of the increased federal funding, the race was on to find one of these cancer-causing mutations. The first oncogene to be isolated from a human cancer cell came from a tumor in the human bladder. In 1981, Robert Weinberg of the Whitehead Institute in Boston found this oncogene, which is called ras. The ras gene occurs not only in humans, but also in rats, mice, and even the lowly yeast.

The only difference between the normal ras gene and the mutated ras gene is a change in just one codon. You'll remember from Chapter 8, "Producing Proteins," that a codon is a group of three nucleotides in the messenger RNA that tells the ribosomes in the cells which amino acid to pick out and add to the protein chain that they are constructing from different amino acids.

The normal gene in humans has the codon GGC, which represents the amino acid glycine (GLY-seen). However, the ras mutation of the gene has the incorrect sequence GTC instead, and this sequence codes for a completely different amino acid called valine (VALL-een or VAY-leen). The amino acid glycine is not produced, and this mutation changes the normal gene into one that can cause a tumor to develop.

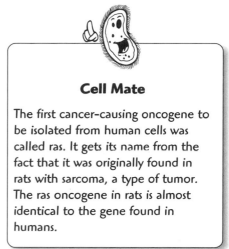

Cell Mate

The first cancer-causing oncogene to be isolated from human cells was called ras. It gets its name from the fact that it was originally found in rats with sarcoma, a type of tumor. The ras oncogene in rats is almost identical to the gene found in humans.

This seemingly minor mistake in the "wording" of the gene means a lot. This particular gene is important in the regulation of cell growth. A normal cell has a normal ras gene that helps it to grow whenever it needs to. A cell with the ras mutation grows without restraint.

A Double Hit

In 1971, a scientist named Alfred Knudson came up with the theory that retinoblastoma (RET-in-o-blast-O-ma), a rare cancer of the retina in the eye, is the result of two different mutational events in the two copies of the retinoblastoma gene.

Knudson already knew that there was an inherited form of the disease as well as a noninherited form. So he hypothesized that in the inherited form, a mutation is passed down from generation to generation, but that it takes another mutation caused by the environment to start the fatal process of unchecked cell division. In the noninherited form, Knudson theorized, because no mutation is passed down from one generation to the next, both the first and second mutations must come from environmental events.

This theory explains why some people have an inborn susceptibility for certain types of cancers but this doesn't mean that they will necessarily develop these rare diseases. There must be outside events which cause other mutations in order for the diseases to develop.

During the 1970s and 1980s, many scientists accepted Knudson's "two-hit" theory of the development of cancer. This theory holds that although cancers may have a genetic basis, they are not necessarily inheritable. Even when inherited, the vast majority of these diseases can still be attributed at least in part to environmental agents.

Four Hits

Sometimes there are more than two mutational "hits" involved in the formation of tumors in certain types of cancers. In the late 1980s, a scientist named Bert Vogelstein discovered that at least four different mutations account for the development of a certain type of colon cancer.

One oncogene may be activated without any problems in cell division. But if a second and then a third and eventually a fourth unfortunate event occur, the result is disastrous. These last three hits are to tumor-suppressor genes, which would normally check the unbridled growth of tumors. These separate events may take years to occur, but eventually, the combination of the four mutations adds up to one lethal combination.

Again, the environment plays a large part in determining whether or not these final mutations will take place in a person's genes. According to a report by the U.S. Office of Technology Assessment in the early 1980s, an estimated 80 percent of all deaths from cancers are attributed to cigarettes, diet, and other lifestyle and environmental factors. Other studies suggest that the risk for colon cancer could be cut in half by the reduction of animal fat in the diet.

So nature works together with nurture here. Epidemiologists, scientists who study diseases that are widespread among the population, have estimated that if people stopped smoking cigarettes, one-third of all cases of cancer in the United States would be avoided. This is because in addition to lung cancer, tobacco has been implicated in cancers of the esophagus, pancreas, kidney, and bladder.

The Least You Need to Know

➤ Genetic diseases are sometimes caused by more than one mutated gene.

➤ The cell cycle is like a timetable for normal cell division.

➤ Scientists are learning about cancer from studying the cell cycle in yeast.

➤ Most cancers have environmental causes and are not passed from generation to generation.

➤ Oncogenes are mutated genes that can cause cancer, and mutated tumor-suppressor genes are unable to stop the development of the disease.

Chapter 12

Mutations and Evolution

In This Chapter

➤ Charles Darwin's theories of evolution and natural selection

➤ Theories about the origin of life on earth

➤ Using mutations to tell evolutionary time

➤ Allan Wilson's theory about Mitochondrial Eve, the theoretical ancestress of all humans

➤ Are we related to Neanderthals?

➤ A British man's reunion with a 9,000-year-old relative

You've learned what mutations are and what can cause them. You've also read about some of the not-so-nice things that mutations can do, such as causing abnormalities and diseases. But mutations aren't always genetic villains. We wouldn't have all the wonderful diversity of life on earth if it weren't for an occasional misspelling of the DNA alphabet.

In this chapter, you'll learn about Charles Darwin, his theory of evolution, and the concept of survival of the fittest. (This does not refer to any natural superiority of a person who goes to the gym every day.) It took a long time for humans to evolve from our long-lost ancestors. In this chapter, you'll also learn how new developments in the study of DNA have given scientists more tools with which to unlock the secrets of our genetic past.

A Natural Selection of Genes

In 1831, a British naturalist named Charles Darwin (1809–1882) embarked upon a five-year voyage as the resident naturalist on a ship called the *H.M.S. Beagle*. While on this voyage, he collected samples of thousands of plants and animals from around the world.

One of his most famous stops on this five-year journey was to the Galápagos (Guh-LOP-a-gose) Islands, which lie in the Pacific Ocean about 600 miles west of Ecuador. On these islands, Darwin noticed an astounding array of unusual animals, such as the marine iguana, the only type of iguana in the world that can regularly take deep-sea dives down to depths of 35 feet. He also saw giant tortoises that could live for a century. But one of the things that impressed Darwin most were the different types of Galápagos finches, which are now called Darwin's finches.

DNA Data

The Galápagos Islands consist of 13 major islands and many smaller islets, which lie on the equator. The word *Galápagos* means tortoise in Spanish, and the islands were named after the giant tortoises that make their home there and are found nowhere else in the world except on an island in the Indian Ocean.

Besides Charles Darwin, another famous visitor to the Galápagos Islands was Herman Melville, the author of *Moby Dick*. These islands continue to be frequented by amazed visitors. They are often referred to as the "living laboratory of evolution."

While wandering through this enchanted destination, Darwin couldn't help but notice that finches on different islands had very different types of beaks. The medium ground finch, for instance, can crack open seeds with its sturdy beak, but another species, the tree finch, would have a hard time with hard seeds because it has a relatively weak beak. It seemed to Darwin that the different types of beaks he observed were best suited to the type of food that could be found on the islands where these different finches lived. Eventually, he came up with a theory that pulled together all the observations he made between bouts of seasickness.

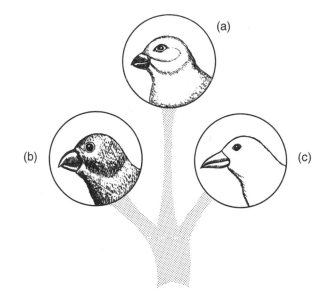

Some of Darwin's finches in the Galápagos Islands. (a) The small tree finch has a dainty beak that's ill-equipped for seed-cracking, but good for grasping and eating insects. (b) The medium ground finch has a stronger beak that can crack open seeds. (c) The cactus finch has a beak that is suited to eating both seeds and insects.

Darwin Versus the Creationists

Maybe he had a bad case of writer's block, or maybe he was just a procrastinator, but it took over 20 years after the voyage of the *H.M.S. Beagle* before Darwin finally published his theory of evolution in *On the Origin of Species* in 1859. The book raised a great debate not only in his native England, but all over the world.

Darwin questioned the going theory at the time, which was that plants and animals had always been the way they are today. His new theory stated that there is a long, slow process by which each species gradually changes or evolves. One animal, for example, might be slightly different from others in that species.

Because there are always natural variations in plants and animals, Darwin reasoned that it made sense to suppose that some of these slightly different types would be better suited to their environment than others. He found other examples of plants and animals that were especially well-matched to their surroundings in the same way that the finches were.

Genetic Jargon

Darwin's theory of **natural selection**, also known as survival of the fittest, states that some variations of a species of plants or animals are better suited to the conditions in their surroundings than others. These variations tend to thrive and reproduce more of their own kind. In the same way, plants and animals that aren't especially suited to their surroundings tend to do poorly and may even die out. As a result, they will reproduce less than the life forms that fare better in a specific place under specific conditions.

137

It seemed obvious to Darwin that if one type of plant or animal fared better in a specific environment, then it would tend to live longer and, as a result, reproduce more, so there would be more of that kind to populate an island or other area. On the other hand, if a certain type of animal or plant did not adapt well to its surroundings, it would tend to die off. Consequently, there would be less and less of its type to mate and bear young and populate an area.

Darwin called this theory *natural selection*. It is also called survival of the fittest. This theory caused quite a ruckus with the Creationists, who believed that God created every plant and animal perfectly and that they had always been in the same form then as they are today. According to the beliefs of Creationism, there were no gradual changes and no evolution.

Darwin and DNA

Darwin did not have a strong background in molecular biology or genetics because there was no such thing as molecular biology or genetics at the time. Darwin didn't know about DNA or genes or exactly how mutations are caused, but he did observe that the different species of Galápagos finches, for example, all seemed to be variations on one theme. He knew that all these types of finches were very much like each other and very much like the finches of mainland South America, yet they had specific variations that made some of them better adapted to the conditions on specific islands and, in some cases, even to different spots on the same island.

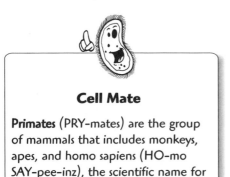

Cell Mate

Primates (PRY-mates) are the group of mammals that includes monkeys, apes, and homo sapiens (HO-mo SAY-pee-inz), the scientific name for us humans.

Now that scientists know more about genes and mutations and the role they play in the long, slow process of bringing about the new, improved versions of people, plants, and animals, they can use DNA to help unlock even more mysteries and see exactly how different species are related and how they diverged. Scientists are now in a position to use DNA samples to understand how humans branched off from a common ancient ancestor that we share with chimpanzees and other *primates*, a group which includes monkeys, apes, and humans.

The Origins of Life

If you just look around, you can't help but notice the immense diversity of life forms. Think of all the species of flowers and trees and animals that you see on a regular basis. Now imagine how many different types of plants and animals must exist in other parts of the world. Consider all the life forms that can be seen only through a microscope. Think of all the unusual life forms that have taken up residence in your refrigerator or under your bed. And what about all the odd life forms that you've dated?

The amazing thing that scientists have learned is that despite obvious differences, there's actually an underlying unity to all of life. At the molecular level, all living creatures share the same As, Ts, Cs, and Gs in their DNA. We all have cells that manufacture proteins, adhering strictly to the same genetic code. The differences are basically just a matter of which of the four letters in the genetic alphabet are arranged in which ways and how often.

So it's generally assumed that all life forms arose from one common ancestor. But when and where and how did life as we know it begin? If all of this evolution has been going on, what did everything evolve from?

The Big Bang and Beyond

One widely accepted theory states that countless years ago, the universe was at first very small and very hot. Then there was an explosion of unimaginable power, which even scientists call the Big Bang. One result of this explosion was that matter got crushed together, forming the first elements. Clouds of dust and gas eventually formed into stars and planets.

DNA Data

You've undoubtedly noticed that the Big Bang doesn't sound very scientific. The term was coined by a scientist named Fred Hoyle, who wasn't totally serious when he came up with the name. However, the name was descriptive as well as funny, and it's been widely accepted as the "real" name for the event.

Eventually our own planet came into being from this explosive mess. You're lucky you weren't around when the earth first formed. It wasn't exactly hospitable to human life. There were all sorts of volcanic eruptions going on everywhere, so it would've been difficult to settle down. Instead of the stuff we breathe, like oxygen, poisonous gases circulated around. There was nothing good to drink, either. Forget wine or whiskey; there's wasn't even any good water. Most of the water that was around was in the form of steam.

Eventually, the earth cooled, water vapor turned into clouds, and those clouds in turn produced the first rainfall. Some scientists think that early life forms could have developed in the earth's primitive atmosphere once it contained chemicals such as ammonia, hydrogen, and water.

The Miller Experiment

A chemist named Harold C. Urey, who had won a Nobel prize in 1934, came to the conclusion that earth's early atmosphere probably contained a mixture of gases such as hydrogen, ammonia, methane, and water. He investigated whether this combination of chemicals could give rise to and support life. He concluded that there would have to be some kind of energy to get things going. Maybe, he reasoned, lightning or ultraviolet light from the sun could have provided the spark of energy that eventually led to the development of life forms on earth.

In 1950, Urey was teaching a class at the University of Chicago on the origins of the solar system. He suggested that it would make an interesting experiment if someone took the gaseous ingredients he thought were present in the earth's early atmosphere and put them together in the lab. Then, he said, that person could add some kind of energy like sunlight or lightning and see whether anything alive came out of the mix.

One of his students, Stanley L. Miller, was fascinated by this suggestion, which planted the seed for what is now called the Miller experiment. Although the Miller experiment sounds like the scientific investigation of a bunch of guys getting together in the frat house and trying to determine exactly how many beers they can drink before passing out, Stanley Miller had something else in mind.

He got himself some laboratory flasks made out of glass, which were connected and sealed. Into this apparatus, he put ammonia, hydrogen, methane, and water. Because waiting for lightning to strike the flask would have taken too long, he scrapped the idea of using this as his source of energy. Instead he decided to use something similar that he had more control over. He zapped his chemicals with 60,000 volts of electricity to simulate lightning.

Cell Mate

Can scientists create life? In 1950, a student at the University of Chicago named Stanley L. Miller recreated the gases of the Earth's early atmosphere in a flask, charged the gases with electricity, and ended up with some amino acids.

The way things were set up in Miller's apparatus, the jolt of energy would propel the gases from one glass flask into another. They would eventually cool off in a section of glass tubing, but then they would pass through the "lightning" again. This process would occur again and again. After a week, Miller decided to check what, if anything, had developed in his energized cocktail of gases.

To his great excitement, some of the molecules in the chemicals were attracted to one another, and bigger and better chemicals had formed. He even found four amino acids. As you read in Chapter 8, "Producing Proteins," there are 20 different amino acids, and they are the subunits of proteins, which are necessary for many life functions. So Miller was quite pleased with the results of his mix of chemicals and energy.

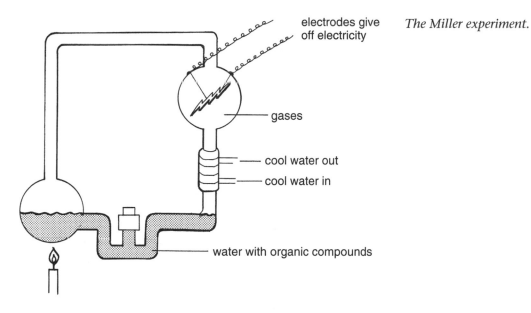

electrodes give
off electricity

The Miller experiment.

gases

cool water out

cool water in

water with organic compounds

When he did the experiment over again, he was delighted to find even more of these amino acids in the mix. Another time, he even dispensed with his electric "lightning" and found that ultraviolet light alone worked just as well. Eventually Miller discovered some bases in the mix—the As, Ts, Cs, and Gs that are the sub-units of the DNA molecule.

As exciting as Miller's experiments were, they still did not conclusively prove that you could manufacture life in a glass tube if only you had the right ingredients. It's a far cry from a few amino acids to actual proteins. And even though Miller found some bases, DNA in living organisms is much longer than anything that was found in his glass flasks. Even the simplest life forms have about one million base pairs of DNA.

Another scientist, Sidney Fox, showed that a mixture of amino acids can eventually form proteins. He applied heat to a combination of several different amino acids in his laboratory and found that some proteins formed from this mixture of raw ingredients and energy in the form of heat.

A Crystal Clear Theory?

Another interesting, but not widely accepted, theory as to how life began dispenses with Miller's entire premise that with the right conditions, earth's gases and water could develop into simple life forms that might eventually evolve into higher life forms. This other theory was put forth by A.G. Cairns-Smith, a chemist at the University of Glasgow in Scotland. His theory is that life could have arisen from clay.

One of clay's ingredients is silica, the most abundant substance on earth. Silica is known to form crystals. Cairns-Smith theorized that as these crystals formed, some mistakes might have happened, and this could've led to different crystal structures. Eventually, the mistakes could have become more and more elaborate.

DNA Data

Historically, the Bible and science haven't been the best of friends. But there are exceptions. A chemist at the University of Glasgow in Scotland, A.G. Cairns-Smith, theorized that life arose from clay—just like Adam.

According to this theory, gene-like crystals can evolve. This theory has never been proved, however, because no one has ever found crystal genes. But the fact that it all starts with clay calls to mind the description of how Adam was made as described in the Bible.

Your Molecular Clock Is Ticking

If DNA carries the secrets of the past, just how do scientists get clues from it to figure out how our species, homo sapiens, evolved from our earliest ancestors? Traditionally, scientists like paleontologists (PALE-ee-un-TAHL-a-jists), who study fossils and early life forms, looked to fossils for all of the answers to the mysteries of evolution. By comparing visual clues such as bone structure, paleontologists attempted to determine which organisms were more related and which were less related to each other.

But now that scientists know more about genes and DNA, they can pick a particular gene or two that's shared by many different organisms and study it. Remember, nature usually sticks with a mechanism that's worked well in the past. It might create some variations, but the basic idea will remain the same. As we discussed in Chapter 11, "Gene Gangs," even lowly yeast cells divide pretty much the same way that human cells do. The genes that regulate cell division are similar pretty much across the board in yeast cells, cats, canaries, and humans. They're so similar, that a human gene could function in a yeast cell, and vice versa.

Genetic Jargon

Once scientists were able to figure out the rate at which DNA mutates, they were able to come up with the **molecular clock**, a model that shows when different species branched off from common ancestors on the tree of evolution.

By studying tried-and-true genes that are shared by so many different species, scientists can figure out how closely species are related to one another. The more genetic variation there is between the DNA of two organisms' genes, the further apart in time these species must have evolved.

Scientists think that as evolution plodded its slow, steady course all the way to the development of present-day life forms, there were several junctures at which some species formed their own branches. Sometimes these branches themselves got split into even smaller divisions of species that broke off from their original ancestors. The totality of these branches makes up what is called the tree of evolution.

Scientists figured out that they could measure the rate at which DNA mutates. So then they reasoned that by studying mutations, they might be able to come up with some kind of *molecular clock*, a time schedule that shows when and how different species evolved from a common ancestor. There is evidence that humans and chimpanzees have a common ancestor, but that the two species went their separate ways about five or six million years ago (although some humans still bear a striking resemblance to our chimp relatives).

All About Mitochondrial Eve

In the January 1987 issue of *Nature* magazine (the same prestigious British journal that published Watson and Crick's paper on the structure of the DNA molecule), an article created a great deal of excitement that spread around the world. An American scientist named Allan Wilson and his research partner, Rebecca Cann, had taken samples of mitochondrial DNA from 182 people of different ethnic backgrounds. The scientists came to some important conclusions about the origin of homo sapiens.

Wilson and Cann chose to use mitochondrial DNA instead of DNA from the nucleus of the cell for several reasons. As you know, there's DNA in the nucleus of the cells of all organisms (except bacteria, which have DNA but lack a nucleus). But DNA can also be found inside the mitochondria. As you read in Chapter 4, "Me, My Cell, and I," mitochondria are the organelles that are considered the powerhouses of cells because they provide energy that's needed for various tasks.

One unusual feature of mitochondria is the fact that they have their own DNA, which contains fewer and different genes than the "regular" DNA contained in the nucleus of the cell. Every time a cell makes copies of itself, all of its DNA is duplicated, including the DNA in the mitochondria.

Another thing that makes mitochondrial DNA unique is the fact that it can only be passed from mother to child, not from father to child. This means that your mitochondrial DNA is exactly the same as your mom's. She got her mitochondrial DNA from her mother, your grandmother, and so on back through time to some distant ancestress who must have passed her mitochondrial DNA down to all of her descendants.

Even though mitochondria are located in a sheath surrounding the tails of sperm, the latest research suggests that the sperm's mitochondria will be destroyed in the egg if they enter during fertilization.

Another reason why Wilson studied mitochondrial DNA instead of nuclear DNA is the fact that it mutates about 10 times faster. This is because its repair systems aren't as sophisticated as the ones in nuclear DNA.

143

Delving Through DNA

Wilson's idea was to study the same specific stretch of mitochondrial DNA in different individuals from various ethnic backgrounds. He hoped to understand which groups were genetically close to one another and which groups diverged from the common ancestor and when. Wilson worked on the assumption that the number of DNA sequence differences would increase with the length of time from the last common ancestor. By making these comparisons, Wilson created a chart that he felt would show which current ethnic groups were most closely related to our earliest ancestors.

Wilson and Cann gathered their data and came to some conclusions. To begin with, there seemed to be relatively few differences between the mitochondrial DNA of even the most widely divergent ethnic groups in human populations. In other words, whether you're a Watusi, Cherokee, or an Australian aborigine, your mitochondrial sequences are very similar to those found in humans in the rest of the world.

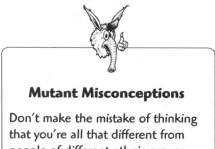

Mutant Misconceptions

Don't make the mistake of thinking that you're all that different from people of different ethnic groups. Genetically speaking, you're almost exactly the same.

A Tree of Humanity

Wilson took his findings and made up a type of tree with two main branches. The first branch included only Africans, and the other branch was made up of some Africans and the other ethnic groups of the world. Noting that Africans seemed to have more mutations in their mitochondrial DNA than other races, Wilson suggested that Africans had been around longer than other races. The extra time they had on this earth allowed more mutations to occur in their mitochondrial sequences.

Wilson concluded from his data that this DNA tree could potentially be traced back to one woman, who may have lived in Africa about 200,000 years ago. Newspaper and magazine writers were fascinated with this idea, and eventually the names Mitochondrial Eve and African Eve were created to describe the ancestress of all humans. But some scientists think that Wilson's assumptions may not be correct. They contend that it might make more sense to assume that there were maybe thousands of women from the group homo sapiens at the time who passed that type of mitochondrial DNA to their descendants.

Wilson's idea was embraced by some, but others pointed out some flaws in this theory. To begin with, Wilson's findings depend on the assumption that the rate of mutation is the same now as it was hundreds of

Genetic Jargon

Dating methods doesn't mean going up to someone attractive and asking, "Haven't I seen you somewhere before?" This term refers to scientific ways of determining age, for example, by studying the composition of a skull or other body part to determine the age of a fossil.

thousands of years ago. Using mutations as a kind of molecular clock might not be as reliable as other *dating methods* used by scientists for this reason.

Another theory holds that it's possible that homo sapiens evolved somewhere in Africa, but that this species could have also popped up in several other places all at the same time. For instance, if homo sapiens evolved in Africa, they might also have spontaneously evolved in Asia and Australia around the same time. But the weight of the evidence suggests that Wilson had it right—modern human beings most likely arose south of the Sahara and then migrated throughout the world.

It's All Relative

Added to the mystery of when and where we humans came about is the question of who's related to whom. Scientists can use DNA testing to figure out who's on which branch of the evolutionary tree. They can even trace the long-buried descendants of living people through DNA comparisons.

Our Infamous Ancestor?

For a long time, scientists have had questions about Neanderthals. This extinct species resembled homo sapiens, and scientists have long wondered whether the Neanderthals were a kind of missing link between modern day humans and a remote ancestor or whether they were on their own separate branch of the evolutionary tree that eventually withered away. If they were a separate species, then what caused them to die out?

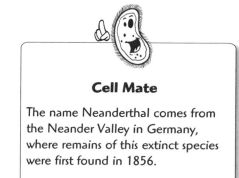

Cell Mate

The name Neanderthal comes from the Neander Valley in Germany, where remains of this extinct species were first found in 1856.

The traditional way to figure out the answer to this question would be to compare the skulls of the Neanderthals and present-day homo sapiens, but another way to go about this would be to use the new tools provided by recent advances in genetics and molecular biology to examine the DNA of these fossils.

Scientists don't need a whole lot of DNA from a Neanderthal bone to be able to study it. Using a technique called PCR (polymerase chain reaction), which was used in the O.J. Simpson trial, they can take tiny bits of DNA and multiply them so there's more to study. You'll read a lot more about this technique in Chapter 17, "Stories Genes Can Tell."

When scientists compared the DNA of Neanderthals to modern-day humans, the result was that the differences between them and us are the same, regardless of which living race they were compared to. Suppose someone believes in the theory that humans originally came from Africa. If Neanderthals were our direct ancestors, then it would follow that they would be more closely related to Africans than, say, to Europeans. But this is not the case, so it stands to reason that the Neanderthals were not our direct ancestors. This suggests that Neanderthals may have been on a completely different branch of the evolutionary tree than homo sapiens. However, some scientists remain unconvinced.

A Cheesy Story

DNA tests on the remains of ancient bones can also show if they are related to anyone today. Sometimes, there are some pretty unexpected living relatives. Take the 1997 case of a 42-year-old history teacher in England named Adrian Targett. Through DNA tests, he was amazed to learn that he shares a common ancestor with a hunter who lived in the Cheddar Gorge 9,000 years ago.

Scientists at the Institute of Molecular Medicine at Oxford University conducted a study by taking DNA from the tooth of Cheddar Man, as the ancient skeleton found in the Cheddar Gorge has been called. The researchers then took samples of DNA from teachers and students at a nearby school. Mr. Targett, who works at the school, has DNA that showed that he and Cheddar Man are distant relatives. Coincidentally, the teacher and his wife live close to Cheddar Caves, where his prehistoric relative's remains were found in 1903.

Mr. Targett was astonished to find out that he was a distant descendant of Cheddar Man. His wife joked with reporters from a British newspaper, saying that maybe her husband's kinship with a primitive man might account for her husband's preference for eating his steaks on the rare side.

The Least You Need to Know

➤ Charles Darwin, a British naturalist who lived from 1809 to 1892, proposed the theory of natural selection or survival of the fittest, which states that variations of organisms that are better adapted to their environment will thrive and reproduce more of their kind, whereas those less suited will die out.

➤ Stanley Miller performed an experiment that suggested that life forms could have sprung up from the gases found in Earth's early atmosphere that were jolted with energy from lightning or ultraviolet rays from the sun.

➤ Scientists can use the rate of mutations as a molecular clock to figure out approximately when different species branched off from each other.

➤ DNA in the cell's mitochondria—the cell's tiny powerhouses—can give a clue to how humans evolved.

➤ Scientists have tested the DNA of the remains of Neanderthals and most believe they are not an ancestor of humans, as previously thought.

➤ In 1997, scientists conducted DNA tests on the remains of Cheddar Man, a hunter who lived in England 9,000 years ago, and linked him to a distant descendant living in the same area today.

Part 4
Manipulative Scientists: Genetic Engineering and Cloning

You've probably heard of Dolly the cloned sheep, and you might have heard about the multiplied mice. Now you can learn the science behind the headlines. Just what do scientists do to make identical copies of plants or animals?

This section tells you the difference between genetic engineering and cloning. You'll be able to understand how scientists can create new types of plants and animals by recycling genes from different organisms.

Chapter 13

Cut-Down Genes

> ### In This Chapter
>
> ➤ Cut-and-paste DNA
>
> ➤ The first genetic engineers: Stanley Cohen and Herbert Boyer
>
> ➤ Deciding on safety measures at a conference in California
>
> ➤ Biotechnology becomes big business
>
> ➤ Lean, mean gene machines

Since 1953, when James Watson and Francis Crick figured out the structure of the DNA molecule and how it makes copies of itself, a lot has happened on the genetic front. In the 1970s, two scientists came up with a revolutionary idea that built upon the new knowledge of how DNA works. The idea was based on a simple question: If genes code for certain proteins which could translate into specific traits, then what would happen if researchers cut out a gene from one organism and pasted it into another, completely different organism?

This idea became a reality in the 1970s, and since then, scientists have embarked upon incredible experiments that previously had been impossible. In this chapter, you'll read about how scientists created a new technology called genetic engineering, which can combine DNA from animals, plants, and even humans and bacteria. In Part 7, "Playing God? Ethical Issues," you can read about some of the ethical concerns that are growing along with the new knowledge of how to manipulate genes.

The Splice of Life

In November 1972, two scientists who were attending a conference in Hawaii decided to go out for a snack one night. They were Stanley Cohen of Stanford University and Herbert Boyer of the University of California at San Francisco. Boyer had spoken at the conference that day, and Cohen wanted to know more about his research. The two discussed their work over sandwiches at a deli that evening. Boyer was doing work on substances called *restriction enzymes*.

Genetic Jargon

Restriction enzymes are special enzymes that occur naturally in some bacteria. When the bacteria are attacked by invaders such as viruses, restriction enzymes cut the DNA of the attacker at very specific sequences of As, Cs, Ts, and Gs to protect the bacteria.

Restriction enzymes naturally occur in bacteria. These substances help the bacteria to defend themselves against foreign invasion. When a molecular villain like a virus attacks a bacterium, these special enzymes literally rip the invader's DNA to shreds. So these substances are called restriction enzymes because they restrict the growth of foreign invaders. They're something like an extremely small pair of molecular "scissors."

Boyer wasn't the first person to do research with these molecular scissors. Restriction enzymes were discovered in the 1960s, and in 1978, three scientists shared the Nobel prize for Physiology or Medicine for the discovery of restriction enzymes, which can cut DNA in very specific spots. These three men were Hamilton Smith and Daniel Nathans from the United States and Werner Arber, a scientist from Switzerland.

The scissors-like enzymes do their job by recognizing certain patterns of DNA's bases, its As, Ts, Cs, and Gs. Usually these sequences are from four to eight bases long. The enzymes will snip the DNA at that point, but nowhere else. One restriction enzyme might go to work whenever it meets the sequence TTCG, for example. That means that wherever it recognizes this sequence, it will snip the DNA at that point. There are well over 1,000 restriction enzymes, each of which reacts to different DNA sequences.

Circles of DNA

While munching away at his sandwich, Boyer learned that Cohen was doing research on how some bacteria can exchange small circular pieces of DNA called plasmids. He experimented by taking plasmids from one type of bacteria and putting them into other bacteria.

Bacteria, a lower form of life, normally have only one chromosome, which contains lots of genes. But bacteria also have some small, free-floating rings of DNA that contain genes other than the ones in the chromosome.

These miniature DNA circles are called *plasmids*. Some of these genetic extras help bacteria to do things that some of their fellow bacteria can't do. For instance, there could be a gene that allows the bacteria to make proteins that enable them to eat different kinds of food. Other plasmids can make the bacteria resistant to certain antibiotics. When the bacteria multiply, their chromosomes will of course get duplicated, and so does the DNA in their plasmids.

Sometimes bacteria can give these plasmids with useful genes to one another. You read in Chapter 4, "Me, My Cell, and I," that even though bacteria don't have sex, they sometimes engage in conjugation—a less exciting activity that's like the bacterial version of foreplay. It's probably not much fun for the bacteria, but it does transfer some of these plasmids from one bacterium to the other.

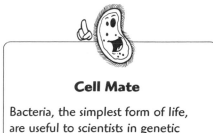

Cell Mate

Bacteria, the simplest form of life, are useful to scientists in genetic engineering precisely because they're so simple. It's easy for scientists to understand what goes on with the bacteria's genes.

Scientists first discovered these little DNA rings in bacteria that cause a type of dysentery, a severe infection of the intestines. At first, antibiotics stopped these bacteria from growing, but eventually, some strains became resistant to the antibiotic. Some of the genes for antibiotic resistance were in the plasmids, and these were being transferred by way of the bacterial sex you just read about.

Not only can plasmids protect some nice and some not-so-nice bacteria from death by antibiotics, but they can also protect bacteria from some heavy metals. This does not refer to the kind of loud music your neighbor's teenager plays at high decibels. It means that the plasmids protect against poisonous metals such as lead or mercury, which can be spread by industrial pollution. Other types of nifty plasmid genes can break down poisonous herbicides or weed killers. A frightening fact is that some plasmid genes can even destroy penicillin, which we generally depend upon to get rid of the tiny bad guys.

Mutant Misconceptions

Don't get the strain of E. coli bacteria that's used in genetic engineering confused with its evil cousin. You've probably read in the newspapers about how some people got sick at fast food restaurants where they were unfortunate enough to be served meat contaminated with the dangerous strain of E. coli. However, the kind that's used in the laboratory does not cause illness. It's a tame strain that doesn't harm humans.

Juggling Genes

Both Cohen and Boyer were eager to learn more about each other's research, so by the time they finished their sandwiches in the deli in Hawaii, they had made plans to get together when they were back in California. They wanted to see whether they could put together what they'd learned in their separate experiments. They met in the spring of 1973 and, with their assistants Annie Chang and Robert Helling, carried out some experiments that involved both restriction enzymes and plasmids.

At first, the researchers took two different strains of E. coli bacteria. (There are many types of E. coli, and as you'll recall, some of them live in the human digestive tract and help us with digestion.) Using Boyer's knowledge of restriction enzymes, they cut some plasmids in the E. coli bacteria that Cohen had been doing research on. After plasmids are cut open, they have what are called *sticky ends*. This means that they can easily attach to other plasmid parts that also have sticky ends.

Genetic engineering in bacteria.
(a) A plasmid is taken from a bacterium. Restriction enzymes cut it in a specific place.
(b) DNA is cut from another plasmid and then "pasted" into the first one.

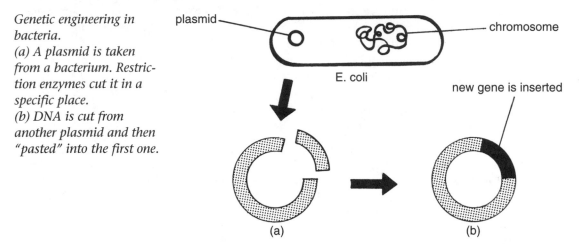

Then Cohen and Boyer cut out specific genes from another type of bacteria. When the cut-down genes from one strain of bacteria met with the sticky ends of the other strain, they joined. Then, using an enzyme called *ligase*, which is like molecular glue, Cohen and Boyer and their assistants pasted the new genes into the cut-open plasmids. In this way, they made a batch of E. coli bacteria resistant to certain antibiotics.

Cohen and Boyer took one type of plasmid containing a gene that made bacteria resistant to an antibiotic called tetracycline (TEHT-ra-SY-clean). Then they took another type of plasmid that gives bacteria resistance to another antibiotic called kanamycin (CAN-a-MY-sin). By cutting these DNA circles in the same places using restriction enzymes, the scientists hoped that the sticky ends of the tetracycline-resistant and the kanamycin-resistant plasmids would team up and form one recombined plasmid that would give the bacteria resistance to both types of antibiotic.

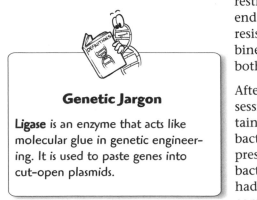

Genetic Jargon

Ligase is an enzyme that acts like molecular glue in genetic engineering. It is used to paste genes into cut-open plasmids.

After what they hoped was a successful cut-and-paste session, Cohen and Boyer put the bacteria into a container with tetracycline and kanamycin. Most of the bacteria died, as they normally would have in the presence of these powerful antibiotics. But some of the bacteria survived. Cohen and Boyer realized that they had succeeded in creating the plasmids with a new combination of antibiotic resistance.

Of Tadpoles and E. Coli Bacteria

After other experiments with bacteria, the two scientists decided to go for the big time. What would happen, they wondered, if they tried to mix genes from very different life forms? Could they cross species boundaries by using this molecular cut-and-paste technique? Or was the DNA from one species so foreign from other species that the experiment would fail?

Using restriction enzymes again, Cohen and Boyer cut out a gene from a toad and experimented with pasting it into E. coli bacteria. Could a gene from an animal like a toad be transferred into the DNA rings of simple organisms like E. coli bacteria? When the bacteria began to multiply, the scientists got their answer. With every new generation of E. coli bacteria, the toad gene was present in the plasmids.

Cohen and Boyer's landmark experiment ushered in a new age of genetics with infinite possibilities. Now species boundaries could be crossed. Human genes could be spliced into pigs and tobacco, and a zebra gene could be pasted into a flower. Although this discovery brought the promise of new medicines and possible preventive cures for diseases, some scientists and laypersons were concerned about the potential for mistakes. You'll read about these issues in Part 7, "Playing God? Ethical Issues."

Cohen and Boyer called the new technology that they had devised recombinant (re-COM-bin-ent) DNA technology, because it *recombines* DNA from different sources. The media picked up on this tremendous innovation and dubbed it genetic engineering. It's also called gene cloning.

The Tiniest Workhorse

As more and more researchers picked up on Cohen and Boyer's genetic engineering experiment, it became clear that E. coli bacteria would remain the microscopic stars of this technology. Even today, E. coli is known as the workhorse of genetic engineering. There are several reasons for this.

In addition to carrying the circular bits of DNA called plasmids, E. coli, like most bacteria, reproduces rapidly. Because it takes only about 20

Mutant Misconceptions

Don't confuse gene cloning with making clones, which refers to making two organisms with exactly the same DNA. An example of cloning in nature is identical twins, because they have the same DNA. You'll read more about this type of cloning in Chapter 16, "Send in the Clones."

Cell Mate

E. coli bacteria that are genetically altered to produce large amounts of certain proteins, such as Human Growth Hormone or insulin, are put in huge vats to reproduce. The temperature is strictly controlled so the bacteria will multiply rapidly and produce the needed substances. The procedure is a lot like the fermentation process that brewers use to make beer. Only beer tastes a lot better than bacteria.

minutes for a new generation of E. coli to be born, scientists can quickly and easily check the results of their experiments.

When researchers have successfully inserted a gene for a protein that they want to produce in large quantities, all they have to do is to be extra nice to the bacteria. Give them a good, warm home with lots of goodies to snack on, and E. coli will grow and grow, producing the protein that the gene codes for. Ungrateful as it may seem, the researchers then kill off the bacteria and extract the protein that was made by these miniature living factories.

Biotechnology or Cloning?

If you read newspaper reports about all the new advances in genetic engineering, you might be confused by all the terms reporters throw around. You could be wondering whether biotechnology is the same as genetic engineering. And is gene cloning the same as gene splicing?

The word *biotechnology* comes from the Greek (what else?) word *bios*, which means life and from which we get the word *biology*, the study of life. In other words, biotechnology is life technology. What does that mean? It's basically the use of any life form to serve the needs of humankind. Thus, a very simple, old form of biotechnology is breadmaking. When bakers use yeast to help the bread rise, they are employing a living thing, the yeast, to help them make a loaf of bread that is lighter and more palatable than bread baked without yeast.

Yeast is also used in making wine. Every time you drink red wine with your spaghetti or white wine to complement your salmon, you're drinking a product that was made possible by putting a living thing to work to help humans quench their thirsts.

Cheesemaking is another type of simple biotechnology. The whole process of turning milk into cheese usually starts with an enzyme. In the past, this enzyme was almost always taken from the stomachs of calves. Nowadays, thanks to genetically engineered copies of this calf enzyme, most cheeses are made with enzymes produced by genetically engineered bacteria that act as factories to crank out copies of the calf enzyme. The enzyme was spliced into the bacteria. The next time you read the ingredients in the cheese or cheese product you're about to buy, you might notice something called *microbial enzymes*. This term refers to the enzyme that was genetically engineered using bacteria.

Genetic Jargon

The phrase **microbial** (my-CROW-bee-ull) **enzymes** refers to genetically engineered enzymes. One type of microbial enzyme is commonly used in the manufacture of cheese.

But biotechnology is much more than making wine and cheese. The term encompasses a number of different technologies, such as genetic engineering and cloning. So although genetic engineering is a type of biotechnology, it doesn't mean that biotechnology is only genetic engineering.

What's What in Biotechnology: A Guide for the Genetically Confused

Term	Meaning
Biotechnology	The technological use of living things to serve the wants and needs of humankind.
Cloning	Making an identical copy of **all** the DNA in an organism, as opposed to changing around just a gene or two as in genetic engineering. So identical twins, which formed from the same fertilized egg that split into two separate embryos, are considered clones. All of their DNA and genes are exactly the same.
Gene cloning	Same as genetic engineering.
Gene splicing	Same as genetic engineering.
Genetic engineering	The manipulation of genes. Using restriction enzymes, which can cut DNA in specific places, researchers remove a specific sequence of DNA, a gene. Then they cut open the DNA of another organism the same way and paste the new, foreign gene into this organism's DNA. To make sure that the ends stick together, they add an enzyme called ligase, which is like molecular glue. When the new gene is successfully cut and pasted into the new DNA strand, it will be duplicated along with the rest of the DNA every time the cell that it's contained in divides.
rDNA	Same as genetic engineering.
Recombinant DNA	Same as genetic engineering.

Monkeying Around with Viruses

Stanley Cohen wasn't the only molecular biologist working at Stanford University in California in the early 1970s. Around this time, another molecular biologist named Paul Berg was embarking on experiments in which he spliced DNA from a virus into E. coli bacteria. Before he started his experiments, however, everything was changed by a chance encounter between someone who worked in Berg's laboratory and a scientist named Robert Pollock.

Pollock was teaching a course at the Cold Spring Harbor Laboratory on Long Island, New York, when he coincidentally learned about Berg. One of the students in Pollock's class was going to do the

Cell Mate

The Cold Spring Harbor Laboratory on Long Island, New York was under the direction of Dr. James Watson, who was awarded the Nobel prize along with Francis Crick for their discovery of the structure of the DNA molecule.

experiments that Berg was planning to perform and happened to mention this to Pollock. When Pollock heard that Berg intended to splice a gene from a virus into E. coli, he was alarmed. The virus, called SV40, is found in the kidneys of African monkeys. Pollock was worried because he knew that the SV40 virus could cause cancer to develop in certain animals, such as mice.

Pollock also knew that the strain of E. coli that Berg intended to work with can survive in the human digestive system. What would happen, Pollock worried, if Berg or one of his workers accidentally ingested some of this E. coli bacteria containing genes from the SV40 virus? Would it cause cancer in these people? Would it spread a new type of disease? Would this cause a new, lethal epidemic?

Pollock decided to call Berg in California to share his thoughts on these unsettling possibilities. It turned out that Berg was open to Pollock's concerns. As a result of the phone call, Berg made the decision to hold off on his research until he was assured that the experiments would be harmless.

Assessing the Risks

At Berg's prompting, the National Institutes of Health, a government agency, conducted a series of tests and, after more than a year of serious consideration of the possibilities, concluded that the engineered organisms would not harm anyone. Eventually, Paul Berg went through with his planned experiments and later won a Nobel prize for his research.

Pollock and Berg had raised some serious questions about genetic engineering. Other scientists started thinking about the repercussions, too. So in 1975, about 150 scientists gathered at the Asilomar (a-SILL-o-mar) Conference Center in California to discuss the possible risks involved in conducting genetic engineering research.

They discussed the potential risks and came up with suggested guidelines for carrying out this kind of work. They hoped that following these guidelines would minimize the potential dangers of recombinant experiments, such as the accidental escape of harmful organisms.

Measures for Safety

The scientists at the Asilomar conference divided genetic engineering experiments into four categories and ranked them according to the degree of possible danger that they might entail. Then they called for specific procedures for different experiments to ensure safety.

For instance, the Asilomar participants suggested conducting experiments that involved disease-causing bacteria or viruses that could threaten the lives of laboratory workers and others under strictly controlled conditions. Laboratory personnel, for example, might need to wear special protective suits to keep out harmful bacteria or other dangerous microscopic organisms. Another suggestion was to weaken any harmful bacteria that were used. That way, even if the harmful bacteria escaped from the laboratory, they wouldn't be able to survive outside of the site of the experiments.

As a result of the 1975 conference in Asilomar, the National Insitutes of Health (NIH) formed a Recombinant DNA Molecule Program Advisory Committee (RAC). In addition, scientists who received funding from the government were required to follow the guidelines that were issued.

Genes for the Stock Market

The experiments of Cohen and Boyer resulted in more than the Asilomar conference and scientific safety guidelines. In 1976, Herbert Boyer teamed up with investment broker Robert Swanson to form a company that would get into the business of marketing the products of genetic engineering. The company was named Genentech (from GENetic ENgineering TECHnology). You could say that companies like Genentech recombined biology with finances.

DNA Data

When biotechnology was spliced into the stock market, the result was recombinant high finances. On October 14, 1980, Genentech, the company formed by Herbert Boyer and Robert Swanson, offered its stock for public sale. Their original plan was to sell one million shares of the stock for $35 each. Considering that the company was brand new, in an industry that didn't exist, and hadn't even come up with any products yet, this was considered a high price. But the demand was great, so the company decided to offer more shares than originally planned. On the same day, the price of shares rose to an incredible $89, which broke the previous record for an initial public offering of stocks. Later that day, the price went down to $71.50, still a nice day's haul for an investor.

Relying on the workhorse of genetic engineering, E. coli bacteria, to act as living factories, Genentech came out with its first commercial product: bioengineered insulin. People with the disease diabetes need insulin because the specialized cells in their pancreases, the organ that is responsible for manufacturing this hormone, can't produce enough of it. A lack of insulin results in an excess of sugar in the blood and other health problems.

In the past, insulin was obtained from the pancreases of animals such as pigs and cows. However, even though the insulin of these animals is nearly identical to human insulin, one amino acid is different. (You'll remember from Chapter 8, "Producing Proteins," that amino acids are the building blocks of proteins and that proteins are necessary for many of the body's functions.) To produce genetically engineered

Cell Mate

Other genetic engineering companies were founded around the same time as Genentech. Companies such as the Cetus Corporation and Genetic Systems were formed in the United States, and Biogen was formed in Switzerland. By 1998, there were close to 900 biotechnology companies in the United States.

Genetic Jargon

A **gene machine** can construct fragments of a desired gene from scratch. The researcher has to know the sequence of the gene and type it into the machine's computer. Then the machine mixes the sequence of As, Cs, Ts, and Gs to make the desired gene fragment.

insulin, E. coli had the human gene that codes for insulin pasted inside their DNA. So when they produced insulin, it was according to the genetic code that comes from humans.

Other medical substances, such as Human Growth Hormone and interferon, are now manufactured by bacteria. Vaccines are also made by E. coli bacteria that have had the human genes for these substances inserted into their DNA. You can read more about these biomedicines in Chapter 23, "Molecular Medicines."

The Incredible Gene Machines

One development in the field of genetic engineering that sounds like something out of science fiction is the creation of *gene machines*. These machines can build small parts of a desired gene from scratch. As early as 1981, there were two companies—Vega Biochemicals in Tucson, Arizona and Bio Logicals in Toronto, Canada—that offered these amazing devices.

How do these incredible devices work? They have containers filled with each of the four bases (those As, Ts, Cs, and Gs that we keep coming back to in this book). As you've learned, these bases form the four-letter alphabet that makes up DNA. In addition, the gene machines have other containers with chemicals that ensure that the four letters are put together in the right order and stick together once they're assembled.

In order to obtain the desired gene part, researchers need to know its sequence. In other words, they have to know the exact order of the As, Cs, Ts, and Gs in the gene fragment that they want to make. Then they type this information into a computer in the gene machine, and one by one, the bases attach to one another according to the instructions.

After the researchers have typed in the genetic sequence, they can walk away, have a bite to eat, take a snooze. They just leave the rest of the work up to the machine. After a number of hours, they can come back, and voilá! There's the gene part that they wanted, all ready for future use. These gene fragments are used in some types of DNA testing, which you'll read all about in Chapter 17, "Stories Genes Can Tell," and Chapter 18, "DNA's Day in Court." They're also used to help discover new genes for the Human Genome Project, which you'll read about in Chapter 21, "The Human Genome Project."

The Least You Need to Know

➤ Some bacteria have structures called plasmids, which are small circles of DNA, in addition to one main chromosome of DNA.

➤ Restriction enzymes, which occur naturally in bacteria, can cut DNA at specific points.

➤ In 1973, scientists Herbert Boyer and Stanley Cohen used restriction enzymes to cut out a gene from a toad and then paste the gene into the DNA of E. coli bacteria. With this experiment, Cohen and Boyer created recombinant DNA technology, popularly known as genetic engineering.

➤ In 1975, about 150 scientists met at the Asilomar Conference Center to discuss ways of keeping genetic engineering experiments safe.

➤ Devices called gene machines can construct a desired fragment of a gene from scratch.

Chapter 14

Seedy Science: Engineering Plants

In This Chapter

➤ Tumor-inducing plasmids, gene guns, and other scary-sounding ways that scientists inject foreign DNA into plants

➤ Pumping up plants to resist pests and disease

➤ The tomato that made history

➤ Putting plants to work making vaccines, fertilizers, and other useful stuff

Mary, Mary, quite contrary, how does your garden grow? There's a lot more springing up in the garden since Mary planted her pretty maids all in a row. Now that genetic engineering makes it possible to mix genes from different species into plants, you're likely to see greenery that contains DNA from animals and even humans in the future. So far, there are no tiger lilies with genes from tigers or cauliflowers with DNA from collies, but the future possibilities are endless. In this chapter, you'll learn about the genetically altered plants that have been developed so far, and you'll get a glimpse of what might be sprouting from the soil in the near future.

Foods of the Future

What's for dinner in the 21st century? You might be surprised by the offerings on the menus of the future. Suppose you went to a cozy little place called the DNA Cafe with a very special date. You sit down at a table decorated with a double helix border, and the waitress hands you a menu. You read the following specials of the day:

Dinner at the DNA Cafe

Appetizers
Spiced Potato with Waxmoth Gene
Juice of Tomatoes with Flounder Gene

Entrees
Blackened Catfish with Trout Gene
Pork Chops with Human Gene
Scalloped Potatoes with Chicken Gene
Cornbread with Firefly Gene

Dessert
Rice Pudding with Pea Gene

Beverage
Milk with Genetically Engineered Bovine Growth Hormone

National Wildlife Federation, "A Dinner of Transgenic Foods," The Gene Exchange, December 1991.

If you think that the food du jour listed here sounds like something out of Aldous Huxley's futuristic novel *Brave New World*, think again. This mock menu from the DNA Cafe was created by the National Wildlife Federation for their newsletter, "The Gene Exchange," and was meant to give the public an idea of some of the unusual combinations of plant, animal, and human genes possible using the new genetic technologies. Every item on the menu has already been developed, and most of the foods listed are already on the market. And there's plenty more where they came from.

Cell Mate

A number of foods that are equipped with a gene from an entirely different species have already been developed, and some of them are even for sale at your local market.

You may well wonder about these novel products. Isn't crossing a potato with a waxmoth gene something like Mr. Potato Head meets Mothra? Would these potatoes produce french fries that flutter out of the frying pan because they're attracted to the flames beneath them? Would pigs with human genes eat more entrees at a sitting than your father-in-law at a fast-food restaurant? And is a potato with a chicken gene likely to be the potato that crossed the species boundary to get to the other side? Will you need to debone and defeather your mashed potatoes in the future? None of these scenarios have happened, but the combinations leave some people wondering.

Beyond Selective Breeding

Farmers have traditionally relied on a method called selective breeding, or artificial selection, to help produce improvements in their crops. They took the hardiest, tastiest, most nutritious varieties of a type of plant and crossbred them (like Mendel, who crossbred pea plants with the traits he wanted). Using this procedure, it can take years for a farmer to get exactly the kind of crop that he wants. The process takes so long because along with the desired characteristics of a chosen plant, the breeder is also likely to get some unwanted traits that are passed along in the genetic shuffle.

But even though selective breeding may seem slow and short on excitement, it has had some profound effects. For example, the domestication of corn over the centuries has produced a hybrid plant that looks very different from the one originally found in nature, which had very small ears. Modern hybrid corn is totally dependent on humans for its continued propagation.

It could take five years or longer of interbreeding different plants to get a bumper crop of apples with tastier fruits that are more resistant to pesky worms or pears that resist colder temperatures and taste especially sweet. This crossbreeding approach might work after a short time, after a few years, or possibly not at all. It might work better for some traits, for example taste, and less effectively for other characteristics, such as resistance to certain insects or kinds of plant fungus.

Of course, breeders using artificial selection are limited to varieties of plants that are in the same species, or at least closely related to one another. Before genetic engineering, breeders couldn't get plants to cross species boundaries. They certainly couldn't imagine taking a trait from an animal such as a zebra and adding it to a plant's DNA—for instance, a tomato—to get red-and-white striped salad tomatoes. (We're just joking. This has never been done—not yet, at any rate.)

Some researchers feel that the new genetic technologies are the way to go to produce tastier, hardier types of plants in the future. This approach might take a fraction of the time that selective breeding does, and researchers hope that they can zero in on just one particular trait and not develop plants that inherit other undesirable traits from the preceding generation.

Cell Mate

Before the development of genetic engineering, farmers had to rely on traditional plant breeding methods to get plants with desired characteristics, such as better taste or better durability. This traditional method *is* called selective breeding or artificial selection.

Genetic Jargon

A **transgenic plant** is one that has been genetically altered. It may have a gene or two from an animal, a human, or another plant spliced into its genome, which is the totality of its DNA. A transgenic plant can also have a gene or two deleted from its genome.

How Do They Do It?

You can't get a plant to mate with an animal or a human, so how do researchers cross species boundaries and get DNA from walking, talking, or howling organisms into growing greenery? The answer is genetic engineering, or recombinant DNA technology, which you read about in the last chapter.

By cutting and splicing DNA, researchers are now trying to engineer *transgenic plants* to taste better, have superior nutrition, produce their own herbicides or pesticides, and withstand inclement weather or adverse environmental conditions, such as poor or salty soil. Hopes for the future of agriculture include increases in plant yields and lower production costs for food crops.

Scientists can engineer plants to have different traits, such as herbicide resistance.
(1) Gene for herbicide resistance is spliced into a plasmid (a small ring of DNA) in bacteria.
(2) The engineered plasmid is put into a type of soil bacteria.
(3) The soil bacteria gets mixed with plant cells, and the DNA with herbicide resistance gets into the plant's chromosomes.
(4) The plant cells are grown into whole plants that can withstand certain herbicides.

gene for herbicide resistance

(1)

(2)

(3)

(4)

Microscopic Soldiers Without Weapons

As you learned in Chapter 13, "Cut-Down Genes," in order to get a gene into an organism, the researcher first has to use a vector, which is something that will get the foreign DNA into the plant DNA in the first place. This can be an organism, such as a naturally occurring plant virus, that has no problem invading the chromosomes of a tomato or a potato.

There are several ways of achieving this goal of delivering foreign DNA. One way is to take what's known as a *Ti plasmid* from bacteria and use it as a genetic delivery service to the plant. A plasmid, as you learned in the previous chapter, is a circular ring of DNA that floats around inside bacteria.

Scientists use bacteria that would normally cause a type of plant cancer. First, they insert the foreign gene into the bacteria's plasmids. Then they disable the genes in the bacteria that cause harm to the plant. But the bacteria can still invade the plant's cells. This is something like sending soldiers to invade a country, but taking away their guns and ammunition. They can get inside enemy lines, but they can't do any harm once they're there. Once inside the plant cells, the foreign DNA that was spliced into the bacteria's plasmid now becomes a permanent part of the plant's genetic material.

One drawback with this method is that it can only be used on certain types of plants, such as tobacco. Important food crops like rice, corn, and sugar cane don't respond to this method, so researchers have to use other techniques to implant new DNA into these plants.

Genetic Jargon

The *Ti* in **Ti plasmid** stands for tumor-inducing. Genetic engineers sometimes splice a new gene into a Ti plasmid because it is naturally able to invade plant cells. However, the scientists just want to deliver the gene; they don't want the plant to develop a tumor. So they disarm the plasmid by taking out its harmful genes.

Annie, Get Your Gene Gun!

Another method of getting new DNA into plants is the use of *gene guns*. Although you can't shoot to kill with one of these devices, you can change the gene structure of a plant with it. Although similar to a regular gun, it's not loaded with bullets that actually go through a plant or it would shoot the plant to smithereens.

Instead of bullets, researchers use tiny round metal spheres that are made of either tungsten or gold.

Genetic Jargon

A **gene gun** is a device that shoots tiny metal spheres coated with foreign DNA, which gets into plant cells. Some of the foreign DNA is incorporated into the plant's permanent genetic makeup.

These spheres are 10 times smaller than the period at the end of this sentence. They're coated with the DNA that the researchers are trying to put into a plant's cells.

When one of these spheres goes down the barrel of the gene gun, it's stopped and does not exit the gene gun the way a regular bullet would exit a regular gun. Although the sphere is stopped, a tiny hole in the gene gun allows particles of the DNA-coated spheres to get sprayed around. Some of these particles physically penetrate the cells of the plant.

Once the foreign DNA has been shot into the plant, some of it is incorporated into the plant's genetic structure. This procedure is hit-or-miss, because not all of the DNA finds a new home in the plant cells. If too many particles bombard the cells, the cells themselves can die. Of the particles that make it to the plant cells, some overshoot their mark and pass completely through. The success rate for this procedure is usually very low.

You Can Go Back Again

Before being shot with the gene gun, the plant has to be specially prepared. This preparation can be done in several ways. For instance, small squares can be cut out of a leaf. These small squares are then soaked in a Petri dish, a small dish that scientists often use in experiments.

The small squares are then taken back to the time in their leafy little lives when they were so young that they weren't any particular part of the plant yet. They were what are called *undifferentiated cells*. This means that they were waiting for instructions from the plant to turn into, for example, a root, together with lots of other cells that were instructed to do the same thing. At this early stage in the life of a plant cell, it could have also been given instructions to turn into a stem or a flower cell.

Scientists turn these adults into infants again by breaking them up with special chemicals. Now the cells that were in groups as a leaf or a root are no longer connected to each other. As a result, they regress and, if need be, can form an entire new plant if given the right foods and a comfortable environment.

If some of the new DNA shot from the gene gun successfully makes it inside these regressed plant cells, the researchers take these plant pieces and feed them with the right nutrients and plant growth hormones in the culture dish. Roots begin to form, and eventually a small seedling will shoot up. Then the seedling is transplanted into a tiny flower pot, where it continues to grow.

Genetic Jargon

When plant cells are very young, or when they've been artificially regressed to a young state, they are called **undifferentiated** or callus (KALL-us) **tissue**. At this point, they can be given an assignment to turn into a root or a stem or a flower.

Shocking Plants

Yet another method for delivering new genes into plant cells is called electroporation (ee-LECK-tro-pour-AY-shun). In this process, the plant cells are given a big shock, one that people couldn't withstand. The shock is that the scientist tells the plant that it was adopted. (Not really—just making sure you're paying attention.) Actually, about 200 to 600 volts of electricity are passed into the cells for a short amount of time.

As a result, little holes appear in the plant cells, and this provides a brief window of opportunity for the foreign genes to be inserted. These genes are floating around in a solution in a culture dish that the plant cells are using as a swimming pool.

The cells will suffer damage from the electric shock (you wouldn't look too good if someone threw a toaster into your bathtub as you relaxed after a hard day), but the cells can repair themselves fairly quickly. The whole point was to get the cells to open up a bit to allow the delivery of the foreign genes, and some of them do indeed make it into the plant's genetic makeup.

Defense Plants

One reason why researchers are looking for quicker ways to develop new plants is that the earth's population is on a steep and rapid rise, and more and more food will be needed just to make sure no one goes hungry. If the world's population continues growing at its present rate, by the year 2035, you can expect the number of people living on this earth to have doubled.

That means there will be somewhere in the vicinity of over a whopping 11 billion people sharing the limited resources of our planet. If you think it's hard getting a seat on the bus or train these days, just wait. And all those long lines at the video store will double in size, too. But the worst part of all this will be the 11 billion mouths to feed.

The population explosion is of great concern to people involved in the agricultural, food producing industries. Some researchers hope that if they can insert genes into plants that can enable them to grow quicker or under inclement weather conditions such as severe cold, or get them to adapt to poor soil conditions, or to enable them to survive droughts, or even to thrive when irrigated with salt water, that this might be a step in the right direction toward providing food for the world's ever increasing numbers.

B.t.—Death from Within

Farmers are all too familiar with insect pests such as caterpillars and beetles, which can ravage food crops. Unfortunately, food growers get to know these critters intimately because they can often be found inside the farmer's crops, which the insects view as huge smorgasbords, ripe for the taking. As a result, many farmers spray toxic insecticides on their food crops. Environmentalists worry that these poisons will have adverse long-term and short-term effects on people, plants, and animals and may upset the ecology of the areas where they're used.

For many years now, organic farmers—farmers who do not use toxic chemicals in the soil or on their plants—have been using a type of naturally occurring bacteria that keeps insect pests in check. The bacteria are known as *B.t.* These microscopic organisms work something like biological time bombs in insect pests, but they are harmless to the environment because they eventually dissipate and are harmless to people and animals as well.

Genetic Jargon

B.t. is the abbreviated name of a harmless type of naturally occurring soil bacteria that has been used by organic farmers for many years to keep away insect pests. The long and unwieldy full name is *bacillus thuringiensis* (ba-SILL-us THOOR-in-JEN-sis).

B.t. works like this: When a hungry insect takes a bite of a mouth-watering plant that has been sprayed with B.t., it doesn't die on the spot. When the B.t. gets inside the digestive system of the unsuspecting insect, a poison is formed. Eventually, the insect dies of a meal that's worse than your mother-in-law's cooking.

So the insect dies, but the B.t.-sprayed plant is nevertheless safe for everyone and everything else. Neither the spray itself nor eating the sprayed crops will cause any harm to humans or animals. And any run-off into nearby streams or other bodies of water does not harm fish, either. These qualities make B.t. the harmless pesticide of choice.

Some biotech companies have taken the gene from B.t. that codes for the production of the substance that poisons insect pests and transplanted it into food crops. Instead of being sprayed on the plant, the biotech approach splices the poison-inducing gene into every cell of the new transgenic plants. For instance, the Monsanto company has created genetically modified potatoes that carry the B.t. gene. This was done in an effort to save crops from Colorado potato beetles, which consider farmers' crops to be an open-air potato salad bar.

Monsanto has also genetically engineered corn plants to be resistant to the corn borer, an insect that does much of its damage inside the stalk of corn plants and isn't reached by the usual pesticides. (Yes, like people, some insects can accurately be described as boring.) Each cell of these modified plants contains the B.t. gene, and as a result, the plants can fend off these voracious insects.

Disease Resistance

Another reason why researchers are developing modified plants is to enhance a crop's ability to resist disease. For example, in the United States, the Asgrow Seed Company has developed a type of bio-squash engineered to be resistant to certain viruses that naturally prey on this type of plant.

The United States isn't the only country to utilize genetic technology to make new, improved plants. In China, for example, some tobacco has been altered to resist the Tobacco Mosaic Virus, which can decimate crops. Other plants that have been engineered to have built-in virus disease resistance are African sweet potatoes, melons, and cucumbers.

One Hot Tomato

In 1994, the FDA (United States Food and Drug Administration) approved the first genetically altered plant to be marketed to consumers. Calgene, a biotech company in California, came up with the FLAVR SAVR™ tomato. The intention was to create a genetically altered plant that could withstand the long, bumpy ride from farm to market to consumers' dinner tables.

Ordinarily, when tomatoes are picked off the vine after they ripen, the long journey to the market can prove too much for these sensitive individuals of the plant world, and the result can be soggy tomatoes that have been bruised and abused. To get around this problem, tomatoes are usually picked when they are still green and unripe. They're harder then and can endure long rides and bumping around in crates, but this creates a new problem. Some consumers complain that often store-bought tomatoes that are picked when they're unripe don't meet their standards for taste and freshness.

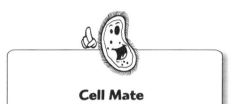

Cell Mate

The first genetically altered plant approved by the FDA for marketing to consumers was the FLAVR SAVR™ tomato, engineered for greater durability.

Researchers at Calgene decided to use genetic technologies to get around this durability versus taste standoff. They spent millions to find the gene that codes for the protein that causes tomatoes to go soft and watery and overripe.

They succeeded in taking this gene out and inserted a gene with the opposite letters. This is known as an antisense gene. If you'll remember what you learned about DNA's As, Ts, Cs, and Gs and how they code for specific proteins, then think of what would happen if you put in the opposite letters. This will cancel out the activity of the original gene. In this case, the gene that makes a tomato get overripe was deactivated. This is something like shutting off the switch for only one light bulb in your home.

The arrival of the FLAVR SAVR tomato, the first genetically modified food on the market, was met with a wide range of reactions. Some labeled the altered plants "Frankenfoods;" others viewed them as new, improved versions of the food that we normally eat. At any rate, although the shelf life of the FLAVR SAVR was long, its popularity was short. According to an article in *The Wall Street Journal* in 1995, a year after the tomato first came out, Calgene was having problems shipping the altered tomatoes because they apparently didn't always stand up to the trip to market.

Since 1986, over 2,000 transgenic plants have been developed around the world, and it seems that this trend will continue in years to come. However, there are a number of unanswered questions about the use of these altered plants. There are so many controversial issues that this book contains an entire chapter devoted to talking about the related questions of safety, ethics, ecology, and economics.

You can read about all these controversial topics in Chapter 26, "Food for Thought: Are Biofoods Safe?" We'll present both sides of any arguments, and you can weigh the facts and judge for yourself whether this new technology is a blessing or a bringer of future dangers.

DNA Data

Scientists have inserted genes from viruses, bacteria, chicken, fish, humans, and other organisms into plants. One company spliced a gene that codes for resistance to cold from an Arctic flounder into tomatoes. Some consumers might've wondered whether the tomato was very flat and capable of migrating to the bottom of the salad bowl!

Green-Collar Workers

Scientists are also developing plants that can produce substances that are needed by humans, agricultural animals, or even by the plant itself. The researchers are developing novel crops to increase productivity and growth, with the hope that these new crops may lead to cutting costs in the production of food.

Cell Mate

In the future, your doctor might prescribe a banana dessert for you when he wants you to get vaccinated. Some researchers are probably well aware that most people would prefer to eat a banana split with chocolate sauce, whipped cream, and a vaccine than to get an injection in the arm or the bottom.

Plants as Factories

Researchers have managed to put genes into a variety of plants to make them function as little factories or, well, plants. On the horizon are canola plants engineered to manufacture their own low-fat margarine or to be the source of raw materials for nylon, which is presently made synthetically from petroleum.

Researchers at Texas A&M University's Institute for Biosciences and Technology in Houston and Tulane University in New Orleans announced in 1995 that they were developing plants that they hoped could one day be used to deliver vaccines to people. Other researchers at the Boyce Thompson Institute for Plant Research in Ithaca, New York are working on producing bananas that contain a vaccine for hepatitis B in their DNA. Researchers at AgriStar Inc. in Texas are working to develop vaccines in lettuce and tomato plants.

Built-in Fertilizers

One of the many ways that scientists are using biotechnology to alter plants is by putting different organisms together to make the plant more efficient. For instance, scientists are trying to develop built-in fertilizers in popular food crops that need them.

Plants called legumes have a special relationship with friendly bacteria that naturally live in their roots. Plants like beans, peanuts, and peas are legumes. What these bacteria do is to take nitrogen from the air and make it readily available to the plant, which needs nitrogen for the growth of its green, leafy parts.

Plants such as wheat, rice, and corn are not legumes, so farmers generally add fertilizers with nitrogen to the soil so these plants can thrive. One drawback to the use of nitrogen fertilizers, however, is that it contributes pollution to rivers, which can result in the growth of unwanted plants in the bodies of water and the poisoning of fish. Researchers hope to employ engineered bacteria to turn plants that normally would not host the nitrogen-fixing bacteria into nitrogen-providing varieties.

The Least You Need to Know

➤ Researchers are genetically altering plants by adding genes from animals and humans to them.

➤ Using DNA delivered through bacteria, gene guns, and other tools, researchers can insert foreign genes into plant cells.

➤ In 1994, the Food and Drug Administration approved the first edible transgenic plant, the FLAVR SAVR™ tomato.

➤ Plants are engineered to resist diseases and pests and in the future may be able to produce built-in vaccines and fertilizers.

MOOOO....

Chapter 15

Down on the Pharm: Engineering Animals

In This Chapter

➤ The magic of microinjection

➤ Human protein–packed sheep milk

➤ Mice as models for disease and genetic research

➤ Putting people parts into pigs and pig parts into people

A lot has been going on down on the farm since Old MacDonald cared for his chickens and cows and other barnyard animals. Since the advent of genetic engineering, researchers have found ways to put animals to new uses that Old MacDonald never could have imagined. Since the early 1980s, scientists have been creating human genes in sheep's clothing. They've also succeeded in splicing human genes into a variety of other animals. In this chapter, you'll learn how and why pigs, sheep, cattle, and other creatures are being genetically altered to produce human proteins that are in short supply, and how "designer" mice have developed to aid researchers in their quest for cures for human diseases.

Molecular Pharming

You learned in Chapter 13, "Cut-Down Genes," that genetic engineering, or recombinant DNA technology, works by taking one or more genes from one organism and inserting them into the DNA of another organism. By doing this, you can cross species boundaries and join together organisms such as bacteria and humans or fish and tomatoes, which normally would never be seen together on a cozy date in a romantic restaurant.

DNA Data

An animal that is the result of splicing DNA from two different sources is a *transgenic* animal or a *chimera* (kih-MEE-ra). Its DNA is called chimeric DNA. A transgenic animal has had one or more genes from another organism added to its DNA or one or more genes of its own genes taken away from its DNA through genetic engineering. The term was coined by J.W. Gordon and F.H. Ruddle in 1981. A transgenic animal is also sometimes called a *chimera*. This word comes from the name of a monster in Greek mythology that had the head of a lion, the body of a goat, and the tail of a dragon. Appropriately, it was killed by a hero riding on a horse with wings.

Scientists are using this technology to alter animals to make biological products that humans need. For instance, transgenic pigs produce human insulin for use by diabetes patients. Human growth hormone is being produced by sheep. One reason for doing this is that using animals as factories is much cheaper than producing these pharmaceuticals in traditional ways. This type of genetic engineering of animals is called *pharming* or *molecular farming*.

Genetic Jargon

Pharming, or **molecular farming,** refers to the genetic altering of animals to produce human proteins that would otherwise be in scarce supply.

Waiter, There's a Gene in My Egg

When researchers want to alter animals to produce human hormones, the most common way for them to achieve this is using a process called microinjection. A glass needle thinner than a human hair is loaded up with a few hundred copies of a gene or two from human DNA, and it's ever-so-gently injected into newly fertilized eggs, for instance, sheep eggs. While looking at the eggs under a microscope, the researcher guides the needle straight into the nucleus of each egg, so the human gene or genes can hopefully recombine with the animal's DNA.

This procedure is delicate, and not every egg survives the needle's invasion. Even those that aren't destroyed by the microscopic operation may not take up the foreign gene into their DNA. Some eggs, however, will pick up the genetic hitchhiker. The genes are incorporated at random spots in the egg's chromosomes. Sometimes a gene ends up where the researchers intended, and other times, it winds up in a spot that hinders its ability to be expressed (or put into action). The success rate of this procedure is only one percent.

When a successfully altered egg develops into an embryo, it is put into an adult female sheep and allowed to develop normally. This is just like in-vitro fertilization for women. After birth, the offspring becomes a full-grown sheep that has to be tested to see if it produces the human protein that the inserted gene codes for.

The Milk of Human Kindness

If the microinjection procedure is successful, the human protein will be present in the full-grown sheep's milk. It's not exactly what Shakespeare had in mind when he wrote about "the milk of human kindness," but this sheep-made protein is identical in every way to the one that is produced in the human body. The protein is then extracted from the sheep's milk and can be used by an ailing person whose body is unable to make enough of it.

Animals are engineered to produce a human protein in their milk because the animals simply have to be milked and the human protein purified. This also means not having to kill the goose that laid the genetic "golden egg" to reap the protein harvest.

Cashing In

A number of companies are vigorously pursuing research into pharming. The Genzyme Transgenics Corporation in Massachusetts, for instance, has been developing goats to produce a human protein that controls blood clotting. Genzyme is also working on transferring human genes to goats in the hope of finding cures for maladies such as multiple sclerosis, skin diseases, and some cancers.

The United States isn't the only place where scientists have been adding human genes to the DNA of animals to produce human proteins. For many years, PPL (Pharmaceutical Proteins Ltd.), a company in Scotland, has created genetically modified sheep that produce, among other things, a human protein used to treat emphysema.

DNA Data

PPL, a company in Scotland that genetically alters sheep to produce human proteins in their milk, is the same company that donated one-third of the funding for research toward the development of Dolly, the famous cloned sheep that graced the covers of so many newspapers and magazines in 1997. Cloning is completely different from genetic engineering (recombinant DNA technology), so you'll have to wait until the next chapter to learn about "another ewe."

In the early 1990s, GenPharm, an international biotechnology company, genetically altered a bull named Hermann in the Netherlands. This black-and-white answer to every marriageable cow's dreams has a little extra something in his DNA. Hermann is the proud, if unaware, owner of a human gene that codes for lactoferrin (lack-toe-FEH-rin), a protein found in human milk. In 1994, Hermann became the father of at least eight calves, and they all inherited the human gene for lactoferrin.

In the future, companies hope to produce designer milks from cows that have been impregnated by bulls like Hermann. This humanized milk might be closer in composition to human milk than regular cow's milk, or might be genetically programmed to be fat free, or might be digested with ease by people who are lactose-intolerant. Who knows? Maybe someone will come up with instant chocolate milk or ready-made vanilla shakes straight from the cow.

Supermice and Friends

There's an old saying that if you build a better mousetrap, the world will beat a path to your door. We're still waiting for the new, improved mousetrap, but in the meantime, a scientist named Dr. Ralph Brinster decided to build a better mouse. In 1982, Dr. Brinster, working at the University of Pennsylvania's School of Veterinary Medicine, spliced the gene for Human Growth Hormone into mouse embryos.

If you're wondering what *Human Growth Hormone* (HGH, for short) does in humans, its name is a dead giveaway. It's basically responsible for the fact that you're not as tall as Shaquille O'Neal (and it's responsible for the fact that he's as tall as he is). An overabundance of the substance makes you tall, and a shortage of the stuff makes you short (go figure).

When Dr. Brinster succeeded in inserting the HGH gene into the mouse genome, the result was mice that grew twice as fast and almost twice as large as normal. They will continue to pass this gene on to their offspring, because this gene is now a permanent part of their DNA.

Genetic Jargon

Human Growth Hormone (HGH) is the substance responsible for normal growth in humans. Insufficient HGH results in dwarfism. The hormone is normally secreted by the pituitary (pi-TOO-i-terry) gland, which is found at the base of the brain.

The AIDS Mouse

Another famous mouse experiment was performed in 1987 by Dr. Malcolm Martin, working with a team of researchers at the National Institutes of Health (NIH) in Bethesda, Maryland. Dr. Martin was searching for a laboratory mouse that could help suggest a cure for AIDS. Normally, mice don't come down with the disease, so the researcher decided to design his own living research tool.

Using microinjection, he added the AIDS virus genome into the nuclei of some fertilized mouse eggs. When these fertilized eggs continued to grow, they were put into other mice who served as surrogate mothers. Not all of the mice that were born in this experiment carried the AIDS virus in every cell of their bodies, but about 10 percent of them did. These mice appeared normal, but when they were mated with mice without the AIDS virus, their offspring showed signs of having AIDS, such as skin conditions and pneumonia.

DNA Data

AIDS stands for Acquired Immune Deficiency Syndrome and is caused by HIV (Human Immunodeficiency Virus). In 1983, the virus was discovered simultaneously by two researchers working independently. They were Dr. Robert Gallo, working at the U.S. National Institutes of Health (NIH), and Luc Montagnier, working at the Pasteur Institute in France.

After the results of the experiment were announced, a number of critics, including scientists, expressed concern that if these AIDS mice ever escaped from the lab, they could mate with normal mice and possibly spread a new kind of AIDS. No one will ever know if this would have happened, because in 1988, an accident occurred in the laboratory, and the air supply to the mice was cut off. All but three of the AIDS mice died.

Two years later, in 1990, an article appeared in *Science*, a prestigious journal. In it, Dr. Robert Gallo, one of the discoverers of the AIDS virus, expressed his concern that using genetically altered mice could have grave consequences. If the human AIDS virus were to interact with other viruses normally occurring in mice, would this result in the creation of a new strain of the AIDS virus? Dr. Gallo and his team of researchers cautioned other scientists about the use of mice as AIDS research tools. They also pointed out that mice might make poor tools for human AIDS research, because if the human virus undergoes changes in the mice's bodies, then the researchers' conclusions would be inaccurate.

Designer Mice

Although the AIDS mice didn't turn out to be the laboratory tools that some hoped they would be, scientists have had better luck designing other mice to develop various human diseases. Researchers hope to use these mice to search for cures for these diseases. Some biotech companies have engineered mice that develop human ailments such as Alzheimer's disease, a degenerative disease of the brain, and rheumatoid arthritis, a painful disease of the joints. In 1988, the U.S. Patent and Trade Office issued a patent to Harvard University and the Du Pont Chemical Company for the Oncomouse, which was genetically altered to be prone to develop breast cancer and other tumors.

In 1992, the same Harvard team, led by Dr. Philip Leder, received a patent on another type of designer mice, in which the males develop enlargement of the prostate gland. Another type of mice netted a patent for two Dutch scientists and GenPharm International of California. These mice can receive tissue and bone marrow transplants without rejecting them because their immune systems don't function. Yet a third patent was granted in the same year to the University of Ohio for a mouse with the gene for human *interferon*, a protein that helps improve the body's natural response to disease.

Genetic Jargon

Interferon (in-ter-FEAR-on) refers to three proteins that help improve the body's natural defenses against disease. Interferons fight viruses and can slow the growth of tumors. The three types of interferon are alpha, beta, and gamma interferons. For example, when a virus invades the body, white blood cells produce alpha interferon, the main type, and this substance helps other cells that are not infected to become immune to the attack.

What a Knockout!

Other types of transgenic animals are developed when scientists delete a gene in a normal animal, for example, in mice. When scientists are studying how genes work together, they can determine the function of a specific gene by observing what happens to the animal once that gene isn't there.

In 1989, scientists found a new way of producing mutations called *gene targeting*. They found that if they introduced a new gene that was similar to one that was already on the chromosome, the original gene might swap places with the new gene. This phenomenon made it possible for researchers to trade a new, defective DNA sequence for a specific gene that would normally be expressed. In this way, they can knock out the gene that they want to study.

In 1995, a group of researchers working at Athena Neurosciences in San Francisco, California produced a type of knockout mice that exhibited symptoms resembling those of human Alzheimer's disease patients. But this experiment doesn't prove that the gene that was knocked out causes the disease. The gene that the researchers were studying was given an artificial *promoter* gene, which may have caused the gene to code for more of the protein than it normally would induce in cells. This change, instead of the normal workings of the gene, may be what caused the mice to exhibit Alzheimer's-like symptoms. Another problem is the fact that the mice don't behave exactly like human patients with Alzheimer's. It is hard to ask a mouse if it has forgotten something. Alzheimer's is also likely caused by alterations in more than one gene.

Genetic Jargon

A **promoter** is not a boss known for giving employees raises and new positions. It's a region of DNA in front of a gene that promotes the expression of that gene. In other words, it's like the on/off switch on a lamp or an electrical appliance.

Other types of knockout mice have been developed to study human ailments such as heart disease, hypertension, and other maladies that scientists would like to understand better on the molecular level.

A Certain Time and a Certain Place

In 1997, another development made it easier for researchers to zero in on the mysteries of some of the genes in the brain. Susumu Tonegawa, a Nobel prize winner and the director of Massachusetts Institute of Technology's Center for Learning and Memory, created a special type of knockout mouse. What makes this type of mouse so special is that the gene was deleted only in certain cells in the mouse's brain, and only after the brain had developed to a specific stage.

The way this living research tool was developed was with the addition of a promoter, a stretch of DNA that triggers a specific gene to do its thing. If the gene were a light bulb, then the promoter would be the on/off switch on the wall. The gene (the light bulb) is always capable of working, but it won't be expressed (lit up) until the switch is turned to the on position. In this case, the switch goes on only in certain cells in the brain and at a particular time in development.

Scientists hope to use these so-called regional knockout mice to study more about the brain. They hope to learn more about human ailments such as Parkinson's disease, or to discover more about how human sight works, or even to develop ways of enhancing a person's failing memory in the future.

Creature Commodities

In addition to studying how genes function using animal models, scientists are also interested in imitating or replicating some of nature's well-designed products for human use. For instance, the Army Research Laboratory in Massachusetts announced in 1990 that they had taken the gene that produces spider silk from the Golden Orb weaver spider and spliced it into bacteria. These microscopic creatures were tricked into reproducing more and more of the substance every time they reproduced (and bacteria reproduce quickly and often). This spider silk is much stronger than the kind we normally wear, which is made by silkworms.

In Australia, where there are about as many sheep as people, government scientists have already engineered sheep with enhanced wool growth. Researchers have also developed a genetically engineered hormone that can be injected into sheep that causes them to shed their fleece. This hormone may not be the "shear fantasy" that the researchers had hoped for, because one side effect is that some pregnant sheep abort their young when they are injected with this hormone. Other Australian researchers spliced a gene from tobacco into the sweat glands of some sheep. They hoped this would kill off any fly larvae that might burrow into the sheep's skin.

In one unique experiment, researchers from the University of California at San Diego managed to combine genes from plants and animals when they isolated the gene that causes fireflies' tails to glow in the dark. They took this gene and spliced it into tobacco and created a plant that glows in the dark! You might call this "light tobacco." Some observers joked that such a tobacco plant would come in handy to produce cigarettes that would be easy to find at night.

DNA Data

The reason why the abdomens of fireflies glow in the dark is because this helps them to find mates. But it's not likely that people smoking glowing cigarettes will attract mates in the dark, too.

Another fantastic genetic combination was created when researchers at the University of Davis in California combined the embryo cells of a goat and a sheep to produce a *geep* which has the body of a sheep, but the distinctive horns of a goat, and a face that looks more like a goat's than Mary's little lamb. This type of animal was first created in England and then later in Germany.

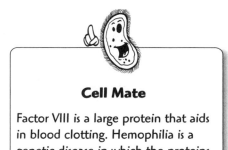

Cell Mate

Factor VIII is a large protein that aids in blood clotting. Hemophilia is a genetic disease in which the proteins responsible for blood clotting are not present. The hemophiliac therefore bleeds excessively.

Other animals that have been genetically altered include fish such as trout, Atlantic salmon, carp, and catfish. Many of these fish have received growth hormone genes from various other organisms, such as chickens, mice, other fish, and humans. Researchers have even spliced genes from cows into fish. Hopefully, you'll never hear of Mad Fish disease.

Popular Porkers

Pigs are also popular with genetic engineers. For example, William Velander of Virginia Polytechnic and State University in Blacksburg and William Drohan of the American Red Cross in Rockville, Maryland have developed "designer pigs" that produce constituents of human blood such as Factor VIII, which is needed by people who have hemophilia.

Humanized Pigs

Possibly inspired by Dr. Ralph Brinster's experiments with "supermice," Dr. Vernon Pursel, working at the U.S. Department of Agriculture (USDA) research center in Beltsville, Maryland, used microinjection techniques to splice Human Growth

Hormone genes into the egg cells of pigs. The animals were expected to grow quickly and to much larger sizes than usual.

But what works for one species does not necessarily work for another. Although the mice grew to nearly twice their normal size with the addition of human DNA, the pigs didn't fare so well. Only one out of every 200 engineered embryos survived. None of the transgenic pigs that were born was as healthy and robust as the mice that Dr. Brinster had engineered earlier. The pigs that lived were arthritic, had vision problems, and apparently had problems with their immune systems, because they were prone to pneumonia.

Since the days of the so-called Beltsville pigs, Dr. Pursel and USDA researchers have tried their luck at splicing genes from a chicken into pig embryos in the hopes of creating large-shouldered porkers. Dr. Pursel referred to these animals as "Arnie Schwarzenegger pigs."

Animal Organ Donors?

One proposed use of transgenic pigs is the future possibility of using them as potential organ donors for humans in need of hearts, livers, lungs, and the like. This use of animal organs is called *xenotransplantation.*

Ordinarily, the human immune system would reject an organ as foreign to the body as one that came from an animal like a pig. Upon being transplanted, the pig organ would turn black because the human immune system would recognize it as an invader and proceed to choke its blood supply. But researchers are hoping that by inserting some human genes into the pig donors, the human immune system will not recognize the pig parts as being foreign, and there will be a lesser chance of them being rejected.

Imutran, a company in Great Britain, views pig organs as a promising possibility. In the United States, Nextran, a corporate collaboration between DNX of Princeton, New Jersey and Baxter Healthcare of Illinois, has spent a long time researching the possibility of obtaining organs from humanized pigs and transferring them to people. Another U.S. company, Alexion Pharmaceuticals, is also working on developing animal organ donors for the future.

Genetic Jargon

Xenotransplantation (ZEE-no-TRANS-plant-AY-shun) is the use of animal organs to replace needed human parts. This term comes from the Greek work *xenos*, which means foreign.

To date, pig organs have not been successfully implanted in humans, but Novartis Pharmaceuticals Corporation has transplanted genetically altered pigs' organs, such as hearts and kidneys, into monkeys. They've met with limited success in keeping the monkeys alive. The longest a monkey recipient of a pig organ has lasted after such a transplant has been about two months.

You may be wondering why we haven't gone into controversial questions that naturally arise concerning whether researchers should proceed to mix and match genes from humans and animals. We'll get into all the ethical issues in Chapter 27, "Creature Concerns." If you can't wait to answer questions like, "Is it safe to have an organ transplant from a humanized animal?" and "What happens if these animals escape and mate with others in the wild?" and "Is this fair to the animals involved?" then skip to Chapter 27. See you later.

The Least You Need to Know

➤ Obtaining human proteins from genetically altered animals is known as pharming, or molecular farming.

➤ Human DNA is most commonly inserted into animals using a process called microinjection, which uses very tiny needles to inject the DNA into an animal cell's nucleus.

➤ Researchers have created designer mice that develop human diseases for use as lab tools to search for cures for these diseases in the future.

➤ Some biotech companies are adding human genes to pigs with the hope of transplanting their organs into humans in the future.

DID YOU SAY CLONES?

Chapter 16

Send in the Clones

In This Chapter

➤ What is cloning?

➤ How Dolly the sheep was cloned

➤ How mice and cows have been cloned

➤ How endangered species might be saved by cloning

➤ How human embryos might be cloned in the future

Everyone's heard about Dolly, the cloned sheep. But do you really understand how she was created? In this chapter, you'll learn about early attempts at cloning, and then read about how Ian Wilmut and his team of researchers in Scotland started counting cloned sheep. You'll also learn about developments that might lead to the cloning of humans. Of course, all of this technology has tremendous ethical implications; you can read more about the ethics of cloning in Chapter 29, "Should We Clone Around?"

Nature's Clones

Did you ever notice that sometimes nature seems to favor certain species with amazing abilities and to skimp on other species? For instance, if you remove one of a starfish's five arms, it will grow back. But don't try this on your cat. You could, however, pull off this trick with a tiny animal called a hydra (HI-drah). If a small part of a hydra is cut off, it regenerates into a whole new hydra. Some plants have this ability as well.

The Great Divide

Clones occur in nature. A *clone* is an identical copy of another organism; all of its DNA is the same as that of the original. An interesting case is the sea anemone (a-NEHM-a-NEE), which is an animal that looks like an underwater flower. Generally,

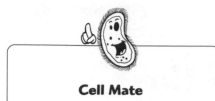

Cell Mate

The tiny hydra, which can regenerate cut-off parts and whose cut-off parts regenerate into a new hydra, was named after the many-headed monster in Greek mythology of the same name. If one of its heads was cut off, two grew back. Hercules, the Greek hero, finally did the monster in.

sea anemones reproduce by the union of eggs and sperm, but a sea anemone can clone itself. It creates an identical twin by gradually pulling itself in two opposite directions until it comes apart. Each side eventually grows into a separate individual that's exactly like the other.

Natural human clones are called identical twins. Generally, when a sperm cell fertilizes an egg cell, the fertilized egg cell begins to divide. More often than not, it will remain one embryo and develop into one child. But sometimes the fertilized egg will split into two identical halves that then develop into two separate babies. These babies are identical twins. Because they originally came from the same egg fertilized by the same sperm, they will, of course, have the same DNA. They look alike and have all the same genes.

DNA Data

Identical twins form when one egg that was fertilized by one sperm splits into two identical halves that develop into two separate babies. They have exactly the same DNA. However, there are also fraternal twins which form from two separate eggs. Each egg was fertilized by a different sperm cell. These eggs develop at the same time, and the children are born at the same time, but their DNA is different. That's why you can have twins that are one boy and one girl. Identical twins are always the same gender, of course.

Genetic Jargon

Clones are organisms that have exactly the same DNA. Identical twins are human clones, for example. The term comes from the Greek word that means twig.

Twice as Nice

In the past, scientists succeeded in creating clones of certain kinds of animals. In 1952, Robert Briggs and Thomas King cloned frogs from the cells of tadpoles. In 1962, John Gurdon replicated this feat, but he used cells from tadpoles that were older than the ones Briggs and King used.

One fairly simple method of cloning animals is called *embryo twinning*. In this procedure, researchers split an animal embryo in half at an early stage in its development, and the split halves sometimes develop into separate embryos, much like natural twins.

Another method for cloning animals has been used with cows. First, a researcher takes single cells from a developing calf embryo. Then the genetic material from one of these cells is put into an unfertilized cow's egg that has had its own nucleus removed. Finally, the egg with the new DNA is implanted into a surrogate mother cow. One drawback with this method of cloning is that sometimes—for reasons that are not yet understood—the calves come out weighing much more at birth than those bred the natural way. Sometimes the calves are so beefy that they have to be delivered by cesarean section.

Hall of Fame

In 1993, an amazing experiment was successfully performed. It was a breakthrough that many people viewed as the promise of great things to come, while others were frightened by its potential. In October 1993, the American Fertility Society held a meeting in Montreal, and two researchers from George Washington University in Washington D.C. reported that they had taken 17 human embryos and divided them into 48.

The two scientists were Robert Stillman and Jerry Hall (not to be confused with the beautiful model formerly married to Mick Jagger of the Rolling Stones). They never imagined that their experiment would be noticed by anyone other than those at the conference in Canada, but it hit the front page of *The New York Times*. The headline announced that for the first time ever, human embryos had been cloned.

Genetic Jargon

Embryo twinning is one method used to clone animals. Researchers take a developing animal embryo at an early stage and split it in two. This splitting usually results in two separate embryos, which develop into identical animals.

Cell Mate

Aside from scientific interest, the cloning of human embryos generated a vigorous ethical debate. If you want to learn more about the ethics of human cloning, then be sure to read Chapter 29, "Should We Clone Around?"

Brave New Clones?

Hall and Stillman weren't trying to produce human clones to be born and raised as identical copies of each other like something straight out of Aldous Huxley's 1958 futuristic novel, *Brave New World*. Rather, they were looking for alternatives to help childless couples become parents. Hall was the director of the in vitro laboratory, and Stillman was the head of the in vitro fertilization program at George Washington University.

In vitro fertilization is a technique used to help couples who are having difficulties conceiving a child. It is popularly known as making test-tube babies. Specialists at in vitro clinics take eggs and sperm from prospective parents and mix them in Petri dishes. Resulting embryos are implanted into the mother's uterus. (You can read more about in vitro fertilization in Chapter 24, "Sex Techs: Reproductive Technologies.")

Because Hall and Stillman performed their experiments in the hope that someday they might lead to an alternative for people who cannot conceive children naturally, they did not want to end up with full-grown human clones. Therefore, they deliberately experimented with abnormal embryos that could not develop for the required nine months to become a child. Some of the eggs had been fertilized by more than one sperm, which usually leads to a spontaneous abortion.

Double Take

The way the procedure worked is relatively simple. When a fertilized egg divided into a two-cell embryo, Hall and Stillman removed the outside coating of the cells and separated them. As you read earlier in this chapter, this occasionally happens naturally. When the two cells separate, they form two embryos, and nine months later, they are a pair of identical twins.

Then Hall and Stillman experimented with creating a new outside coating for each embryo made from a seaweed gel. This artificial outer coating seemed to do the trick, and some of the cells continued to develop. However, none of these clones grew for more than six days. None of them grew to be 32-cell embryos, the size at which they could theoretically be implanted into a woman's uterus to eventually develop into a child.

Ewe Again?

Although scientists had some success in splitting embryos and adding DNA from developing cells into animals, some scientists thought that cloning from an adult cell in a higher organism was impossible. The reason for their skepticism was the fact that DNA in adult cells contains many genes that are switched off.

In the early stages of the development of an embryo, all of its cells are undifferentiated. An undifferentiated cell can grow into any other type of cell in the body. It can turn into hair cells, heart cells, brain cells, or anything else that's programmed in the genes. But as the embryo continues to develop, most of the genes in the cells switch off, so the undifferentiated cell can become whatever type of adult cell it's destined to become. So even though all the genes of an organism are in every single cell that it contains, different sets of genes are turned on in different cells.

There's a practical reason for genes to switch off. You'd have a heck of a time if the genes for your toenails turned on in the cells in your forehead. You'd look like a unicorn. If the cells in your back suddenly turned on the genes to grow a foot, you might give yourself an unwanted kick in the seat of the pants. So it's necessary that most of the genes in your cells aren't activated at all times. But they're still there.

Some scientists theorized that these cells were switched off for good and couldn't be turned back on again. But that theory was proved wrong when a little lamb named Dolly was born quietly in Scotland on July 5, 1996. Her birth wasn't announced until

February 1997, when *Nature*, the same magazine that heralded Watson and Crick's discovery of the structure of DNA, published a paper with the sheepish news.

A Nuclear Reaction

Working at the Roslin Institute in a town near Edinburgh, Scotland, embryologist Ian Wilmut succeeded in creating the woolly clone using a process called *nuclear transfer*. To do this, he needed two cells: one to provide the DNA and the other an egg cell with its own DNA removed. Wilmut took the egg cell from a Scottish Blackface sheep. This ewe, or female sheep, had her egg cell removed shortly after ovulation because this is when egg cells are ready and waiting to be fertilized.

Using a microscope to view the procedure, a researcher at the institute held the egg with an instrument called a pipette (pie-PET). Then another type of pipette, many times thinner than a human hair, was used to suction out the nucleus of the cell. This is something like putting a straw into a gelatin dessert and sucking out a small piece of fruit. This nucleus, containing the DNA of the Blackface ewe, was discarded. What was left was an unfertilized egg without any genetic programming.

Genetic Jargon

Nuclear transfer is the process that was used to clone Dolly the sheep. The DNA from an adult cell was put into an egg cell that had its nucleus removed. After this egg was stimulated to grow using electricity, it was implanted into another sheep that later gave birth to Dolly.

Teaching an Old Cell New Tricks

Next, the researchers took a cell from the mammary gland in the udder of a six-year-old Finn Dorset sheep that had been pregnant for three-and-a-half months. Wilmut's team worked with cells from a pregnant ewe's udder because these cells grow rapidly during pregnancy. Finn Dorset sheep are completely white, as opposed to the Blackface sheep, who have black faces (go figure). The reason Wilmut chose to work with two different breeds was so he could be sure that a clone was produced and that the resulting lamb wasn't produced from the DNA of the Blackface sheep.

Wilmut and his team put these mammary cells into a petri dish. At this point, instead of giving the cells the nutrients that would enable them to divide and grow, the scientists starved the cells. This starvation caused a change in the cell cycle, which you read about in Chapter 11, "Gene Gangs." The cell cycle is something like a time schedule that a cell follows when it divides. Without the proper nutrition, the mammary cells slowed down and went into a kind of sleep-like stage. They stopped dividing.

Next, the "empty" egg cell without a nucleus, which was taken from a Blackface ewe, was put next to one of the sleeping mammary cells. The researchers used pulses of electricity to fuse the two different cells together into one new cell. After this, another electric pulse caused the new fused cell to start dividing. When an egg cell is fertilized by a sperm cell, there is a natural burst of energy that starts the cell dividing. This division progressed the way it would in a normal embryo.

How Dolly was cloned.
(a) Mammary cells were taken from a Finn Dorset sheep and placed in a Petri dish. They were deprived of nutrients to stop cell division.
(b) An unfertilized egg was taken from a Scottish Blackface sheep, and its own nucleus was removed.
(c) A mammary cell and the "empty" egg cell without a nucleus were put next to each other, fused together with a pulse of electricity, and induced to grow into an embryo with another pulse.
(d) About one week later, the embryo was implanted into a second Blackface sheep, who later gave birth to Dolly, a Finn Dorset lamb.

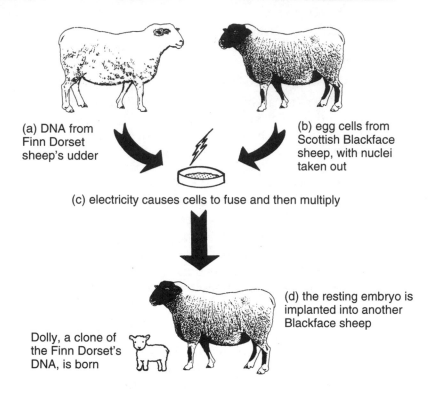

(a) DNA from Finn Dorset sheep's udder

(b) egg cells from Scottish Blackface sheep, with nuclei taken out

(c) electricity causes cells to fuse and then multiply

(d) the resting embryo is implanted into another Blackface sheep

Dolly, a clone of the Finn Dorset's DNA, is born

This new embryo was allowed to develop for about a week and then was implanted into a third sheep. This surrogate sheep mom was also a Blackface ewe. If Wilmut experimented with three sheep of the same breed, the results could have been confusing or misleading, but because the DNA from the mammary cells came from a completely white Finn Dorset ewe, there could be no mistake as to whose DNA the resulting lamb had grown from. The all-white lamb, Dolly, could not have been created from the DNA of either the egg-donating Blackface ewe or the surrogate mother Blackface.

DNA Data

The first lamb to be cloned from the DNA of an adult cell was named Dolly. Because the DNA came from a mammary cell, the lamb was named in honor of Dolly Parton—for obvious reasons.

Hello, Dolly

The creation of Dolly the sheep was a landmark in genetic science. The birth of the ewe put to rest the theory that once genes in cells are shut off, they can't be switched back on again. Dolly's origin in the DNA of an adult sheep cell brings both promise and concern to scientists and laypersons alike.

Scientists at the Roslin Institute hope that this cloning procedure can lead to more efficient use of genetic engineering. Companies like PPL, a company that donated one-third of the funding for the research that led to the cloning of Dolly, have already genetically altered sheep to produce human substances such as Factor IX, a blood-clotting protein. Some hemophilia patients have been treated with this substance. But genetic engineering can be a hit-or-miss proposition. With cloning, when a researcher successfully alters an animal, such as a sheep that produces Factor IX, cloning that animal can produce large flocks of these sheep with human genes in a shorter amount of time than just trying to genetically alter one animal at a time.

Ian Wilmut hopes that this technology will produce animals that can be used to find cures for human diseases. For instance, sheep could be genetically altered to develop cystic fibrosis, a genetic disease that causes breathing difficulties because of a severe accumulation of mucus in the lungs. Then, using cloning techniques, whole flocks of these altered sheep could be raised as laboratory animals.

Cell Mate

Although some people look forward to the applications of new cloning technologies, others contend that the use of living animals for research is unethical. You can read about these issues in Chapter 27, "Creature Concerns."

Another possible use of this technology might be to complement the genetic engineering of pigs that are altered to contain human genes. Cloning could advance research into the use of these pigs as organ donors for people.

Cloning using Wilmut's procedure is not all that easy and reliable, however. Ian Wilmut and his team at the Roslin Institute had to count a lot of sheep's eggs before they succeeded in creating Dolly. They tried to get embryos to grow through the nuclear transfer technique 277 times. Of these attempts, only 29 embryos developed that lived for more than six days. Of these 29 embryos, Dolly was the sole survivor.

If you saw the 1996 film *Multiplicity*, with Michael Keaton and Andie MacDowell, you might have envied the character who had himself cloned several times over so he could work all day and still have time left over for his family. Don't get your hopes up yet. There's still a lot of research that has to be done before you can start making genetic photocopies of yourself.

One concern with cloning Dolly from the DNA of a six-year-old ewe was the possibility that the DNA might have been too old or that it might cause Dolly to age rapidly or grow up in poor health. But Dolly proved to be a healthy and fertile sheep. She had a romance with a Welsh mountain ram, and nature took its course. After mating the old-fashioned way, Dolly gave birth to a lamb named Bonnie in April 1998.

DNA Data

In March 1998, Dolly was fleeced; the sheepish celebrity was shorn and her wool was made into a sweater. The Cystic Fibrosis Trust in the United Kingdom sponsored a competition to design a sweater that would be knit from the cloned animal's wool. The finished sweater, which depicts lots of happy sheep on the front and the back, was exhibited at London's Science Museum.

Multiplying Mammals

Dolly the sheep was the sole survivor of 277 egg cells that Ian Wilmut and his team at the Roslin Institute in Scotland attempted to clone. But since then, other researchers have topped Wilmut's success.

Mighty Mice

In 1997, a Japanese postdoctoral student at the University of Hawaii named Teruhiko Wakayama had success with his own cloning experiment. Wakayma studied with laboratory director Ryuzo Yanagimachi, also at the University of Hawaii, and thought of cloning as a hobby.

Instead of using cells from the udder of a sheep like Ian Wilmut and his team, Wakayama took what are called cumulus (KYOOM-you-luss) cells, which are cells that surround the egg in the ovaries, from female mice. Wakayama's method also differed from the cloning of Dolly because he did not use electric pulses to fuse the two cells or to jump-start cell division, as Wilmut had done. Instead, he injected the cumulus cell nucleus into another mouse cell that had its own nucleus removed.

The first cloned mouse, which Wakayama named Cumulina (after the cumulus cells), was later joined by 49 brothers and sisters, all copies of the same identical DNA. They were in good health and fertile. Wakayama thus learned that clones could be made from clones like photocopies of a photocopy. All together, the 50 furry clones represented three generations of mice. Also, Wakayama's success rate of 2 clones for every 100 tries was much greater than that of Wilmut's team. This was a major advance for biomedical research because so much genetics research is done using mice.

Big and Beefy

1998 was also a big year for the cloning of ewes and the possibility of cloning humans. Scientists at Kinki University in Nara, Japan reported in the December 11, 1998 issue of *Science* magazine that eight calves were cloned from an adult cow. The technique used was similar to that employed in the cloning of Dolly, but the scientists used the cumulus cells that surround the eggs in the ovary that Teruhiko Wakayama favored in his cloning experiments with mice at the University of Hawaii.

The researchers at Kinki University used cumulus cells from one adult cow. The DNA from her cumulus cells was put into egg cells that had their nuclei removed, and much like Wilmut's procedure, electricity was applied to start the cell division. The embryos grew, and then ten of them were implanted into five cows.

All of the cows were successfully impregnated and gave birth to calves. Unfortunately, though, four of these calves died either at birth or shortly afterward. But so far, the Kinki team's success rate has been the highest for cloning large mammals. Japanese have been known to pay as much as $100 per pound for high-quality gourmet beef. The Kinki University cloning technique could be used to produce more of this high-class cattle.

Holy Cow!

Another beefy advance in cloning came from a company in Boston called Advanced Cell Technology, which announced in 1998 that it had developed a method to produce *stem cells*. These cells, found in embryos, have all of their genes turned on, which means that they still have the potential to grow into any type of cell that the body needs, such as heart cells, liver cells, or skin cells. The procedure for generating these cells is similar to the nuclear transfer process that produced Dolly. The aspect of this procedure that some people find a bit unsettling is the fact that it involves fusing a human cell together with a cow's egg cell that has had its nucleus removed.

Researchers are hoping to use this technology to produce a wide range of cells that are needed by humans to alleviate or to cure certain illnesses. For instance, if someone needed a bone marrow transplant or suffered from a disease such as diabetes, the stem cells might be used to grow the specific cells that they need to restore their health. Because the DNA used to produce these cells would come from the same person who would be receiving them, there would be a lower chance of the new cells being rejected by the body as foreign invaders.

Genetic Jargon

Stem cells are undifferentiated cells in an embryo. They can grow into any type of cell in the body, such as heart cells, blood cells, or brain cells.

Saving Species

Another promising application of cloning is the possibility of saving species that are in danger of becoming extinct. Using the genetic material from one or a few endangered animals, scientists might be able to generate clones that could then mate normally and preserve the species.

Rescuing a Lady

In 1998, cloning came to the rescue of the Enderby Island breed of cattle, when scientists at the Ruakura Research Center in Hamilton, New Zealand successfully cloned a very special cow named Lady, the last female of this rare breed. Before turning to cloning, scientists tried artificial insemination, but Lady only gave birth to one bull as a result. Concerned that Lady was getting older and that time was running out, the scientists decided to try cloning her using a procedure similar to the one used by Ian Wilmut to clone Dolly the sheep. The experiment was a success.

Lady's clone is of course a female, because she has the same exact DNA as her mother. The calf's name is Elsie, and when she matures, scientists at the research center in New Zealand intend to artificially inseminate her with the sperm of an Enderby Island bull. The sperm has already been frozen and is being kept in storage.

DNA Data

Need a compatible friend for your collie or your Siamese cat? At a loss for what to do when your guinea pig bites the dust? You can always try cloning your pet. There are rumors that an American millionaire paid to have his dog cloned, and supposedly there's a company on the Internet called Clonapet, which advertises its animal cloning services.

Panda-monium

Another animal that might be saved from extinction by cloning is the giant panda. Less than one thousand of these animals are left in the wild. There are some in captivity, but attempts at artificially inseminating them have not met with much success.

In July 1998, China's Academy of Sciences announced the start of an official project to use cloning to try to increase the dwindling number of wild giant pandas. Professor Chen Dayuan is heading the panda project. Chinese scientists will attempt to take DNA from panda cells and transfer it into the eggs of other animals. These eggs will have their own nuclei removed.

When the scientists decide on the type of animal to use as the egg donor, they will also put these animals to work as surrogate mothers that will hopefully give birth to pandas. Although the scientists haven't decided which animal is the best to use for this purpose, they have suggested that dogs might work as possible egg donors and surrogate mothers because their pregnancies are similar to those of pandas. The Chinese scientists hope that the panda project can be completed within the next three to five years, and the world will see the regrowth of this species.

Be a Clone

A scientific development in South Korea in December 1998 added fuel to the ongoing debate over human cloning. Researchers at the Kyunghee University Hospital took DNA from the cells of an infertile woman. Then, they removed the nucleus from one of her own egg cells, and substituted the nucleus from her adult cell containing her DNA. By now, this process should sound pretty familiar to you. It's nuclear transfer, and it's basically the same procedure that was used in Dolly's cloning.

The embryo was not allowed to grow, however, because South Korea has laws that restrict experimentation on human embryos. When the cloned embryo divided twice, researchers were legally bound to destroy it before the embryo grew any larger. Some people viewed this research as a giant step forward that eventually might result in the cloning of humans.

But aside from the ethical concerns that have always surrounded the possibility of cloning humans, there are also some scientific critiques of this experiment. Some researchers believe that because the cells in the South Korean experiment were not allowed to undergo cell division more than twice, it hasn't been proven that embryos cloned using this method can develop into healthy humans. They feel that just because the embryos were off to a good start doesn't mean that they wouldn't encounter difficulties in developing into more advanced stages.

The Least You Need to Know

➤ Clones are two or more organisms that have exactly the same DNA.

➤ In 1996, Dolly the sheep was cloned using a process called nuclear transfer, in which DNA from an adult cell is put into an egg cell that has had its own nucleus taken out.

➤ In 1998, Terihuko Wakayama cloned 50 mice at the University of Hawaii.

➤ Cloning might be a way to save endangered species, such as the giant panda in China.

➤ In 1998, scientists in South Korea cloned an early-stage human embryo from an infertile woman's body cells, but they terminated their experiment before the embryo grew any larger because of legal restrictions.

Part 5
DNA Detectives

The information in your genes tells your body how to make substances that will keep you alive and healthy. But it can also tell stories about who you are and what family you come from.

Scientists are currently using DNA testing to prove things like paternity or maternity, to solve crimes, and to look out for genes that might cause diseases in a person. This section tells you the nuts and bolts of these rapidly expanding technologies.

Chapter 17

Stories Genes Can Tell

There's no one quite like you. You look and act differently than anyone else, and your DNA is yours and yours alone—unless you have an identical twin or you've been cloned (in which case you're a very well-kept secret).

Scientists have learned to take advantage of the fact that everyone has unique DNA to devise techniques to identify people, to determine whether they're in the same family, or to see whether hair, blood, or other human material left at the scene of a crime matches up with that of a suspect. Read on to learn how these DNA tests, called *DNA fingerprinting*, *DNA profiling*, or *DNA typing*, have reunited families, gotten some dead-beat dads to take responsibility for their children, and solved some long-standing mysteries. (If you want to learn more about DNA in criminal cases, be sure to read Chapter 18, "DNA's Day in Court." You'll learn more about the ethical issues surrounding DNA testing in Chapter 28, "People Problems.")

Your Genetic ID Card

Scientists have discovered the amazing fact that we're all alike yet different. If you take a bus ride and glance at the people sitting around you, you wouldn't think that you

Cell Mate

DNA profiling, also called **DNA typing** or **DNA fingerprinting,** uses genetic testing to compare DNA samples and determine identity. For instance, it can be used to prove maternity or paternity, to positively identify a deceased person, or to determine whether the DNA contained in blood or hair found at the scene of a crime matches that of a suspect.

had anything in common with some of them, especially not the guy who's talking to himself or the woman who's making weird noises with her bubble gum.

But the fact is that we share 99.9 percent of the same order of DNA in almost every cell in our bodies. Yet there's a reason why we're so different. If you consider that there are approximately three billion base pairs of As, Cs, Ts, and Gs that spell out the sequence of our DNA, then that 0.1 percent difference takes on a new meaning. It can translate into several million unique spelling differences in each individual.

So where do most of these spelling variations occur in our DNA? The DNA that codes for proteins can't vary a whole lot, or else we'd be in trouble. As you read in Chapter 9, "Irregular Genes," nucleotide misspellings that occur in the wrong places can wreak havoc with your health. For instance, a misspelling of one particular gene leads to sickled red blood cells, which can cause tremendous problems for a person with this mutation.

What about the DNA that doesn't code for proteins? Mutations or spelling variations in the noncoding areas of DNA—the parts that don't direct the cell to make proteins—probably won't affect a person's health in a bad way. With this in mind, scientists examined many areas of noncoding DNA and observed that some of them were more likely to have variations than others.

You'll remember reading in Chapter 7, "S, M, L, and XL Genes," that in some areas of "junk" DNA—the DNA that doesn't code for proteins—can have special sequences of repetitious letters. Some of these repetitions occur over and over again. They're called *tandem repeats*. The repeating sequences are found right next to each other like a genetic stutter.

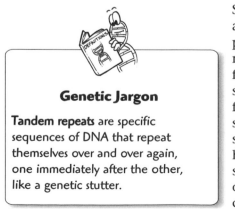

Genetic Jargon

Tandem repeats are specific sequences of DNA that repeat themselves over and over again, one immediately after the other, like a genetic stutter.

Suppose a scientist examines one of your chromosomes and finds one of these tandem repeats. You have a particular spelling variation, and this nonsense "word" repeats seven times, with each repeat immediately following the other. At one particular site on the chromosome, you might have 10 repeats of this genetic stutter from your mother and 15 repeats from your father. But someone else might have, say, 12 repeats of this same sequence inherited from his mother and 14 repeats from his father at the same site on his copies of that chromosome. This difference in the number of repeats would be one way that your DNA and that of the other person could easily be distinguished from each other.

Invisible "Fingerprints"

Some genetic differences are so individual that they have been compared to fingerprints. In the same way that everyone has different fingerprints on the tips of their hands, these DNA variations are also unique. Some people object to the term *genetic fingerprints*, however. They prefer the term *genetic profiles* because only a small portion of the total DNA is compared to a sample in this process.

Although some people in a population may have similar repeat patterns, the odds are extremely high that the DNA sequences in two different people will not be the same provided you look at a sufficient number of different patterns. Usually, scientists compare between five and ten different areas of a person's DNA to a sample to decrease the chances of a random match.

Testing DNA with RFLP

In 1984, a British geneticist named Alec Jeffreys came up with a method for comparing DNA samples to prove or disprove identity. He found areas of DNA that were variable enough to be used in identity testing. His method for creating a DNA profile used an existing technique called a *Southern blot*.

This test works by analyzing a sample of hair, blood, saliva, semen, or other bodily substance obtained from a person. This will be compared with the DNA of another person if the scientist wants to establish that the two are relatives. This comparison can also prove paternity or maternity. Or if blood or hair is found at the scene of a crime, a suspect's DNA can be matched against it. (See Chapter 18, "DNA's Day in Court," for more about some notable cases that have used this technology.)

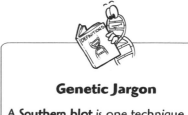

Genetic Jargon

A **Southern blot** is one technique used in creating DNA profiles. It was named after Edwin Southern, who published the technique in 1975.

DNA Data

A British company bought the patent for Alec Jeffreys' method of DNA analysis and, in 1987, established Cellmark Diagnostics, a DNA testing company in England and the United States. One of Cellmark's most famous assignments was the DNA testing used in the O.J. Simpson trial.

The test is highly accurate, but a fair amount of the bodily substance is required. For instance, a blood stain the size of a dime is enough for use in the test. A scientist might need several thousand cells for this procedure. And this method can take a long time as well, sometimes several weeks, before the result is known.

Dividing DNA

The first step in the Southern blot technique is extracting DNA from the blood or saliva or other source. Then a restriction enzyme is mixed in. You'll remember from Chapter 13, "Cut-Down Genes," that restriction enzymes act as molecular scissors.

Therefore, the DNA in the sample is cut at specific sequences. Because the enzymes cut the DNA only when they encounter a certain sequence of nucleotides, for example CAATGC, the result is many, many fragments of DNA of different lengths.

The next step is putting the DNA fragments in a container filled with agarose gel, a gelatin-like substance that is derived from seaweed. Electricity is run through the gel, causing the fragments of DNA to move to one end of the gel. Short fragments of DNA move faster than longer fragments, because littler fragments can make their way through the dense gel in less time. This part of the process is called *electrophoresis*.

Now the different lengths of DNA have been separated and form little bands. But they're invisible at this point. In order to see where they are, a piece of nylon is used like a blotter to pick up the patterns of DNA pieces.

Genetic Jargon

One of the steps involved in DNA profiling is called **electrophoresis** (ee–LECK–tro–for–EE–sis). In this process, electricity is passed through a gel containing DNA fragments. This process moves the fragments from one end of the gel to the other, and the smaller fragments move through the gel quicker than the longer fragments.

Now something called a probe is added. This is like a radioactive "tag" or "label." Each probe is made in such a way that it only connects to a certain sequence of DNA. You'll read more about DNA probes in Chapter 19, "DNA Disease Diagnosis." Finally, the nylon blot carrying the DNA fragments is placed on top of regular X-ray film. The radiation from the tags on the DNA slowly exposes the film. This exposure makes a picture of different bands on the X-ray film, and these bands show where the different DNA fragments that bound with DNA probes wound up. These pictures look a lot like the bar codes that you see in the supermarket, but they tell you a whole lot more than the price of dog food or granola. These DNA pictures can be scanned into a computer and compared with other samples of DNA. If the samples don't match, then they don't come from the same person.

Mama's Boy

Alec Jeffreys devised his method of DNA testing in 1984, and just months later, in 1985, he was approached by lawyers in England who were involved in a highly debated immigration case. They wanted Jeffreys to prove that a young boy was the son of a woman from Ghana.

The woman, Christiana Sarbah, had immigrated to England and established residency there. She later gave birth to her son Andrew in England. When she and her husband subsequently divorced, Christiana remained in England, but her estranged husband returned to Ghana. Andrew eventually went to live with his father in Ghana, but afterwards he decided to return to his mother, brother, and two sisters.

When Andrew returned to England, British officials were not convinced that he was Christiana's son. They suspected that he might have been her nephew or a totally unrelated child from Ghana posing as her son and trying to establish residency in England. As a result, Andrew was about to be deported.

Christiana went to the Hammersmith Law Centre, which gives legal aid to the disadvantaged. The lawyers tried to prove that Andrew was the woman's son by obtaining family photographs and having blood tests taken to use as evidence. Immigration authorities were not convinced. They countered that even though Andrew and Christiana's blood tests showed that they were related, there was no conclusive proof that he was her son and not her nephew.

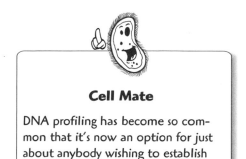

Cell Mate

DNA profiling has become so common that it's now an option for just about anybody wishing to establish that they are (or aren't) related.

Christiana's lawyers had read about Alec Jeffreys and his DNA tests and asked him to help prove that Andrew was Christiana's son. One challenge for the scientist was the fact that Christiana was not completely sure of Andrew's paternity, although it was probable that her estranged husband in Ghana was the father. In addition, none of Christiana's sisters were available for testing to prove that Andrew was not their son instead.

What Jeffreys did have to go on, however, was the fact that Andrew's brother David and two sisters, Joyce and Diana, were in England and could be tested. Jeffreys compared their DNA test results with Andrew's, Christiana's, and an unrelated person. When the "bar code" patterns were compared, Jeffreys determined from their similarities that Andrew was indeed Christiana's son. (Jeffreys also determined that Andrew and his siblings all had the same father.) DNA testing had solved the first of many mysteries that could not be cleared up by blood tests or other means.

A DNA paternity test done by the LIFECODES Corporation. The first alleged father turns out to be the true father, because his DNA matches up with the child's. The test also shows that the second alleged father is not related to the child. Photo courtesy of LIFECODES Corporation

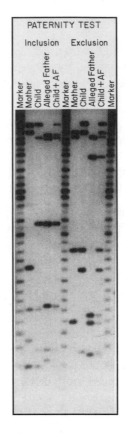

PCR: The More the Merrier

The method that Alec Jeffreys had developed still continues to be used. One small drawback to this method, however, is that it requires relatively large amounts of human substances. But what happens when there's just a drop of blood at the scene of a crime or a small stain on the dress of a White House intern?

Also, what if the DNA is old and broken down? When a living thing dies, its DNA begins to degrade due to the action of enzymes within the organism. Other factors contribute to the destruction of DNA as well, including heat, ultraviolet rays from the sun, and bacteria. As a result, there's less DNA left for analysis.

Copy Machine

The problem of how to analyze small amounts of DNA was solved by a technique called *PCR*. The technique

Genetic Jargon

PCR stands for Polymerase (puh–LIM-a-race) Chain Reaction. This technique is used in DNA testing to produce millions of copies of a short sequence of DNA, even if it has broken down considerably.

Polymerase is a type of enzyme that helps out in the process of copying DNA.

was devised by Kary Mullis, a scientist who was working for the Cetus Corporation in California at the time. One night in 1983, Mullis was driving down a mountain road with a friend when he suddenly got a flash of inspiration. He instantaneously realized how to multiply small amounts of DNA.

The way PCR works is like this. The scientist decides on a specific sequence of DNA to study. The scientist places the DNA sample—from blood or saliva, for example—into a test tube. Along with the sample, he or she adds DNA polymerase—an enzyme that helps out in the process of copying DNA—plus a supply of As, Cs, Ts, and Gs, plus two primers.

A primer is a short sequence of DNA. One primer binds to the left side of the selected DNA sequence to be copied, and the other primer binds to the right side. In other words, the DNA sequence to be copied is "fenced in" by the two primers. This sets the stage for millions of copies to be produced. The polymerase will copy only the DNA that's been "fenced off" by the primers.

The enzyme used in PCR is called Taq polymerase, and it comes from a type of bacteria that lives in hot springs and can withstand high temperatures. This heat resistance comes in handy in this procedure, because high temperatures are used in the PCR method.

Cell Mate

The name Taq polymerase comes from *Thermus aquaticus*, a type of bacteria found in the hot springs of Yellowstone National Park. Taq polymerase is used in the PCR method of creating millions of copies of a small amount of DNA. The bacteria can withstand the high temperatures that are used in the PCR process.

DNA Data

Kary Mullis, the scientist who invented the PCR method for making copies of DNA to aid in DNA testing, worked for the Cetus Corporation, a biotechnology company in California. His employers were so pleased with his invention that they gave him a $10,000 bonus. They later sold the patent for Mullis' invention to Hoffmann-La Roche, a pharmaceutical company, for $300 million. Although Mullis didn't get the big bucks for his discovery, he was awarded the Nobel prize for it in 1993.

The PCR method is so effective that scientists can make about one million copies of the desired DNA in about three hours. Usually after all these copies are made, or *amplified* as it is called, they go through electrophoresis.

In the early days of DNA testing, PCR tests were considered less precise than the technique developed by Alec Jefferys, but they have now been improved and are considered reliable. The great advantage to PCR tests is that they can be conducted quickly. Often they are done in just a few days as opposed to weeks. In addition, PCR is sometimes the only way to go when there's only a small quantity of original DNA or when the DNA being analyzed isn't in the best condition.

From PCR to STR

Currently, there is a trend toward using a kind of combination of the PCR and Alec Jeffreys' methods called *STR* testing. STR testing has the advantages of both methods. Instead of studying long repeated sequences, the STR method uses shorter ones. The PCR technique is then used on these segments to multiply them. The STR technique combines the precision of Jeffreys' method with the speed of PCR.

Genetic Jargon

STR stands for short tandem repeats. STR testing analyzes small segments of repeated DNA. It combines the advantages of Jeffreys' method and PCR.

And there's more on the horizon. In January 1999, a report in a scientific journal stated that Stephen R. Quake and a research team at the California Institute of Technology in Pasadena used a special microchip to analyze DNA. The chip performs the analysis many times faster than conventional technology. Molecules go through channels in the microchip and pass through a laser beam. In this way, the researchers "fingerprinted" the DNA of a virus in just 10 minutes. You can read more about biochips like this in Chapter 19, "DNA Disease Diagnosis."

From Daddies to Mummies

PCR has other uses besides identifying people through their DNA profiles. The process can also detect mutations in a person's DNA that can cause diseases. You read in Chapter 7, "S, M, L, and XL Genes," that the largest gene ever identified is the one for Duchenne muscular dystrophy, a degenerative disease that occurs mainly in males. This gene stretches out over an immense two million base pairs. The invention of PCR has made it easier for scientists to sample this enormous gene along its length to check for mutations.

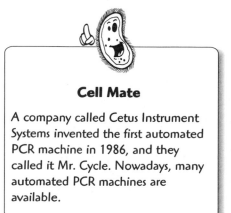

Cell Mate

A company called Cetus Instrument Systems invented the first automated PCR machine in 1986, and they called it Mr. Cycle. Nowadays, many automated PCR machines are available.

The PCR technique is so sensitive that it can be used to study broken-down DNA samples or even DNA obtained from individual cells. Scientists can now glean genetic secrets from such exotic sources as an Egyptian mummy or a zebra-like animal called a quagga, which went extinct in the late nineteenth century.

Some scientists even claimed to amplify DNA from a wasp that was trapped in the sticky sap of a pine tree and died 90 million years ago. When the sap eventually hardened into a golden piece of amber, the intact body of the wasp remained inside.

Deadbeat Dads and DNA

As you learned in the story of the Ghana immigration case, DNA testing is more conclusive than conventional blood tests. Nowadays, if a woman isn't sure who fathered her child, or if she knows but the father needs some reassurance that the child is really his, then DNA testing can help.

An Actor's Genes

DNA testing can even follow a person into the grave to determine paternity. When French singer and actor Yves Montand died in 1991, he left behind an estate worth more than $3 million. He was survived by his live-in partner Carole Amiel and their young son, the only child he ever recognized as his biological child. He also had an adopted daughter with actress Simone Signoret, the only woman he ever married.

But during his lifetime, a beautiful young woman named Aurore Drossard claimed that the seductive French actor was her father. Montand admitted that he had an affair with Drossard's mother in the 1970s, but he denied that he was Aurore's father. Montand also refused to take a DNA test to disprove Drossard's claim.

In 1994, three years after Montand's death, Drossard continued to push for recognition as the actor's daughter. She went to court, and a judge ruled in her favor. One reason was the fact that Drossard's face strikingly resembles Montand's. To be absolutely sure, the court eventually decided to exhume Montand's body for DNA testing to prove once and for all whether Drossard should inherit part of the estate.

When the DNA test results came back in December 1998, the ruling that Drossard was Montand's daughter was overturned. The young woman was shocked to learn that the DNA tests indicated that the actor was not her father. The decision was appealed by Drossard's mother, who insists that Aurore was her love child with Montand.

A Founding Father?

In 1998, another dramatic paternity case was solved, and then challenged, using DNA testing. Unlike Yves Montand, the suspected father in this case wasn't dead for just a few years; the man in question had passed away in 1826. He was Thomas Jefferson, the third president of the United States.

Among Jefferson's slaves was a young mulatto named Sally Hemings. She was reputed to be the illegitimate child of Jefferson's father-in-law and one of his slaves. In other words, Sally Hemings was the half-sister of Martha Jefferson, Thomas' wife, who died in 1782. Hemings became the "property" of Thomas Jefferson when he was given part of his father-in-law's estate.

In 1787, when Hemings was only a teenager, she was sent to Paris to work for Jefferson and watch over his youngest daughter, Mary. Years later, one of Sally's children, Madison Hemings, revealed that his mother was Jefferson's mistress in Paris. Sally later went back to Monticello, Jefferson's estate in Virginia, where she stayed until her death.

As early as 1802, there were rumors that Jefferson, who publicly denounced interracial affairs, was very different in private than he was in the public eye. A nineteenth-century journalist named James Callender accused Jefferson of being the father of a number of Hemings' children. The third president neither admitted nor denied the journalist's accusations.

Following a DNA Trail

In 1998, Dr. Eugene Foster, a retired professor from the Tufts University School of Medicine, decided to perform DNA tests to determine whether the allegations concerning Hemings and Jefferson were true. Instead of digging up the body of the late president, Foster took DNA samples from five people who were definitely related to Jefferson. They were descendants of Field Jefferson, Thomas' uncle on his father's side. Foster also gathered samples from five living relatives of Thomas Woodson, Hemings' first-born son. In addition, Foster obtained DNA from a living descendant of Eston Hemings Jefferson, her youngest son.

DNA Data

For almost 200 years, there have been questions about whether the third President of the United States, Thomas Jefferson, fathered any of his slave Sally Hemings' children. Recently, DNA testing has been performed to establish or refute his paternity once and for all. Unfortunately, nothing conclusive has been proven, and the debate rages on.

Dr. Foster sent the DNA samples to geneticists in England, and the conclusion was that Woodson was not Thomas Jefferson's son, but Eston probably was the child of the president. The scientists came to this conclusion because the five descendants of Field Jefferson, Thomas' uncle, carried unique DNA sequences that are extremely rare in the rest of the population. These sequences were shared by the descendant of Hemings' son Eston. Dr. Foster and others published these findings in November 1998 in *Nature*, the prestigious British science journal. The article stated that there was only a remote possibility that Eston was not Jefferson's child.

The revelation in *Nature* resulted in a flurry of articles in publications across the country. Some Jefferson historians admitted that they needed to reevaluate their glorified views of the third president. In the light of these new findings, Jefferson's stated views against interracial mixing appeared to be hypocrisy on the part of the late president.

But in the January 1999 issue of the journal, the editors admitted that Dr. Foster's article might have been exaggerated. They noted that the article did not mention the fact that there were other possible conclusions than that Thomas Jefferson had fathered Eston. There were eight other Jefferson relatives who might have fathered Eston and passed along the distinctive DNA sequences that occur in the men in the Jefferson family.

One Jefferson historian points a finger at Thomas' younger brother Randolph. The younger Jefferson often visited his brother's estate in Virginia. The president's slaves recounted that Randolph was fond of playing his fiddle and dancing with the slaves at Monticello. Others think that one of Thomas' maternal cousins might have fathered Eston. But then again, there is still the possibility that it was Thomas Jefferson.

Solved Mysteries

DNA testing has also been useful in identifying the remains of people who died many years ago and in answering questions about missing persons. These tests were also responsible for the reunion of the children of murdered parents with their grandparents in Argentina.

The Known Soldier

In 1998, DNA testing cleared up the mystery of the unidentified remains of a Vietnam War veteran. The Defense Department made the decision to dig up the body of a man buried beneath the inscription "an American soldier known but to God" in the Tomb of the Unknowns in Arlington National Ceremony. DNA testing identified the man as Air Force Lieutenant Michael Blassie, a pilot whose plane crashed in Vietnam in 1972.

This technology has also been used to identify other soldiers who had been listed as Missing in Action. As you've learned, DNA begins to decay after a person's death. However, DNA in the bones' mitochondria—those small powerhouse structures in the cell that you read about in Chapter 4, "Me, My Cell, and I"—tends to last longer.

The Armed Forces DNA Identification Laboratory in Maryland uses a technique that can help determine a person's identity through mitochondrial DNA analysis. The scientists take samples from the mother and maternal relatives of a soldier missing in action. As we discussed in Chapter 12, "Mutations and Evolution," mitochondrial DNA is passed only from mother to child, never from the father, so only blood samples from maternal relatives are useful in these cases. The scientists then compare the maternal relatives' mitochondrial DNA with that from the bones of the deceased.

Mitochondrial DNA analysis can't positively identify a specific individual, but it can tell if that person is related to the others tested. Therefore, these tests can be helpful when used together with other means of identification, such as comparing dental records with the teeth of the deceased.

The Mystery of Anastasia

DNA testing using mitochondrial DNA also cleared up a long-standing mystery involving the family of Nicholas II, the last czar of Russia. Nicholas II, his wife Alexandra, and their five children were assassinated with several servants during the Russian Revolution. For many years, no one knew exactly where they were buried.

Finally in 1992, Dr. Peter Gill and a team of researchers from the United Kingdom Science Service were asked to identify the remains of bodies found in a mass grave. Gill and his team compared DNA from the bones with samples from living relatives and concluded that the remains belonged to the czar, the czarina, and three of their five children.

Part of the mystery had been solved, but one question still needed to be answered. After the death of the royal Romanov family, several women came forward claiming to be Anastasia, the daughter of the czar. The most famous of these was a woman who called herself Anna Anderson.

Anna, who was married to an American named John E. Monahan in 1968, said that she escaped execution by hiding behind the body of one of her dead sisters. Although living relatives of the Romanovs viewed her as an impostor, she insisted that she was the royal daughter and, as such, was the legal heiress to the Romanov family fortune, which was held in Swiss banks. But in 1970, West German courts dismissed her claim, and she did not receive a share of the czar's fortune.

Anna Anderson inspired the 1954 movie *Anastasia*, for which Ingrid Bergman netted an Academy Award. Anderson died of pneumonia in Virginia in 1984. By the time Dr. Gill and his associates were conducting DNA tests on the royal Russian family's remains, Anna Anderson was dead, and her body had been cremated. However, Dr. Gill managed to get samples of the woman's hair, and a long-standing mystery was solved. He and his team concluded that Anderson's DNA did not match that of the Romanovs.

DNA Data

Some people believe that the woman who called herself Anna Anderson and claimed to be Anastasia, the daughter of the last Russian czar, was Franziska Schanzkowska, a woman from Poland, who lied about her link to the royal family.

The Disappeared

PCR testing of mitochondrial DNA was also responsible for the reunion of kidnapped children in Argentina with their families. In 1975, a military junta in that country overthrew the Perón government. From that time until 1983, the military killed between 9,000 and 20,000 persons. These political victims were referred to as the disappeared.

After the fall of the military junta in 1983, many of the parents of murdered adults began looking for their grandchildren. They had reason to believe that these grandchildren were still alive. There were occasional stories of pregnant women brought to military prisons who were killed after they gave birth. Their infants were either adopted by people who didn't know their origin or the prison guards themselves.

A large group of Argentine women contacted the American Association for the Advancement of Science to ask for help in identifying their kidnapped grandchildren. Mary-Claire King, a geneticist and professor at the University of California at Berkeley, went to Argentina to help them. Using mitochondrial DNA techniques, King was able to determine which children were not biologically related to the families that raised them.

Dr. Mary-Claire King.
Photo credit: Mary Levin.
Photo courtesy of the
University of Washington

This pursuit raised difficult emotional issues. Once the children's family background was established, how would knowing their true origins affect the children? The adopted families argued that because they raised the children, they were the only parents that these children ever knew. But the grandmothers of the kidnapped children countered that as difficult as it would be for the children to learn the truth and then to be turned over to their grandparents, it would still be the best course.

Ultimately, it was decided that it would be more emotionally damaging to the children to learn in their adulthood that the families that raised them might have been involved in the murder of their true parents.

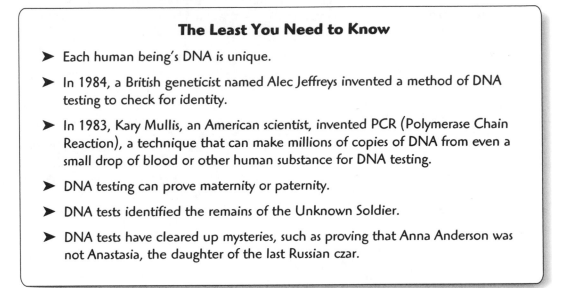

The Least You Need to Know

➤ Each human being's DNA is unique.

➤ In 1984, a British geneticist named Alec Jeffreys invented a method of DNA testing to check for identity.

➤ In 1983, Kary Mullis, an American scientist, invented PCR (Polymerase Chain Reaction), a technique that can make millions of copies of DNA from even a small drop of blood or other human substance for DNA testing.

➤ DNA testing can prove maternity or paternity.

➤ DNA tests identified the remains of the Unknown Soldier.

➤ DNA tests have cleared up mysteries, such as proving that Anna Anderson was not Anastasia, the daughter of the last Russian czar.

Chapter 18

DNA's Day in Court

In This Chapter

➤ DNA testing helps identify killers in England and the United States

➤ How the attorneys challenged DNA testing

➤ DNA evidence comes into question in the O.J. Simpson trial

➤ A dead man's DNA proves he was innocent of a murder charge

Shortly after Sir Alec Jeffreys' DNA testing method was used successfully in an immigration case in England, people began to take notice of this incredible new technology. Soon the scientist was called upon to help find the culprit in two murders in an English village that happened in the early 1980s. In 1987, the technology led to the conviction of a rapist in the United States. From cases as famous as the O.J. Simpson "Trial of the Century" to lesser-known criminal trials where suspects were freed from charges of murder or rape, DNA testing has made its way from scientific circles into the courthouses.

A Double Nightmare

Everyone in the quiet British village of Narborough was shocked to hear about a murder in their neighborhood. A 15-year-old schoolgirl, Lynda Mann, went to visit a friend one evening in November 1983. She never returned home. The next day, the young girl's body was found near a psychiatric hospital. She had been raped and strangled. No one in the area had seen or heard a thing, and there didn't appear to be any evidence left behind that could lead to the murderer.

Three years later, the nightmare was repeated in the village. Another 15-year-old named Dawn Ashworth was reported missing. Her body was found soon after, not far from the area where Lynda Mann's corpse had been lying. Ashworth had also been raped and killed.

The British authorities were stumped. In both brutal murders, the killer had been careful enough not to leave any apparent clues to his identity. Then a detective hit on the idea of calling in Alec Jeffreys, who had already used his DAN analysis technique (which you read about in the last chapter) to convince immigration authorities that a young boy returning to the United Kingdom from Ghana was the son of a woman residing in the United Kingdom. The technology could identify a person by the variations in certain segments of their DNA, which is unique to each individual.

A Genetic Calling Card

The rapist and murderer of the two teenage girls left no visible clues, such as shreds of clothing or a lost wallet that could be traced back to him. But with the new technology, any bodily substance could be transformed into a genetic ID card. The police had collected semen samples from the crime scenes. They turned over these samples to Jeffreys in the hope that a match could be found and the culprit could be brought to justice.

The police already had one suspect, a 17-year-old named Rodney Buckland, who had confessed to one of the murders. But when the semen samples from the crime scenes were brought to Jeffreys along with a blood sample taken from Buckland, the DNA tests yielded some surprising results.

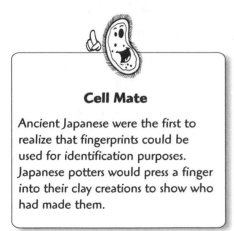

Cell Mate

Ancient Japanese were the first to realize that fingerprints could be used for identification purposes. Japanese potters would press a finger into their clay creations to show who had made them.

Jeffreys discovered that the semen from both crime scenes belonged to the same person. So presumably both girls had been raped and killed by the same man. But according to the tests, Buckland was not that person. When a person's DNA sample does not match up with one found at the scene of a crime, it proves without a doubt that the person was not involved in the crime. Buckland was freed, and now the detectives had to begin their search all over again. But how would they go about looking for a criminal, when all they knew about him was the sequence of some of his DNA?

A Trail of Blood

Going on the assumption that the killer probably lived in the area, the police decided to take blood samples from every male resident between the ages of 17 and 35. Over 4,500 men participated in this mass blood testing, yet none of the subsequent DNA tests showed a match with any of these thousands of men. It seemed as if the murderer would go unpunished.

Then in 1987, the manager of a bakery in the village gave the police a tip that would change everything. While she was talking with some of her co-workers in a pub, she found out that Colin Pitchfork, one of the bakers, never submitted to a blood test. He managed to get out of it by convincing one of his colleagues to take the test for him and to falsely identify himself as Pitchfork.

Following the tip, the police arrested Pitchfork and confronted him with what they knew. He confessed to the two murders. The police had already obtained one false confession, so to prove beyond any doubt that Pitchfork was the culprit, they had a sample of his blood sent to Jeffreys. This time, the DNA sample matched perfectly with the semen samples taken at the scene of the crimes.

DNA in the U.S.A.

The news spread around the world that a new genetic technology could aid crime fighters. This technology could be especially helpful in cases of murder or rape in which there were no witnesses. In that same year, 1987, Cellmark Diagnostics, the British firm that held the patent to Jeffreys' technique, opened a branch in the United States. Another company, the LIFECODES Corporation, at that time based in Valhalla, New York, started performing DNA testing in the same year.

One in Ten Billion

LIFECODES carried out the first DNA tests that led to the conviction of a criminal in the United States. A man named Tommy Lee Andrews was accused of raping and stabbing a woman. His trial began in November 1987 in Orlando, Florida. When LIFECODES compared a DNA sample taken from Andrews with the semen obtained at the scene of the crime, it matched perfectly.

DNA Data

The first crime that was solved using fingerprints to identify the culprit occurred in Paris in 1902. A murderer left the telltale fingerprint at the scene of the crime, and this fingerprint led to his conviction. The idea of using fingerprints to find criminals originated with the Englishman Francis Galton, a cousin of Charles Darwin.

LIFECODES estimated that the probability of finding another person with the same DNA sequences would be about 1 in 10 billion. In February 1988, the jury was convinced by the evidence, and the suspect was found guilty of the crime. Andrews was sentenced to prison, mostly as the result of the telltale DNA test. The use of DNA testing as evidence attracted much attention to the trial, and many crime fighters saw the new technology as a reliable method that could revolutionize criminal investigations in the future.

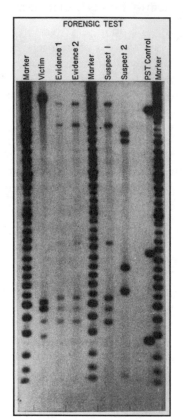

This DNA test done by the LIFECODES Corporation matches suspect 1 with evidence from a crime scene. Suspect 2 is excluded because his DNA profile does not match up. Photo courtesy of the LIFECODES Corporation

Murder in the Bronx

A few years after the conviction of Tommy Lee Andrews, two lawyers began to cast doubts on the reliability of DNA testing as evidence in the courts. They were Barry Scheck and Peter Neufeld, who were later employed by O.J. Simpson in his famous trial.

Vilma Ponce and her two-year-old daughter Natasha were found brutally murdered in their apartment in the Bronx in February 1987. Police suspected Jose Castro, a janitor who worked in a nearby building. A detective noticed a stain on Castro's wristwatch, and suspected that it might be dried blood. He took the suspect's watch for further investigation. Castro was then asked to give a blood sample. Then the LIFECODES Corporation tested the samples, along with blood from the murder victims.

LIFECODES discovered that the stain on the watch was blood, as the detective had surmised. The scientists concluded after testing that the blood on Castro's watch matched up with the sample of Vilma Ponce's blood. They estimated that the chance of a random match—in other words, that the blood on the watch belonged to someone other than the murder victim, was just one in 189,200,000 among the Hispanic population. *Forensic* DNA scientists, scientists who work with DNA testing in court cases, calculate their figures according to the ethnic population a person comes from because a DNA variation might be more common in one ethnic population than in another.

Genetic Jargon

Forensic (for-REN-sic) DNA testing refers to the use of DNA testing as evidence in court cases.

DNA on Trial

Castro's lawyers, Barry Scheck and Peter Neufeld, gathered six expert witnesses on the topic of DNA testing. This case was the first time that experts presented a serious challenge to the admissibility of DNA evidence in court. The media picked up on the story and reported that DNA testing evidence was now on trial. The experts represented both the prosecution and the defense, and they concluded that LIFECODES' performance of DNA testing was flawed in this case.

Among their concerns with the way LIFECODES carried out their DNA analysis was the estimate given for a random match between Ponce's blood and that on Castro's wristwatch. In this case, the scientists made their estimates for an accidental match with Ponce's blood based upon their knowledge of the Hispanic population. But critics such as Scheck and Neufeld, along with their expert witnesses, argued that sometimes these estimates can be very difficult to figure out. As a result, they can sometimes be misleading, they said.

The hypothesis that Scheck and Neufeld gave in an interview was that if you're looking for a tall, blond, blue-eyed, light-skinned person in the general population, the chances of another person having these characteristics might be one in one thousand. But if you took those same traits and looked for a random match in a place like Minnesota, where many of the residents have a Scandinavian background, then the chance of finding someone with these characteristics might be more along the lines of one in three.

It can also be argued that it's difficult to define one specific Hispanic population. There are Hispanic people from North America, South America, the Caribbean, Central America, and Europe, and each group might have distinct genetic differences not found in the others. When people intermarry with different ethnic groups, estimates of the accuracy of the DNA test become even harder to determine. Add to this possible human errors such as the mislabeling of samples or contamination of the samples by bacteria, and the test results may be inaccurate, some people argue.

Striking Out a Match

Another problem cited by the experts in the Castro case was the fact that the bands on the profiles for the blood sample on Castro's watch and the bands from Vilma Ponce's profile didn't match up exactly. The LIFECODES scientists thought that the gels they used may have differed a little, but that the bands of the DNA profiles were still a precise match.

DNA Data

Barry Scheck is the attorney who defended Jose Castro, a Bronx janitor, in a murder trial in 1990. Scheck's defense of Castro centered around the lack of reliability of the DNA tests. Scheck is now co-director of the Innocence Project at the Benjamin N. Cardozo School of Law in New York City. The project uses DNA testing to defend prisoners who have been wrongfully convicted. From 1992 to 1998, the Innocence Project has exonerated 35 inmates, including six who were taken off death row. In 11 of these cases, DNA testing led to the conviction of the actual criminals.

After hearing from the expert witnesses in the pretrial hearing, Judge Gerald Sheindlin ruled that the bloodstain on the watch would not be admissible evidence. He added that although DNA testing can be reliable when done properly, he was not convinced that LIFECODES adhered to the correct scientific techniques required to obtain accurate results. This was the first time in history that DNA tests results were successfully challenged and ruled as inadmissible evidence.

This victory for the defense lawyers was played up by the media. Some people thought that it would set a poor precedent for DNA testing in future trials. Ironically, Jose Castro confessed that he had murdered Vilma Ponce and her young daughter. The bloodstain on his watch did not belong to an unknown Hispanic woman. It was blood from Castro's murder victim.

In Defense of Three Angels

In 1990, Scheck and Neufeld were again acting as defense lawyers in a trial that involved DNA testing, this time in Ohio. Once again, they questioned the validity of the way the DNA testing was carried out. The trial was *United States vs. Yee*.

Three members of the Hell's Angels motorcycle gang, Stephen Wayne Yee, Mark Verdi, and John Ray Bonds, were accused of murdering David Hartlaub, a record store clerk, while he was making a deposit in a bank in Sandusky, Ohio. The three defendants allegedly mistook the record store clerk for a member of a rival gang, called the Outlaws Motorcycle Club. Hartlaub was shot in his own van.

DNA Data

In 1997, a Canadian man was convicted of the murder of his former girlfriend. Evidence in this trial included DNA testing. This case was unusual because it was the first time animal DNA had ever been admitted as evidence in court. When the victim's body was found, her leather jacket was covered with blood. DNA testing pointed the finger at her former boyfriend. In addition, strands of white hair were also found on the body. These strands turned out to belong to the murderer's cat.

Police took samples of the blood in the van and handed them over to the FBI laboratory for DNA testing. Test results implicated John Ray Bonds. Because the blood samples and Bonds' DNA matched, the prosecution reasoned that the Hell's Angel had been injured during the killing, and he lost some blood in Hartlaub's van.

Which Population?

To challenge the results of the DNA tests, Scheck and Neufeld brought in expert witnesses, including *population geneticists* Daniel Hartl, then of Washington University, and Richard Lewontin of Harvard University. Prosecutor James Wooley called in other experts in the field, including Kenneth Kidd, a population geneticist from Yale University, and Thomas Caskey, the director of the Institute for Medical Genetics at the Baylor College of Medicine.

The experts proceeded to debate several points that have often been brought up in criminal cases like this one. Scheck and Neufeld's experts raised the same doubts about the interpretation of the DNA test in the Yee case that had been raised in the Castro case. Did the FBI laboratory carry out the job properly? What exactly was its criteria for concluding that there was a match between the two samples?

The experts went back and forth debating questions like these for 15 weeks. In October 1990,

Genetic Jargon

Population genetics is a branch of genetics that deals with the hereditary makeup of different groups in the population. Population geneticists study the frequency of different genes in these populations.

Judge James Carr made his decision. Although he concluded that the laboratory did not carry out its DNA testing according to the highest scientific standards, Carr ruled that the DNA evidence provided by the FBI laboratory was nonetheless reliable and therefore admissible in the trial.

Testing the Tests

Cases such as the Jose Castro and Hell's Angels murder trials raised serious questions about the reliability of DNA test evidence in court. To better understand the new technology's strengths and weaknesses, the National Academy of Sciences' National Research Council (NRC) was given federal funding to conduct a study to help resolve some of the questions and concerns that scientists and laypersons alike were asking about DNA testing. A panel of 14 experts was assembled, including Daniel Hartl and Richard Lewontin, the two population geneticists who acted as expert witnesses in the Hell's Angels case.

DNA Data

The National Academy of Sciences is a private organization that was founded by an act of Congress. The members of the prestigious organization are noted for their contributions in the fields of science or engineering. The National Academy of Sciences offers advice to the federal government on questions involving science and engineering.

As expected, one of the bones of contention during the writing of the report was the continued debate over how best to calculate the odds of a random match between two samples in a DNA test. Hartl and Lewontin continued to argue that because there are so many subgroups in any ethnic population, the chance of finding a match within a subgroup could be considerably different than the chance of a random match in that entire ethnic population. Other members of the panel countered that DNA test calculations already accounted for the possibility of differences in subgroups of the population. As a result, they contended that the chance for error was minimal.

The NRC report was issued in April 1992. The panel recommended the use of DNA testing as evidence in the courts, but it also recommended that a stricter set of national standards be upheld during DNA testing procedures. Some people criticized the 1992 DNA report for not being thorough enough. The media also initially misinterpreted the report as saying that DNA testing was not valid, although this was not the case.

In 1996, the National Research Council issued another report. Among other points, this new report addressed the question of possible false matches in DNA testing. It addressed ways to ensure that the probability would be more carefully calculated than in the past. For instance, if evidence at the scene of a crime comes from a person whose race is unknown, the study suggested that the odds of a random DNA match should be calculated using different ethnic profiles as a basis. Like the first report, the 1996 report also had its detractors. Some people criticized its lack of specific standards for all DNA testing labs.

DNA Data

In October 1998, police in Florida suspected a man named Charles Peterson of theft and rape, but they lacked samples of his DNA to compare to those obtained at the crime scenes. Sergeant Michael Puetz followed Peterson as he drove through the streets on his motorcycle. When the suspect stopped at a red light, he turned and spat in the street. Puetz took a paper towel and retrieved the saliva sample, which contained the suspect's DNA. It turned out to match the DNA of semen taken from the rape victims, and Peterson was charged with the crimes.

O.J. and DNA

In 1994, Barry Scheck and Peter Neufeld were once again called to represent a defendant in a murder trial involving DNA evidence. They would be part of a team of attorneys in a much-debated murder trial. Unlike the cases involving Jose Castro and the three Hell's Angels members, this time the defendant was a celebrity. All of America was watching TV daily to see what would happen to O.J. Simpson, the popular football player, who was accused of murdering his ex-wife, Nicole Brown Simpson, and her friend Ron Goldman. The two were stabbed to death outside of Nicole's house in Los Angeles on June 12, 1994.

The Trial of the Century

DNA testing was once again put on trial along with O.J. Simpson. Cellmark Diagnostics, a prominent DNA testing company in Germantown, Maryland, was called on to perform the DNA testing in this famous case. Many pieces of physical evidence were allegedly recovered from the crime scene, and some of these were sent to Cellmark's lab. Among them were blood stains from 13 different locations, including the infamous bloody sock allegedly found at the foot of O.J.'s bed. Cellmark Diagnostics concluded that the blood stains found in O.J.'s Bronco belonged to Nicole Brown Simpson and her friend Ron Goldman.

Scheck and Neufeld argued, as they had in the Castro and Hell's Angels trials, that there is room for error in DNA testing if it is not done according to strict standards. If the samples are not collected properly, they can be contaminated. Any type of contamination, they alleged, such as that from bacteria or from another person's DNA, could result in false conclusions about the samples.

In addition, they pointed out that only a small portion of a person's DNA is analyzed in these tests, as opposed to a person's entire range of DNA. They discussed once again that depending upon the population group used to calculate the odds of a random match, the odds can change dramatically.

DNA Data

The world's first national DNA databank was established in the United Kingdom on April 10, 1995. The British take samples from a wide variety of crime scenes, including rapes, murders, and thefts. By 1998, they were making several hundred matches per week between crime scenes and DNA profiles in their databank. Some of these matches linked different crime scenes to the same unknown culprit, and others linked DNA samples from a crime to an offender who had been convicted in the past.

The defense also argued that O.J. might have been framed by investigator Mark Fuhrman, who they described as a rampant racist. They claimed it was possible that O.J.'s blood might have been planted on pieces of evidence. Could the whole scenario have been conjured up because of a racial bias?

Cell Mate

In October 1998, police across the United States gained access to a national DNA database of known criminals. The database was created by the FBI. There were 250,000 DNA profiles already in the database in October 1998, with a backlog of 350,000 others that still needed to be entered by the FBI.

"Beyond a Reasonable Doubt"

The jurors had the obligation to decide whether the prosecution had argued its case beyond a reasonable doubt. As Barry Scheck later wrote in an article in *Newsweek* magazine, this high standard in our legal system protects people who could lose their life or liberty based on the notion that they are "probably" guilty.

Adhering to this legal standard, on October 3, 1995, the jurors voted to acquit O.J. Simpson of the murder charges. Barry Scheck said that the not guilty verdict had many positive aspects. In addition to acquitting his client, Scheck felt that the professional standards of DNA laboratories would be improved, and crime fighters would be better prepared to collect samples of DNA evidence according to stricter protocols to avoid contamination.

Defending "The Fugitive"

DNA can also solve crimes that happened many years ago. In September 1997, the body of Dr. Sam Sheppard was exhumed in Columbus, Ohio, in the hopes that DNA testing would clear his name once and for all. He had been accused of murdering his wife Marilyn 43 years earlier. Before the O.J. Simpson trial, the Sheppard case was known as "The Crime of the Century." The story inspired the popular TV series as well as the 1993 movie *The Fugitive*, although Sheppard never ran away from the law.

The Bushy-Haired Man

On July 4, 1954, Dr. Sam Sheppard fell asleep on the first floor of the home that he shared with his wife, Marilyn, and their son, Sam Jr. At around four o'clock in the morning, he heard his wife scream for him, and when he ran upstairs, he was knocked out by someone who came up behind him. Eventually, Sheppard came to and chased a man he described as bushy-haired into the basement and finally outside. There, the man knocked him out again.

When Sheppard came to, his assailant was gone. Sheppard staggered up the stairs to his bedroom and found his wife dead. After a trial that received the same type of media attention that the O.J. trial received many years later, Sheppard was convicted of the murder of his wife. One reason for the jurors' decision was the fact that there was no sign of forced entry into the Sheppard house.

After Sheppard had served 10 years in prison, the United States Supreme Court ruled that because of the widespread coverage and media hype during the original trial, a retrial was necessary. This time, in 1966, Sheppard was found not guilty. He died four years later of liver failure.

The son of the accused and the murder victim, Sam Reese Sheppard, later wrote a book with author Cynthia L. Cooper called *Mockery of Justice: The True Story of the Sheppard Murder Case*. The younger Sheppard hoped the book would clear his father's name. Sam Jr. suspected that the "bushy-haired man" (who was transformed into the one-armed man in the TV series and the movie *The Fugitive*) was Richard Eberling, the Sheppards' window washer.

In 1959, Eberling was arrested for theft, and when police searched his home, they found Marilyn Sheppard's diamond ring. Eberling later served a life sentence for the 1984 killing of a rich widow. He had changed the woman's will in his favor before murdering her. In 1984, he died in prison at the age of 68. Eberling never confessed to the murder of Marilyn Sheppard.

The Forgotten Report

While writing the book, Sheppard's co-author, Cynthia L. Cooper, gained access to a police report from July 1954. It stated that there was a mark on the Sheppards' door that looked like it had been made with a chisel. Dr. Sheppard's attorneys never received this information, which would have been helpful in corroborating his story that an intruder had broken into the house.

Now there is irrefutable proof that Dr. Sheppard did not commit the crime. Blood found at the murder scene was tested and compared to Sheppard's DNA, which was obtained when his body was exhumed in September 1997. Sheppard's blood did not match the sample from the crime scene. When samples do not match, there is no debate—the two DNA specimens do not belong to the same person. It's only when the samples do match that there have been disagreements as to how the samples were collected and the interpretation of the results.

Earlier DNA tests performed in February 1997 showed that the blood at the crime scene did not belong to Marilyn Sheppard, either. So apparently Sam Sheppard was telling the truth when he said that a bushy-haired intruder fought with him and killed his wife. After more than 40 years, long after his death, Sam Sheppard's DNA told the compelling story of his long-hidden innocence.

Historical Highlights of DNA Testing

Date	Event
1983	Kary Mullis invents PCR, a procedure that can make millions of copies of a small amount of DNA so it can be tested.
1984	Alec Jeffreys invents a method of DNA profiling.
1987	British rapist and murderer Colin Pitchfork is the first person ever convicted of a crime because of DNA evidence.
1987	Jose Castro, a janitor in the Bronx, is accused of a double murder. Attorneys Scheck and Neufeld argue that the DNA evidence in the case is flawed. Judge Gerald Sheindlin rules that the tests are not admissible evidence. Later, Castro confesses to the murders.
1988	DNA testing is used as evidence to convict Tommy Lee Andrews of rape.
1991	Scheck and Neufeld defend three Hell's Angels on murder charges. They challenge the validity of the DNA tests, but Judge James Carr rules the tests admissible. The challenge nonetheless raises doubts about DNA in court.
1995	O.J. Simpson is acquitted of the murder of his ex-wife Nicole Brown Simpson and her friend Ron Goldman, in part due to his attorneys' efforts to discredit the DNA tests submitted as evidence of his guilt.
1997	DNA tests performed on the exhumed body of Dr. Sam Sheppard proved that he did not murder his wife more than 40 years earlier.

The Least You Need to Know

➤ Since 1987, DNA testing has become an increasingly useful tool in identifying criminals, particularly in rape and murder cases where there are no witnesses.

➤ Forensic DNA scientists calculate estimates of the probability of a random match on the DNA being tested according to the ethnic population a person comes from. Certain DNA sequences are more common in one ethnic population than in another.

➤ Attorneys Scheck and Neufeld challenged DNA evidence in several court cases (including the O.J. Simpson case), citing sloppy scientific technique and inaccurate estimates as reasons to question the validity of DNA testing results.

➤ Although a DNA match is not conclusive proof of someone's guilt because of the room for error in the process of collecting, testing, and interpreting the DNA, DNA samples that don't match can always prove someone's innocence.

Chapter 19

DNA Disease Diagnosis

In This Chapter

➤ How DNA tests for diseases can alert you to problems in your genes

➤ How biochips work to test genes

➤ How genetic counselors help patients understand DNA test results

Some people predict that the Genetic Revolution will radically change medicine and health care. In the future, they say, a routine visit to the doctor's office might not just consist of a physical examination and questions about how you feel. Instead of simply saying "Ahhh" and getting your blood pressure checked, you might be asked to give a sample of your DNA for testing. Instead of getting a prescription for a generic medicine, you might get one for a genetic medicine, one that's been uniquely tailored to your specific genes. Instead of having to treat diseases that have already developed, doctors eventually may be able to use new genetic technologies to provide preventive medical care.

Because scientists will learn more and more about genes and how they work, some people feel that scientists will try to mend faulty genes before they cause damage. One step in this direction is the creation of DNA tests for disease. Scientists have already come up with DNA tests to help people determine whether they or their children are at risk for diseases such as Tay-Sachs disease, hemophilia, and muscular dystrophy.

The new tests also raise a number of ethical questions, such as whether they could inadvertently lead to genetic discrimination. There are concerns about the tests' accuracy and other concerns. To learn more about these touchy issues, be sure to read Chapter 28, "People Problems."

How Do DNA Tests Work?

You read in Chapter 17, "Stories Genes Can Tell," and Chapter 18, "DNA's Day in Court," about new advances in DNA technologies, such as PCR, that make it possible to find out about certain segments of DNA. You read how everyone has a unique DNA "profile," and how examining some of the sequences in a person's DNA can lead to positive identification or to the exclusion of a suspect in a criminal trial. These procedures are also used in paternity and maternity cases.

Chapter 2, "Assembling Your Genes," discussed how a nucleotide in the wrong place can sometimes cause a mutation that leads to a disease. Scientists use the same tools—the Southern blot method and PCR—to determine whether a person has a mutation in specific genes known to cause diseases. Once this genetic diagnosis is made, scientists hope preventive care will be possible.

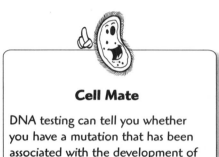

Cell Mate

DNA testing can tell you whether you have a mutation that has been associated with the development of a certain disease, but it won't tell you for sure whether you'll get the disease.

Now that more and more genes are being studied in the Human Genome (JEE-nome) Project, scientists plan on using this knowledge to help create even more DNA disease tests. The goal of the Human Genome Project is to map and sequence all of the approximately three billion base pairs, those rungs on the spiral ladder of DNA. To map the genes means to figure out which chromosome they're on and their specific position on that chromosome. To sequence a gene means to figure out the order of the four nucleotides (A, T, C, and G) that make up its rungs. You will learn more details about this project in Chapter 21, "The Human Genome Project."

Probing Questions

When scientists know the "misspelling" of mutations that can lead to certain diseases, they try to create DNA tests that can detect whether that misspelling is in your DNA. These tests work a lot like the DNA tests used in court, but their ultimate goal is a little different. Instead of trying to figure out whether your DNA matches another person's DNA or DNA left at the scene of a crime, these tests look for specific mutations.

You have about three billion base pairs of DNA. How can scientists wade through all those billions of letters to find the exact gene that they want to study? It's a lot like trying to find a specific sentence in an encyclopedia that's divided up into 46 volumes. The volumes are the 46 chromosomes that humans have, and each one of these volumes has millions of pages. Suppose you wanted to find a certain sentence. How would you know which page to look on, when there are so many pages in those volumes? Even if you found the right page, finding the specific sentence you're looking for would still be a challenge amid all of that text.

Measuring Molecules

Suppose you want to know whether you have the mutated gene that causes sickle cell anemia. In this mutation, just one nucleotide is misspelled, but this one nucleotide is enough to cause the malformation of red blood cells. Instead of being round and looking something like donuts without holes, sickled red blood cells are crescent-shaped. These irregularly shaped cells often clog blood vessels, leading to pain as well as possible organ damage.

Determining the order of letters in a gene associated with a disease is very difficult. As you'll remember from Chapter 7, "S, M, L, and XL Genes," genes can consist of hundreds or thousands of base pairs. To simplify the search for a mutation, scientists examine just a small portion, just 20 or 30 base pairs, for example, of the gene.

To test for the sickle cell mutation, scientists use a procedure called RFLP to analyze the DNA. As covered in Chapter 17, "Stories Genes Can Tell," the first step of this procedure involves using restriction enzymes, those molecular scissors that cut DNA in specific spots. If you have the normal gene, the restriction enzymes that scientists use will cut it in three spots. But if you have the gene with the sickle cell anemia mutation, this mutation will affect how the gene is cut up by the enzymes, and there will be only two cuts.

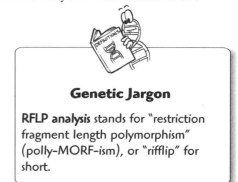

Genetic Jargon

RFLP analysis stands for "restriction fragment length polymorphism" (polly-MORF-ism), or "rifflip" for short.

Imagine this gene is a tape measure that's three billion inches long. Instead of studying the whole gene—the whole three billion inches, a scientist looking for the sickle cell mutation will look at a small portion of the tape measure. Suppose the area from 10 inches to 24 inches will be studied. That's a lot easier than looking at the whole tape measure.

If you have the normal gene, the restriction enzymes will cut it at three sites: at 10 inches, at 20 inches and at 24 inches. However, if you have the mutation, the restriction enzymes will cut the tape measure only at 10 inches and at 24 inches. Thus, the pieces of the tape measure that you end up with are different, depending on whether you have the normal gene or the mutation. Looking at the size and the number of pieces you wind up with tells you whether your tape measure, your gene, was made correctly.

Developing a DNA Picture

In order to make the size and number of DNA pieces in the gene visible, scientists use electrophoresis. As you'll remember from Chapter 17, "Stories Genes Can Tell," in this process, cut-up pieces of DNA are put into a gel. The application of electricity to the gel makes the smaller DNA pieces move faster and the bigger DNA pieces move slower.

Eventually, some of these different-sized fragments are made radioactive using DNA probes (which are described in the following section), and when they're put on a piece of film, the radioactivity exposes a pattern of bands, which look a lot like a supermarket bar code. Depending on how many times the gene segment was cut up—three times in the case of the normal segment or two times in the mutant segment—you will get a different pattern of bands on the film. The normal gene will show a band in a different place on the film than the mutant gene. Using this method, a scientist can visualize this information from just a small section of one gene out of all of your DNA.

DNA Data

RFLP technology was invented in 1980, when several geneticists were trying to come up with ways to track down genes for disease.

A Complementary Affair

Scientists take advantage of the fact that the nucleotides that comprise DNA can pair up only in specific ways. As you read in Chapter 6, "Assembling Your Genes," A can pair only with T to make a complete rung on the twisted rope ladder of DNA, and C can pair only with G. Therefore, if you know the order of the letters on one strand of the double helix, you can easily determine the letters on the other strand. Suppose you're looking for a sequence like AATTCTG. You could use the sequence TTAAGAC to find it, because this is the only possible sequence of letters that can mate with the original sequence.

You could make a whole bunch of TTAAGAC sequences using a gene machine. (You read about these in Chapter 13, "Cut-Down Genes.") You would add radioactive tags to these sequences and then throw them into a test tube. At this point in the process, the sequences are called *DNA probes*, because they search out their complementary sequence. These DNA probe sequences eventually match up with the sequences that read AATTCTG, because they can't mate with any other sequence.

Genetic Jargon

DNA probes are short sequences of As, Ts, Cs, and Gs that have been given tags made of radioactive molecules. The probes join up with their complementary sequence. For instance, AATC will latch onto TTAG, because As are attracted to Ts, and Cs are attracted to Gs. These probes are used in DNA testing.

Many Causes, One Disease

Creating a test for a disease such as cystic fibrosis was more complicated than developing the test for sickle cell anemia because more than one type of mutation can result in cystic fibrosis. You read in Chapter 9, "Irregular Genes," that this disease causes an accumulation of mucus in the lungs, with resulting complications, and people with this disease often do not live past their 20s or 30s.

About 70 percent of the people who suffer from cystic fibrosis have the same mutation. The problem in the genetic spelling is that three nucleotides are missing. You'll remember from Chapter 8, "Producing Proteins," that a three-letter "word" is called a codon, and each codon codes for a specific amino acid, which is a building block of proteins. The fact that there's a misspelling here means that a defective protein will be made.

But in addition to that type of mutation, other mutations can result in the same disease. In fact, over 700 mutations account for the remaining 30 percent of cases of cystic fibrosis. It would be too difficult to include all of the approximately 700 mutations in a genetic test, so scientists developed a test for cystic fibrosis that includes just the most common mutations. Therefore, when someone is tested for the disease, scientists can tell them their prognosis with only 98 or 99 percent accuracy, because not all of the rare forms of the mutations are accounted for in the test.

Breast Cancer and DNA Tests

Only a small percentage of breast cancer is caused by inherited mutations in genes, and there are several genes that have been implicated in the development of some different forms of this disease. These mutated genes fall into the category of *incomplete penetrance* genes, which means that a person who carries these genes will not necessarily develop the disease. Rather, people with these genes carry what is considered an increased risk for the disease, and other factors such as smoking, diet, and environmental influences determine whether the disease will appear.

In recent years, two of these incomplete penetrance genes were identified. Some researchers feel that they carry some risk for the development of breast cancer. In 1994, the BRCAl gene (for Breast Cancer l) was discovered, and in 1996, BRCA2 was discovered. Tests have been developed for each of these genes, and currently, the cost is about $2700 for each test.

There is some controversy surrounding these and other tests. Because scientists still don't know exactly what these genes do, some people wonder about the value of knowing whether they carry

Genetic Jargon

A gene has **incomplete penetrance** when it does not necessarily cause a specific disease, but may carry an increased risk for the development of that disease if other factors, such as diet, smoking, and environmental influences are present.

them. What does it mean if a woman tests positive for one of these genes? What course should she take, if any? Will her employers or insurance company discriminate against her? Questions like these will be discussed in Chapter 28, "People Problems."

A Gene in the Community

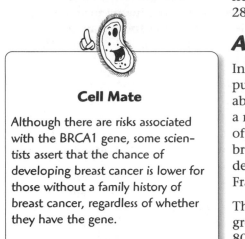

Cell Mate

Although there are risks associated with the BRCA1 gene, some scientists assert that the chance of developing breast cancer is lower for those without a family history of breast cancer, regardless of whether they have the gene.

In 1995, researchers at the National Cancer Institute published the results of a study that determined that about one percent of all Ashkenazi Jewish women carry a mutated copy of the BRCA1 gene, which is suspected of putting women at an increased risk for developing breast cancer. Most American Jews are of Ashkenazi descent, which means that they come from northern France, Poland, Scandinavia, and Russia.

The one percent figure is considered high for an ethnic group. Researchers have estimated that only about 1 in 800 people who are not of Ashkenazi descent carry this mutation.

This discovery raises some ethical concerns. Would some people misinterpret this data to mean that this group of people has flawed genes? Could this lead to discrimination?

To answer this question, scientists are quick to point out the Ashkenazim do not have more genetic diseases than other ethnic groups. It's just that they've been studied more than some ethnic groups, for several reasons. To begin with, they have kept good family records, and secondly, they are very cooperative with people conducting scientific research. The same is true of Mormons, who have extensive family records. In any given population that scientists study, there will always be specific genetic mutations that are more common in one group than in another.

DNA Data

Scientists estimate that every one of us harbors between 5 and 10 lethal mutations. So why don't we just keel over from all these life-threatening mutations? The answer is simple. Many mutations are only lethal when they're present in a double dose. Remember, you get one set of your chromosomes from your mother and another one from your father. So if you have a nasty mutation or two, it won't matter as long as each mutation codes for a different recessive disease—a disease that requires two copies of its gene, one from each parent, to manifest. You'd need to inherit the same lethal mutation from both parents for something terrible to occur, and thankfully, this usually does not happen.

Hunting for the Huntington's Gene

In 1968, a young woman named Nancy Wexler was asked to come to her father's house to celebrate his birthday. She and her sister Alice were gently told the sad news that would forever change the course of their lives. Their mother Leonore, who was divorced from their father, was suffering from Huntington's disease, a genetic illness that slowly destroys sections of the brain that help to control the body's movements. This disease eventually leads to death.

Nancy Wexler and her sister knew that three of their uncles had died of the disease. Their father explained that because their mother was diagnosed with this illness, they each had a 50/50 chance of developing Huntington's. Huntington's is a dominant mutation, which means that even if only one copy of the mutation is inherited from one parent, the disease will develop. The tragic news spurred Nancy Wexler to fight the disease and to try to find a cure.

In 1974, her father, Milton Wexler, created the Hereditary Disease Foundation, which pooled the resources of many people to aid scientific research that might one day lead to a cure for the disease.

Cell Mate

Huntington's disease was named after Dr. George Huntington, who first described the disease in 1872. It was formerly called Huntington's chorea (pronounced ko–REE-a or KORR-ee-a), which comes from the Greek word for dance. This refers to the uncontrolled body movements characteristic of people suffering from this disease.

Dr. Nancy Wexler. Photo courtesy of Columbia University

Collecting Samples

Nancy Wexler knew that in order to find the gene for Huntington's disease, researchers would need to study large families that had the disease. She heard about a small

Venezuelan village near Lake Maracaibo where large groups of the inhabitants suffered from the disease. In 1979 and 1980, she visited the village in Venezuela. In 1981, she started to collect blood samples to create the largest family tree that had ever been compiled for the disease, which included over 13,000 people with the disease.

Marking the Spot

You've already learned that the DNA of most people is 99.9 percent the same. You also read that what makes one person's DNA different from another's is small variations. The scientific name for these variations is *polymorphisms*. These polymorphisms, or variations, are passed down from one generation to the next. Scientists have theorized that some of these polymorphisms might be close to disease-causing genes on a chromosome. If this is the case, the polymorphism might be like a signpost that points to the general location of the disease gene. This type of genetic signpost is known as a *marker*.

Going on this theory, a researcher named James Gusella discovered a marker that indicated that the Huntington's disease gene was somewhere on chromosome 4. (As you know, humans have 46 chromosomes, and most of these are referred to by numbers.) After much research and Nancy Wexler's repeated trips to Venezuela to contribute more data to the search, the gene responsible for the disease was found.

Genetic Jargon

A variation in a short stretch of DNA is called a **polymorphism** (polly-MORF-ism). The name comes from two Greek words meaning many forms.

A **marker** is a genetic variation that sits near a disease gene on a chromosome. The marker is like a signpost that points the way to the general location of a specific gene.

A researcher named Marcy MacDonald noticed a genetic stutter in this gene—the three letters CAG. Although people without the disease might have a few of these repeat sequences, people with Huntington's disease can have up to 86 repeats of this sequence in their DNA. For some reason, the number of repeats of this sequence increases as the mutation is passed from one generation to the next.

The cause of Huntington's disease has been found, but much more research lies ahead. Nancy Wexler still makes visits to the village in Venezuela, and scientists continue to study the gene and its tragic effects in the hope that some day they will understand enough about its workings to discover a cure.

Biochips Off the Old Block

The latest technology which promises to give quick disease diagnoses consists of tiny squares of silicon or glass that contain gene fragments that test for the presence of specific mutated genes in DNA samples. These tiny squares are called *biochips*, and some people think that this invention may revolutionize disease diagnosis in the same way that microchips with electronic circuits revolutionized the computer industry.

*The SpectroCHIP, manu-
factured by Sequenom, Inc.
Measuring less than 1¹/₄
inches in length, the
SpectroCHIP can detect
mutant genes.
Photo courtesy of
Sequenom, Inc.*

Companies such as Affymetrix in Santa Clara, California, Sequenom in San Diego, California, and others are involved in this new technology. Affymetrix pioneered this technology in 1993, and other companies have since done research on this new approach to disease diagnosis.

In the same way that microchips in a computer store huge amounts of information in a tiny amount of space, biochips use many bits of DNA in different spots on these little squares of silicon or glass. They enable scientists to determine whether a person carries a specific mutation of a gene and whether a particular gene is switched on or off. (In Chapter 16, "Send in the Clones," you learned that most genes are switched off after an embryo has developed.)

In some ways, biochips work like DNA tests that use RFLP and PCR, only they're tremendously faster. Scientists have determined the DNA sequences of some mutations that can cause disease. To test for the presence of these sequences in someone's DNA, scientists use DNA probes, which, as you learned earlier in this chapter, are a synthetic set of nucleotides that will match up with a mutated gene sequence. These DNA probes are then put on the biochips. If the DNA probes used in biochips stick to a mutated gene, a specific portion of the chip will glow under a laser beam. By observing which minuscule parts of a biochip light up, scientists can determine whether a person has the gene mutation in question.

Genetic Jargon

A **biochip** is a microchip about the size of a dime that is used in diagnosing genetic diseases or genetic predispositions to diseases. Biochips are also referred to as DNA arrays, chip arrays, DNA chips, and microarrays.

A Glowing Technology

There are different types of biochips because different companies have slightly different systems. Some types of biochips contain DNA probes that use fluorescent dyes that glow when they are held under a laser light. If the sequence matches the probe exactly, the probe will glow brightly. If the sequence is close, but not exact, the glow is more subdued.

In the future, biochips may reveal the reason why some people respond to one type of medicine and not another. Some medicines have never hit the market because they were ineffective for most of the people in a pre-market study. But a minority of people in such studies often respond positively to the new drug. Unfortunately for that minority, these drugs are not marketed, because the percentage of positive responders is too small for the drug companies to recoup their investment. In the future, biochips may make it easier to identify people that these medicines would work on, thereby making it more cost-effective for pharmaceutical companies to create drugs targeted to individual genetic makeups.

DNA Cards

In addition, the use of biochips might lead to the day when each person has an individual DNA card. Some researchers predict that in the future all of a person's genetic information might be contained on something the size of a credit card. When put into a computer, this card would alert the person's health care provider about any pre-existing conditions or help them discover what's wrong in cases of emergencies.

Worming a Way into DNA

Scientists are also discovering that they can transfer human genes to different organisms in order to test how these genes work. You read in Chapter 11, "Gene Gangs," that scientists study yeast cells to see how cells divide and to learn about how similar genes function in humans.

Genetic Jargon

Nematodes (NEEM-a-toads) are a type of worm that have threadlike bodies. They are also called roundworms. The name comes from the Latin word meaning "the threadlike ones."

For instance, NemaPharm, a small biotech company in Massachusetts, has unique plans for testing out new drugs intended for humans. Instead of trying out the drugs on people to see whether they cure or have unexpected adverse effects, the company uses roundworms, also called *nematodes*, that have been genetically altered to carry some human genes.

In the future, plants may even be used to test for the mechanisms of human diseases and to develop possible cures. Gloria Coruzzi, a plant biologist from New York University, contends that plants use a chemical for "memory" that is similar to a chemical found in the human brain. Who knows? This chemical might lead to the diagnosis and cure for diseases such as Alzheimer's.

Guidance Counselors

It's difficult to keep up with all the new advances in genetic testing technologies. Even some doctors have trouble sorting it all out, because most doctors have not been trained in clinical genetics. Trained professionals called genetic counselors help people interpret the results of their DNA disease tests. These counselors try not to make the decisions for a patient, but rather inform them about what the tests results may or may not mean. There are many different factors that come into play that determine whether a person will develop a disease, and if so, how severely they will get the illness.

For instance, if someone is a carrier for a genetic disease like cystic fibrosis but has not developed the disease, a genetic counselor can explain to the person what the chances are of passing this gene on to his or her offspring. If a man and woman who are both carriers of the mutation for cystic fibrosis marry, then the genetic counselor would explain that they have a 25 percent possibility of giving birth to a child who receives both copies and consequently develops the disease.

But what does this mean? Some people might misinterpret this to mean that if they have one child with the disease, then their next three children will not have it. This is incorrect. The genetic counselor will explain that every child has a one-in-four chance of getting both mutated genes, regardless of whether any of his or her siblings has one, two, or more of the mutated genes.

A Risky Business

A genetic counselor will also explain that carrying any mutated gene carries a specific risk for developing the disease associated with that mutation. But every person is different.

If you took a group of 100 people who had the same type of mutated gene for cystic fibrosis, you might expect them all to develop the disease at the same time and in the same way. But the fact is that some of them will live quite a bit longer than the others. This is because aside from a person's genetic makeup, environmental factors also figure into whether a person develops a disease. Where you live and what you eat contributes to if, when, and how you develop the disease. If a person eats only foods that have a low nutritional value, or if they smoke, for example, these factors will certainly contribute to their chances of developing an illness.

What's Available?

Genetic counselors have a lot to explain to their clients. There are a number of DNA diagnostic tests out there, and the list is growing every day, especially because scientists are finding out more and more about genes from the project that is cataloging the entire human genome. You'll read about this scientific effort in Chapter 21, "The Human Genome Project."

There are a large number of DNA diagnostic tests for genetic disorders. Here's a list of just a few of the tests that are currently available:

➤ Down syndrome

➤ Muscular Dystrophy

➤ Cystic Fibrosis

➤ Sickle Cell Anemia

➤ PKU

➤ Fragile-X syndrome

➤ Huntington's Disease

➤ BRCA1 and BRCA2 (implicated in some forms of breast cancer)

➤ Tay-Sachs Disease

➤ Hemophilia

➤ ALS (Lou Gehrig's Disease)

➤ Familial Hypercholesterolemia (results in extremely high cholesterol)

The technological advances we've made so far can determine whether a person is at risk for a disease. But some of the gene mutations that can be discovered in a person's DNA are incurable. In such cases, it's debatable whether these tests do more harm than good. You'll read more about such ethical concerns in Chapter 28, "People Problems."

The Least You Need to Know

➤ Some DNA testing for diseases is a lot like the identity testing used in courts.

➤ Scientists have created tests for the mutations that are linked to certain diseases such as Huntington's disease.

➤ Biochips are silicon or glass microchips carrying specific DNA sequences that can test for the presence of mutations in a person's DNA.

➤ Genetic counselors are trained specialists who explain the risk factors and other issues to patients before and after DNA testing.

Chapter 20

Did Your Genes Make You Do It?

Is there a gene that causes some people to drink too much? To be paranoid? To get up and dance on a table at parties? Is there a stretch of DNA that causes a genetic predisposition to program the VCR correctly? These days, you hear about a gene for this and a gene for that with such frequency that you'd think there was a Gene-of-the-Month Club.

Genes certainly can code for diseases such as cystic fibrosis and sickle cell anemia, but not everyone is in agreement about how much or even whether your genes affect complex forms of your behavior. It's the old nature versus nurture debate, which has been going on for decades. Some people feel that your genes are your destiny. Whatever you're born with, you're stuck with, they say, and it will affect not only how you look, but how you act. Others insist that your genetic endowment means much less than the environment you grow up in. According to this view, your parents, siblings, teachers, and peers affect your personality much more than your DNA. Others believe that a complex combination of genes and environment makes you who you are.

Welcome to the world of *behavioral genetics*, the branch of scientific inquiry that aims to find a genetic cause for all those wild and wacky things you do. Read on to learn how some scientists are trying to find genetic links to mental illnesses, addictions, sexual orientation, shyness, and even thrill-seeking.

Do Twins Behave Identically?

Pamela Burford and Patricia Ryan look alike, and sometimes they even think alike. They're identical twins. Formed from the same fertilized egg that split into two distinct embryos, Pamela and Patricia have identical DNA. Every gene in every cell in their bodies has the same order of As, Cs, Ts, and Gs. So how has this affected their behavior?

Pamela and Patricia, or is it Patricia and Pamela? Identical twins share the same DNA, but do they share the same behaviors?

When they were children, they both liked to sew. They were attracted to the same types of friends, and they both were interested in art. When they grew up, they got married just two months apart. They each have two children, and they both enjoy cooking. They also decided to embark on new careers at the same time. Today, they're both successful romance novelists, and sometimes they even write these books together.

Was it their genes that caused these similarities? Did their identical DNA destine them to have identical courses in life? Were their likes, dislikes, future careers, marriages, and motherhood all predetermined by their DNA, like some inner astrology? Were these romance writers born or bred?

Some people would say that no matter what, Pamela and Patricia would have become writers, been married, and had two children when they did because of their

Genetic Jargon

Behavioral genetics is the branch of genetics that investigates the relationship between the genes you're born with and the behaviors you exhibit.

DNA. But others would point out that not only did the twins have identical genes, but they were also raised by the same parents at the same time in the same place. Their environment, and not their genes, could have caused them to grow up the way they did.

So what would happen if twins like Pamela and Patricia were raised separately? If Pamela and Patricia were raised in different homes and by different families, would one have grown up to become a civil engineer and the other a Las Vegas showgirl?

Separated at Birth

A researcher named Thomas Bouchard decided that if he could find sets of identical twins who were separated at birth and then reunite them, he could study whether nature or nurture played the greater role by comparing their similarities and differences. In an effort that resulted in the Minnesota Study of Twins Reared Apart in 1988, Bouchard ferreted out hundreds of twins who had been put up for adoption at birth. These twins were raised separately by different families, often in completely different parts of the country. Sometimes one was adopted by an American family, and the other was adopted by a foreign family.

Some of the reunions that Bouchard arranged astounded him. Take the case of Oskar Stohr and Jack Yufe. Stohr grew up in Germany and was raised a Catholic; his identical twin grew up in Trinidad and was raised by a Jewish family. But when they met for the study, they both showed up in the same type of clothes, wearing the same type of glasses and sporting identical mustaches. As they spoke about their similarities, they discovered the most unusual one. They both were practical jokers who liked to sneeze in elevators to surprise the people around them.

DNA Data

Sir Francis Galton (1822–1911), a British biologist and a cousin of Charles Darwin, was the first researcher to systematically study pairs of fraternal and identical twins to determine whether nature or nurture played the larger role. His conclusion was that nature had more to do with determining what people become. Unfortunately, he also believed in the theory that "superior" people should be encouraged to mate to improve the population and "inferior" people should be discouraged from reproducing. These ideas, known as eugenics (you-JENN-ics), created some ethical problems. Ideas such as these led to forced sterilization laws in the United States and, at its worst, Hitler's ideas of "purifying" the population. You can read more about eugenics in Chapter 28, "People Problems."

Another reunited pair were identical twins Jim Lewis and Jim Springer. When they met at age 39, many years after their separation as infants, they were exactly the same height and weight. But the similarities went way beyond the physical. They both had dogs they named Toy, and each Jim had been married two times. Their first wives were both named Linda, and their second marriages were to women named Betty. One twin named his son James Alan; the other twin called his son James Allen.

Another pair of twins separated at birth were named Mark Newman and Gerald Levey. When reunited, they both had the same type of mustache and wore the same type of glasses. In addition, each of them wore a belt with a key ring on the right side, and they were both volunteer firemen in New Jersey.

Nature and Nurture

Some people, and especially the media, picked up on these astounding findings and concluded that genes must determine everything, ranging from people's choice of careers to their choice of spouses and pets. But some scientists see flaws in the twins research.

Many scientists and laypersons alike think that the way people act and look has a great deal to do with their genes, but this influence doesn't mean that people are prisoners of their DNA. What you're born with isn't necessarily a life sentence, they say. Personalities are a complex blend of the genes that people are born with as well as the family surroundings that they grow up in, their diet, their friends, when and where they're born and raised, and many other factors.

And complex traits such as whether a person is outgoing or shy, why a person is drawn to one career but not another, or how a person chooses a mate are all determined by much more than the order of As, Cs, Ts, and Gs in that person's DNA. Although each gene codes for a specific protein, there is no protein that controls whether someone will want to study philosophy as opposed to anthropology.

However, because genes do affect everything about our bodies, a certain genetic makeup may predispose someone to act in one way rather than another. But the surroundings that the person grows up in may change all that. If a person craves an illegal drug, for example, a religious family background may alter that person's decision to try it.

Researchers at the George Washington University School of Medicine and Health conducted a study of more than 700 families in various parts of the United States. They found out that even when parents raise twins, they don't always treat them identically, and each twin responds in their own way to these differences in their parental environment. One of the study's conclusions was that a loving environment leads to more well-adjusted children who behave well. So even if a child is born with the best of genes, an abusive environment can take its toll.

Planting a Future

Those who believe in the combined influence of nature and nurture often point out that if a gardener takes the best seeds for roses, plants half of them in fertile soil, waters them sufficiently, and makes sure they have enough sunlight, soon there'll be a garden plot with lovely, sweet-scented red roses. But if the gardener plants the other half of these seeds in poor soil, waters them infrequently, and lets them grow in a dark patch of the garden, you don't need to be Gregor Mendel to guess the poor results.

Some people say the nature and nurture camp holds that the same is true of humans. If a child is not raised in a nurturing, healthy environment, then the promise of the genes will never be fulfilled. Nature and nurture must work together to yield the best results.

Your Gender's in Your Genes

You already know that if you've got two X chromosomes from your parents, then you were born a woman, but if you have one X and one Y chromosome, you're a man. But if genes determine sex, do they determine sexual preference as well?

In the 1990s, a geneticist named Dean Hamer decided to investigate whether a genetic basis for homosexuality existed. He began by studying whether homosexuality runs in families. If it did, Hamer reasoned, he might be able to determine whether it was caused by genes or the family environment.

DNA Data

Hamer's associate, Angela Pattatucci, studied data on female participants in the study and concluded that lesbianism also tended to run in families. However, the patterns were different than those found in homosexual men in the study. Hamer concluded that lesbianism had more to do with cultural factors than with heredity.

Hamer advertised in gay publications for people to participate in the study. He asked the participants if they had gay siblings or cousins, aunts, uncles, or grandparents. When he had collected data from over 100 families, he concluded that the sexual preference seemed to run in many of these families. He realized that this tendency might be an indication that genes were implicated in this preference. However, he knew that this one fact didn't necessarily prove it.

Upon closer examination of his data, Hamer discovered that there were patterns in the distribution of gay relatives in some of the families that he studied. It seemed that most of the gay relatives were on the mother's side of the family. Hamer took this pattern to be particularly significant.

If there were a gene for homosexuality, and if it were passed down on the maternal side of the family, Hamer reasoned, then it would have to be on the X chromosome. You'll remember from Chapter 10, "The X in Sex," that men get their sole X chromosome from their mothers. If homosexuality was connected to a gene located on the X chromosome, Hamer reasoned that there would be more gay relatives on the mother's side of the family and few or none on the father's side of the family. This theory was consistent with the patterns that were observed in the study.

A Genetic Address

The next step was to look for a gene on the X chromosome. But there are several thousand genes on each chromosome, and Hamer didn't know exactly where to find the gene, or what protein it coded for. It could also be possible that a number of genes working together produced this trait, and, in addition, Hamer couldn't rule out environmental influences.

To narrow the search, Hamer looked at large sequences of DNA that acted as markers. You read in Chapter 19, "DNA Disease Diagnosis," that markers are noncoding sequences of DNA that always seem to be in the vicinity of certain genes. You read how following a marker led to the discovery of the gene that causes Huntington's disease.

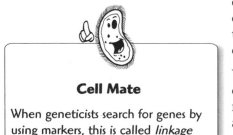

Cell Mate

When geneticists search for genes by using markers, this is called *linkage analysis*, because the marker is linked to the gene.

You could think of a marker as a landmark when you're driving. If you are given directions to a certain restaurant, you might be told to look out for the gas station, and then the restaurant is just up the block. The gas station has nothing to do with the restaurant, and you certainly wouldn't want to eat there, but it's a definite indication that you're going in the right direction to your final destination.

Xq28 Marks the Spot

Concentrating on families with brothers that were both gay, and excluding families that had gays on the father's side of the family, Hamer continued his research. Narrowing down the participants in his study to 40 pairs of gay brothers, Hamer tested their DNA for the presence or absence of 22 different markers. In a section of the X chromosome called Xq28, which is at the end of the long arm of the X chromosome, Hamer came up with some interesting findings.

Out of the 40 pairs of brothers studied, 33 of them had five markers that were the same. Hamer concluded that a gene or genes that might influence sexual preference might near these markers. In July 1993, *Science* magazine published Hamer's paper, "A Linkage Between DNA Markers on the X Chromosome and Male Sexual Orientation."

DNA Data

Even genes have their own "addresses" nowadays. You can't mail letters to genes at your local post office, but there is a system that scientists use to explain exactly which chromosome a particular gene is on and approximately where it lies on that chromosome. For instance, Dean Hamer narrowed down his search for a possible genetic link to homosexuality to a region called Xq28. The *X* means that the section of DNA lies on the X chromosome, the sex chromosome that men and women get from their mother. The *q* refers to the long arm of this particular chromosome (as opposed to the short arm), and the *28* refers to the 28th region on the chromosome.

The study met with an enormous amount of controversy. Some gay activists reacted favorably to the news; others accused Hamer of depicting homosexuality as a genetic flaw. Reputable newspapers such as *The New York Times* reported the finding accurately, but one tabloid ran a headline proclaiming that there would be a supposed "cure" for homosexuality within a few years.

When a research paper is presented, it is more likely to be accepted by the scientific community if other laboratories in addition to the one that initiated the study are able to replicate the study, meaning that they can do the same experiment and come up with the same results. In 1995, two years after Hamer's report was published, George Ebers and George Rice in Canada studied close to 200 families with gay brothers. Like Hamer's research team, they also found gay relatives on the maternal side. But their research into the gay brothers' DNA did not yield the same results as Hamer's. In his book *Living with Our Genes*, written with Peter Copeland, Hamer explains that the Canadian study was conducted differently, so he does not think their findings dispute his own.

Despite the controversy, one fact remains. No one has found a gene or genes for sexual preference. Hamer wrote an article for *Advocate*, a national newsmagazine for gays and lesbians, describing that sexual preference is a behavior too complex to be explained by heredity alone. Despite the media's talk of a possible "gay gene," Hamer admitted that there could not be such a thing. Rather, genes could impart a predisposition to a particular type of behavior, but this predisposition would always be molded by outside forces such as society and family.

The Chemistry of Mood

Genes code for proteins, and some proteins can affect the brain, so does this mean that there are genes for different moods? Scientists know from studies of animal behavior that certain instincts are inborn. For instance, no one has to teach a duck how to swim or a bird how to fly. But do humans have similar behaviors that are caused by genes?

Some scientific investigations in our Age of Genetics have focused on determining whether there are genes for mental disorders or addictions. Since the 1980s, scientists have been hunting for genes that they believe cause some psychological disorders. Yet other scientists hotly dispute the possibility that just one gene may have such an enormous impact on a person's behavior. They think that the way we act is more likely to be influenced by the interaction of many different genes, as opposed to a specific one. So this move from a psychoanalytical approach to mental illness to a biologically-based approach continues to be very controversial among psychologists.

DNA for Delusions?

Psychologists have long tried to discover the cause and a possible cure for a mental illness called *schizophrenia*. This disease often appears in a person's late teenage years or early 20s. This mental ailment is characterized by hallucinations, delusions, disorganized thinking, and in some cases, the inability to tell the difference between fantasy and reality. About one percent of the population has this serious mental illness.

In 1987, a team of researchers at the University of British Columbia in Canada did a study of schizophrenic patients and their families and came to the conclusion that there might be a genetic link to this disease. A 20-year-old man had been brought in to the emergency room in the university's hospital because he had spent weeks in his room, talking and laughing to himself. His anguished mother told the psychiatric resident at the hospital that her afflicted son not only acted like her brother, who had been diagnosed with schizophrenia, but he also looked like her brother.

Genetic Jargon

Schizophrenia (skits-a-FREE-nee-a) is a mental illness characterized by hallucinations, delusions, and disorganized thinking. In some cases, a schizophrenic cannot tell the difference between fantasy and reality. The disease usually appears in a patient's teenage years or early twenties.

The psychiatric worker in charge of the case, Ann Bassett, had DNA testing performed on both the young man and his uncle. It turned out that a part of chromosome 5 was not only duplicated, but was inserted backwards into another chromosome in both men. Then the young man's mother was tested. She, too, had the same mutation on one of her copies of chromosome 5, but not on the other. You'll remember that we all have two copies of each chromosome, one that we inherited from our mother and one from our father. But men only inherit one X chromosome, and this chromosome comes from their mother.

In the following year, 1988, a British team led by Hugh Gurling and Robin Sherrington reported in *Nature* magazine that they obtained similar results when they studied families with a history of schizophrenia from Iceland and England. On chromosome 5, the same one that had a mutation in the young man and his uncle in Canada, the British team found a special segment that appeared in the British and Icelandic family members with this mental disorder. They concluded that the odds of this happening by chance were 50 million to 1.

However, in the very same issue of *Nature*, another article presented data that contradicted Gurling and Sherrington. Kenneth Kidd, a researcher at Yale University, studied a large Swedish family with a widespread incidence of the mental disorder. When Kidd analyzed their DNA, he found no connection between segments on chromosome 5 and the occurrence of schizophrenia.

Later studies on families with a history of schizophrenia implicated genes on other chromosomes as possible candidates involved with the development of the disease. However, there are still doubts about what role genes play in this complex behavior pattern. In 1990, Richard L. Suddath studied pairs of identical twins in which only one twin developed schizophrenia. The researcher and his team reported in the *New England Journal of Medicine* that his research team found abnormalities in the brains of the affected twins. This finding calls into question the idea of genes as the culprit, because although both twins shared the same DNA, only one developed the disease.

A Debated Issue

As you can see by the conflicting conclusions that have come out about schizophrenia, genes and behavior are very hard to study. Here's some of the reasons why:

➤ Scientists can't study generation after generation of humans in the same way that they study lab animals such as mice or fruit flies.

DNA Data

Researchers have also reported finding genetic predispositions for alcoholism, manic depression, and even "thrill-seeking." Some researchers even think there might be a genetic link to whether we enjoy eating alone and whether we change jobs frequently! However, genetic links for any complex behavior such as these are hotly debated by the scientific community.

➤ Behavioral traits such as mental disorders and addictions are highly complex. They're not as simple to study as the instinctive behavior of a duckling that follows its feathery mom around. Even experts frequently disagree on the diagnosis of a mental illness in a patient.

➤ It's nearly impossible to separate the genetic contributions from those that came from the environment.

Is There a Gene for Genius?

You learned in Chapter 9, "Irregular Genes," that mutations can cause mental retardation. For instance, an extra copy of chromosome 21 causes Down syndrome. Fragile-X syndrome, which results from a mutation that causes the tip of the X chromosome to break off, is the second leading cause of mental retardation.

But if genes can cause lower intelligence, does this imply that there may also be genes for higher intelligence? Is there a gene for genius? Once again, you'll get a different answer to this question depending on which scientist you ask. There are behavioral geneticists who feel that intelligence is inborn, but other scientists think that a complex mixture of genes and environmental factors determines how bright a person will be.

Dangerous Curves

In 1994, a storm of controversy was set off by a book called *The Bell Curve* by Charles Murray and Richard Herrnstein. One premise of this book was that intelligence is hereditary. The book went on to assert that certain groups, such as African Americans, are genetically programmed to be less intelligent than other groups.

In addition to its obviously racist overtones, this assertion is dangerous in the fact that it can lead to an inaccurate view of underprivileged children. There are programs such as HeadStart that try to improve the social and educational environments of these children to improve academic performance. But if intelligence is predetermined by genes, as *The Bell Curve* suggested, then why would anyone bother to develop programs to help children from poor households without access to better school programs? According to Murray and Herrnstein's theory, it would just be a waste of time and money.

Most experts think that this simplistic view of nature versus nurture in relation to intelligence does not hold up under closer examination. For instance, a study done by Sandra Scarr and Richard Weinberg in the 1970s showed quite the opposite. They tested the intelligence of 99 African American children who were born into poor families, but had been later adopted by middle-class Caucasians in Minnesota. The researchers found that not only were the IQ scores of these children higher than the average African American children, but they were also higher than those of Caucasian children. The researchers concluded that environment played a highly significant role in shaping the thinking skills of these children.

More Than One Intelligence

Another problem with the theory that genes are the sole determining factor of intelligence is the fact that no one can agree on just what intelligence is. Although IQ (Intelligence Quotient) tests are widely accepted as a valid evaluation of a person's thinking abilities, critics point out that these tests can be culturally biased.

For instance, if there's a question on one of these tests about a regatta (a boat race), a child from an inner city neighborhood would have a hard time answering this question correctly. But this difficulty wouldn't mean that the person taking the test was not intelligent. It would just mean that the question falls outside of that child's life experiences.

Psychologists Howard Gardner, in his book *Frames of Mind,* and Thomas Armstrong, in his book *In Their Own Way,* argue that there are different types of intelligence. Typical intelligence tests are aimed at logical, linear thinking, but there are other kinds. Among these are musical intelligence, creative abilities, social abilities, and other forms of intelligence that are just as valid as linear thought.

A Smart Environment

Sometimes changing our environment can override the "predictions" of our genes. Take the case of a disease called PKU, which is short for *phenylketonuria*. Because of a faulty enzyme in the liver, an amino acid called phenylalanine can't be broken down in the body of a person with this mutation. The result is a buildup of toxic products in the body, and this buildup ultimately causes brain damage, which can lead to severe mental retardation.

However, when infants with this genetic problem are kept away from foods that contain the amino acid they can't break down, there is no damage to the brain. In some cases, when the child becomes a teenager, the child's brain becomes able to withstand the toxic products resulting from this condition, and he or she can consume a normal diet without damage. From this example, you can see how nurture can make up for a genetic flaw in a person's nature.

As scientists continue to find out more about genes, we should cautiously examine any simplistic beliefs that one gene could cause a complex human behavior. More and more, we are finding out that nature and nurture, like the two separate strands of the DNA double helix, are very much intertwined, and one is just as important as the other.

Genetic Jargon

Phenylketonuria (FENN–il–KEET-a–NYOOR-ee-a), or PKU, *is a genetic disease that is characterized by the lack of a liver enzyme that can break down an amino acid called phenylalanine (FENN–il–AL-a-neen). If the disease is not controlled by a special diet, the toxic buildup in the brain of the person with the disease can lead to severe mental retardation.*

The Least You Need to Know

➤ Studies of twins reared apart have shown that they often exhibit similar behaviors as adults.

➤ Geneticist Dean Hamer found segments of DNA that appeared with high frequency in gay brothers.

➤ Studies have been conducted to find genetic links to schizophrenia, but the results have been hotly debated.

➤ Intelligence results from interactions with the environment as well as the genes you inherit.

Part 6

Mending Your Genes: Gene Therapy and New Medicines

Scientists have embarked on the Human Genome Project, a huge effort to find out about all the genes humans have. Eventually, they hope to be able to fix faulty genes that might cause diseases. In the future, they plan to cure diseases by adding genes that are missing or fixing ones that don't work properly.

Researchers also hope to create new medicines specifically tailored to each individual's DNA. Currently, scientists are experimenting with genetically engineered vaccines and other medicines that have been made using the latest discoveries about genes and DNA.

Chapter 21

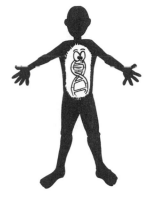

The Human Genome Project

If you were going to visit an immense foreign country that had three billion unmarked streets, you'd have one heck of a time finding your way around. If you wanted to get to a specific address, you'd have to bring along a good map to help you find your way through the long and winding maze of unknown territory.

The human genome, the totality of all the genetic material in people, is similar to that vast, complex country. Suppose you were a researcher and you wanted to find out the location of a certain gene amid all that human DNA. With the estimated three billion base pairs of Ac, Cs, Ts, and Gs, you'd be wandering around in the DNA wilderness for quite a long time if you went about your task in a haphazard way.

In an effort to figure out how human DNA is organized (and some animal and plant DNA, too), scientists formally began an international effort called the Human Genome Project in October 1990. (This was the official starting date, but preparations had been in place for a number of years before that time.) Read on to discover how scientists are trying to find out more and more about DNA and how they feel it will affect the future of medicine. They hope it will transform today's health care from one that treats symptoms and diseases after they've developed to a more preventive approach that can stop diseases before they even begin. To learn more about some of the ethical issues raised by this project, be sure to read Chapter 28, "People Problems."

The Holy Grail

For many years, scientists have wondered where each of the approximately 80,000 human genes are located. Humans have 46 chromosomes (two pairs of 23, one set from each parent), and each chromosome can contain thousands of genes, so making random guesses wouldn't help. Eventually, the idea of mapping and sequencing the human *genome* was discussed more and more in scientific circles. The idea gathered momentum with more and more scientists, and finally, federal funding was granted to researchers in the United States to *map* and *sequence* the human genome.

The United States Genome Project, which was formally begun in October 1990, is divided into the National Institutes of Health (NIH) National Human Genome Research Institute (NGHRI) and the Department of Energy (DOE) Human Genome Program. Francis S. Collins is the head of the NIH division of the Human Genome Project, and Ari Patrinos directs the DOE Human Genome Program. Initially, the total cost of the project was estimated at $3 billion.

The project in the United States had several goals from the start, including the following:

Genetic Jargon

A **genome** (JEEN-ome) *is the totality of DNA in an organism. To* **map** *the human genome means to figure out exactly which chromosome each gene lies on and exactly where each gene sits on that chromosome. To* **sequence** *DNA means to figure out its specific order of As, Cs, Ts, and Gs.*

➤ To make a genetic map of the entire human genome, in other words, to show where the genes are in relation to one another along the chromosome

➤ To make better physical maps of all the human chromosomes, that is, to determine their "address" on the chromosome

➤ To make genetic and physical maps of the DNA of several other organisms besides humans

➤ To sequence (determine the order of the As, Cs, Ts, and Gs) the entire human genome as well as the genomes of several other organisms, including some bacteria, plants, and animals

➤ To develop technologies that make sequencing the genome easier, faster, and cheaper

This mammoth program is being carried out not only by scientists in the United States, but also by scientists in more than 18 other countries who are also involved in research programs that are connected to the U.S. Human Genome Project's goals. Some of the larger international programs are being conducted in the United Kingdom, Germany,

Japan, Australia, Brazil, Canada, Denmark, Sweden, Korea, Mexico, the Netherlands, Israel, Italy, and France. Scientists consider the collection of all this data to be so monumental and significant for the future of biology and medicine that they've dubbed it The Holy Grail of science, referring to the medieval legend of the quest for a cup that Christ used at the Last Supper.

The Inner Country

The Holy Grail isn't the only comparison that's been made for this extensive project. To conceive of the enormity and importance of the human genome, one comparison says that you can think of the genome as the Earth. If the genome is like the Earth, then geneticists can see some of it, but they can't make out all the details. Imagine a group of astronauts up in a space shuttle, looking down on the Earth. They know that everyone and everything is down there, but they can't tell you exactly where their families live by looking out the window of their spacecraft.

DNA Data

The Department of Energy and the National Institutes of Health estimate that the Human Genome Project in the United States will cost about $3 billion. Out of this amount, from three to five percent of the budget will be used to study the ethical, legal, and social issues that will undoubtedly develop as the project continues. This study is called the ELSI project.

When these hypothetical astronauts start their journey back home, they're able to make out the shapes of the continents as they get nearer and nearer to the earth's surface. If scientists got closer to the human DNA inside a cell, then they could start to make out the chromosomes, the 46 units we have inside our cells. These chromosomes are home to our genes.

As the astronauts get even closer to Earth, they can start to make out a particular state. This can be compared to a geneticist looking at a section of a chromosome, such as one of the arms, either the long arm or the short arm. The space shuttle finally brings the astronauts close to a small town. This would be like scientists looking inside a specific segment of a chromosome and examining one stretch of the DNA there. This stretch of DNA is a gene.

Finally, as the astronauts peer out the window, they can see not only the small town with its houses and buildings, but they can also see people walking around, driving cars, and crossing streets. This would be like scientists getting close enough to a gene to identify its base pairs, those rungs on the twisted rope ladder that makes up DNA. At this stage, the scientists can examine the gene in detail to determine the order of its nucleotides.

If a cell is like the world, then a chromosome is like a country, a section of a chromosome is like a state, a gene is like a small town, and base pairs (As, Cs, Ts, and Gs) are like people and houses.

The Encyclopedia of Life

The study of the human genome has also been compared to working with an enormous encyclopedia divided up into many volumes. Think of these volumes as the chromosomes in humans. Within each one of these volumes, there are pages and pages of text.

The funny thing about this encyclopedia is the fact that there are only four letters used in the text. In addition, there are no spaces between the letters, and there isn't any punctuation to show you where the words begin and end. You also have no clue what you're reading because there are no explanatory pictures, charts, or drawings to help you figure out what's written there.

Suppose you can fit 4,000 letters on each page of each volume in the encyclopedia. Then you would need about 750,000 pages to register all of the DNA in our chromosomes. The resulting stack of books would measure about 150 feet high. If for some odd reason you had too much time on your hands and felt like reading all these letters (all three billion bases) aloud, then it would take you about 9.5 years to do it, assuming you didn't take any time off for eating, sleeping, or reading this book.

If it would take you all that time just to say these combinations of letters, think of how different it would be to store all of this information on your home computer. One million bases alone would take up about one megabyte of storage space, and the human genome is estimated to be about three billion base pairs long. That means you'd need three gigabytes of storage space, not including any space you might need for comments or additional information you'd want to throw in with your list of human DNA.

Putting Genes on the Map

In order to learn about DNA, scientists use a number of high-tech tools. Some of these tools help scientists map the genes. Researchers hope the process of mapping the genes will help them to gain a greater understanding of how genes work and what happens when they don't work correctly.

The Missing Linkage

There are several types of what are known as genetic maps. Think of these as maps to a country that can tell you all the different addresses you could find there. To make a genetic map, scientists look at thousands of genetic markers, which, as you learned, are like signposts on a road pointing out the location of your genes.

Although scientists have the tools they need to identify these markers, they also need to study markers in a lot of people who are related to one another. In order to do this, scientists need to study extremely large families. For example, the Mormons, a religious group based in Utah, have kept detailed family records for many generations and therefore make good subjects for genetic study. By studying the DNA of groups like the Mormons, researchers can find some of the markers that are common to their extended families. This in turn helps the scientists find out whether they're in the particular region of the wilderness of DNA that they want to study and understand.

There are two main types of genetic maps. One is called a *genetic linkage map*. This type of map examines the position of one gene or

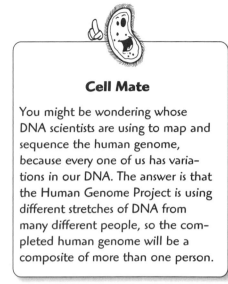

Cell Mate

You might be wondering whose DNA scientists are using to map and sequence the human genome, because every one of us has variations in our DNA. The answer is that the Human Genome Project is using different stretches of DNA from many different people, so the completed human genome will be a composite of more than one person.

Genetic Jargon

A **genetic locus** (LOW-kus) is a specific area of DNA that is being studied in a genetic linkage map. The name comes from the Latin word that means place, which is also where we get the word *location*. A **genetic linkage map** compares the position of one genetic locus with others.

sequence of DNA in relation to another gene or sequence of DNA. The piece of DNA that's being studied is called a *genetic locus*. It is like a map that says New York is halfway between Washington, D.C. and Boston, but doesn't tell you how many miles it is.

Let's Get Physical

This sort of map is pretty useful, but it would be much more informative to have a map that showed the mileage between the cities. This kind of genetic map is called a physical map. A physical map tells you exactly where a specific gene or other sequence of DNA is located on the DNA strand.

To make a physical map, scientists use several procedures. For instance, the scientists decide which specific chromosome they want to study. Then they chop it up into a bunch of pieces using restriction enzymes, which, as you read, act like molecular scissors. Then these different sized pieces are inserted into lower organisms such as yeast or bacteria. Every time these organisms reproduce, more of the sequences are also made. Then researchers go on to the next step, which is figuring out how these pieces fit together.

Delving into DNA

At the National Institutes of Health, the United States Human Genome Project involves three steps. The first step is mapping. One advantage of doing this step first is the fact that it divides the unwieldy three billion base pairs of the human genome into smaller, easier-to-work-with sections. It's like examining separate volumes or pages in an encyclopedia a little at a time instead of trying to read the whole thing at one sitting.

The next step is determining the order of the nucleotides in these DNA segments. It would be difficult, if not impossible, to try to sequence all three billion base pairs at once. So the small sections are cut up into even smaller pieces, and then scientists try to put them back together again to see just how they fit together.

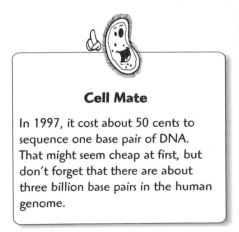

Cell Mate

In 1997, it cost about 50 cents to sequence one base pair of DNA. That might seem cheap at first, but don't forget that there are about three billion base pairs in the human genome.

One way to understand this better is by imagining that you're trying to figure out the order of all the letters in this book. If this book were the human genome, then what you'd do would be to take several copies of *The Complete Idiot's Guide to Decoding Your Genes* and then cut them up into random pieces of about a few paragraphs each.

Now you have a tremendous number of pieces, but you have no idea what order they used to be in. But by analyzing all these randomly cut paragraphs, eventually you'd find phrases or even sentences that match up. Some of these phrases might match up for only a few words, but one of the copies of the cut-up books might have even more words to continue the phrase. So by matching up different overlapping sections, you would sooner or later be able to put together the meaning of the entire text of the book.

The last step in the process is called finishing, which is something like proofreading. In this step, scientists re-examine difficult areas by sequencing them one more time to ensure accurate results.

So far, the Human Genome Project has revealed a great deal about human DNA, as well as the genomes of some other species. All in all, the project seems to be ahead of schedule. Originally, one goal was to complete all of the mapping and sequencing by the year 2005. Now it seems as though the project may be completed ahead of schedule, by 2001.

A Race to the Finish

The Human Genome Project was going along just fine when an unexpected announcement shocked the scientific world. On May 9, 1998, J. Craig Venter, a renowned genome researcher, said that he and the Perkin-Elmer Corporation of Connecticut were going to form a new company. Perkin-Elmer is the world's leading manufacturer of automated DNA sequencing machines.

What Venter and Perkin-Elmer had in mind was a plan to use a new type of automated machine to complete the sequencing of the order of bases in human DNA in just three years. Not only does their expected completion by the year 2001 threaten to pull the rug out from under the official Human Genome Project and its later completion date, but Venter's plan will cost less. He estimated that it could be done for about $200 million dollars, which is less than the government-funded project's estimate.

This announcement has raised quite a debate among scientists. Some think that Venter's methods are the works of a genius; others question whether his unconventional approach will achieve its intended goals. Even if he does succeed, some people wonder whether those results will be as accurate as the ones that the official project will yield.

Messenger Service

Venter's approach differs considerably from the original Human Genome Project's methods. He came to this approach after trying other, much slower ways of isolating genes. In the late 1970s, the technology in this area wasn't as sophisticated as it is today. So Venter spent 10 years to find his first gene in 1986, using available tools.

Eventually, newer methods were devised, and when Venter heard about automated DNA sequencers, he decided to try them out. This was in 1987, when he was the head of a lab at the National Institutes of Health. Employing this new technology, Venter discovered eight genes in just two years.

But Venter wanted to do much more and much faster. By 1991, he found a new way to uncover the mysteries of DNA. He realized that any cell in a person's body can

Cell Mate

In addition to the human genome, the Human Genome Project is also studying the DNA of E. coli and other bacteria, mice, fruit flies, yeast, and several other organisms.

Genetic Jargon

Complementary DNA, also known as cDNA, is a synthetic form of DNA made from messenger RNA. Usually, messenger RNA is used by the tiny ribosomes—"factories" in the cell—as the blueprint for manufacturing a specific protein. But cDNA is used by scientists for purposes such as locating specific genes.

Genetic Jargon

EST stands for expressed sequence tags. This is the name that researcher J. Craig Venter gave to parts of genes that were fished out using a synthetic form of DNA called complementary DNA.

sequence DNA many times faster than the speediest supercomputer. How? Remember from Chapter 8, "Producing Proteins," that when a cell is going to manufacture a specific protein, it first copies the gene that codes for that protein and sends this copy from its nucleus to its ribosomes. This copy, called messenger RNA or mRNA, is a type of molecule that's something like DNA, only it's single-stranded instead of double-stranded like the double helix. This mRNA is like a template of the gene that is used to code for the protein.

The only difference—and it's a big difference—between messenger RNA's readout of a gene and the one that is obtained by painstaking work in the lab is the fact that messenger RNA contains only the coding part of a gene. Genes can have lots and lots of noncoding parts, called introns, separating the coding parts, called exons.

Fishing for Complements

One challenge that Venter came up against was the fact that messenger RNA isn't such a sturdy molecule. But researchers have a remedy for this. They use special enzymes that change the messenger RNA into a more stable synthetic molecule called *complementary DNA* or cDNA.

Although Venter wasn't the inventor of complementary DNA, he found a new way to use it to quickly isolate parts of unknown genes at an unprecedented pace. Venter figured out that complementary DNA can be used like a lure to fish out parts of genes. He called these fished-out parts *ESTs*, or expressed sequence tags. This method not only worked, but it also obtained results much quicker than any other methods used before.

When Venter's employer at the time, the National Institutes of Health, saw the huge successes he had with his new method, they tried to get patents on his ESTs. This attempt brought on a storm of controversy. Among the critics was Dr. James Watson, the Nobel prize winner who was the director of the U.S. Human Genome Project at the time. He was very much against these attempts at patenting and called the idea "sheer lunacy." Watson resigned as director of the project, and some people feel that this was in part in protest of the patenting issues that were raised.

In 1992, Venter founded a company called The Institute for Genomic Research (TIGR). In 1994, Venter left the National Institutes of Health. Then, together with a sister company, Human Genome Sciences (HGS), Venter's company discovered expressed sequence tags (ESTs) for 35,000 genes. In other words, they had tiny bits of sequences of the coding parts of all of these thousands of genes.

Tagging Along

After his success with ESTs, Venter tried his hand at sequencing chromosomes. Characteristically, he went about this quest with a unique approach. Most scientists involved in the Human Genome Project sequenced fragments in order along the chromosome, but Venter used a so-called shotgun approach.

In this approach, the genome is first broken up into millions of random fragments. Then these fragments are sequenced, even though Venter has no idea where exactly they belong on the chromosome. The last step uses supercomputers to figure out where the fragments are located on the chromosome.

In 1995, Venter and his company joined with researchers led by Hamilton Smith at the Johns Hopkins Medical Institutions in Baltimore, Maryland. You'll remember from Chapter 13, "Cut-Down Genes," that Smith shared the Nobel prize in 1978 for the discovery of restriction enzymes, the molecular scissors that can cut DNA in very specific spots. This landmark discovery eventually led the way to Cohen and Boyer's creation of genetically engineered bacteria.

With Smith's team, Venter and his researchers discovered the entire genetic sequence of a type of bacteria that causes ear infections and *meningitis*, a dangerous inflammation of the membranes that surround the brain and the spinal cord. This discovery made Venter and Smith the first team to ever sequence the entire genome of a living organism.

After that, Venter continued his successful experiments. He discovered the genetic sequence of the bacteria that cause syphilis, as well as the kind that cause stomach ulcers. Then, in 1998, he announced his daring plan to finish off the entire human genome in just three years.

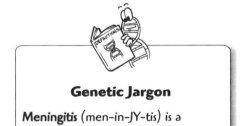

Genetic Jargon

Meningitis (men–in–JY–tis) is a serious disease, characterized by an inflammation of the membranes that surround the brain and the spinal cord.

But the scientific community is divided in their opinion about the accuracy of Venter's nontraditional approach to sequencing. Some scientists point out that the human genome is much larger than that of the bacteria that Venter worked with in the past. They point out that human DNA has lots of repeated sequences which make the task more difficult than sequencing bacterial DNA, which does not contain these repeated sequences.

Critics assert that Venter and his colleagues will obtain only incomplete results. Venter's answer is that the sophisticated computer programs that he works with are more than capable of filling in a lot of the gaps in the sequencing. He's also said that the repetitive, noncoding DNA doesn't interest him all that much. It's the coding parts, the genes, that he intends to continue working on. The controversy over patenting remains as well. Also, Venter's company releases their findings every three months or so, whereas the Genome Projects throughout the world release their new information within 24 hours.

Although Venter's methods and his proposal to finish sequencing the human genome are still hotly debated, some people feel that his work and intentions will spur on the team of the official U.S. Human Genome Project.

The Future of Genome Research

In an article in *Science* magazine's October 23, 1998 issue, Francis S. Collins, the director of the NIH Human Genome Program, Ari Patrinos, the director of the DOE Human Genome Program, and others involved in the effort discussed the new goals for the U.S. Human Genome Project for the years from 1998 until the project's completion. So far, the Human Genome Project has yielded a wealth of information on the workings of human DNA, as well as the genomes of other organisms.

As of January 1999, the project was responsible for the mapping of about nine percent of the entire human genome. The U.S. National Institutes of Health and the U.S. Department of Energy predicted that they will sequence the entire human genome by the end of the year 2000.

One definite aim of the project is to make all the information it obtains available to researchers and the public for free. The NIH and DOE directors feel that this information is such a valuable resource that medical researchers and others need to be guaranteed free access to this knowledge so they can utilize it to come up with possible cures for diseases. Another aim is to develop even more efficient technology to sequence DNA. The directors hope that existing technologies can be improved to speed up the process of finding out the order of all the bases of DNA.

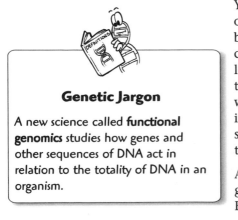

Genetic Jargon

A new science called **functional genomics** studies how genes and other sequences of DNA act in relation to the totality of DNA in an organism.

Yet another goal is to find out more about the function of genes and the genome. The project's findings will bring about a revolution in the way biology and medicine will develop in the future. But just knowing the location and sequence of genes isn't enough for scientists to understand everything about the way genes work. A new science called *functional genomics* is evolving from all of this new information. This science is the study of the function of DNA sequences in relation to the totality of DNA in an organism.

At present, scientists can study the functions of human genes by comparing them to genes in other organisms. But one goal for the future is to find new ways to understand how all the different parts of DNA in an organism function. New research methods must be found to make this study easier and more complete.

In addition to studying genes, their structure, and how they work, the Human Genome Project planners would like to see more work done on the so-called "junk DNA" sequences. Many scientists feel that although these segments of DNA don't code for specific proteins, they probably serve some purpose that hasn't been discovered yet.

Much more needs to be done in the field of evolutionary genetics as well. You read in Chapter 12, "Mutations and Evolution," that scientists have determined that all life descended from a common ancestor at the beginning of a vast evolutionary tree. By comparing the sequenced genomes of different organisms, scientists can understand more about how this evolution took place.

Comparisons between the genes of widely differing species can make it easier for scientists to discover which genes are similar. For instance, the entire genome of E. coli bacteria has already been mapped and sequenced. By seeing how bacterial genes function, scientists can learn more about the genes of more evolved organisms.

In December 1998, researchers at Washington University in St. Louis, Missouri and the Sanger Centre in England completed the sequencing of the first entire animal genome. This genome consists of 97 million base pairs and belongs to the *nematode*, a type of roundworm. The genomes of organisms like the nematode, along with those of the fruit fly and even mice when they are sequenced, will be compared to those of humans and other animals. Then scientists will have a clear picture of how genes function and where all these organisms belong on the branches of the evolutionary tree.

The Least You Need to Know

➤ The Human Genome Project is an international effort to map and sequence the totality of human DNA.

➤ Mapping means locating genes on chromosomes, and sequencing means finding out the order of their As, Ts, Cs, and Gs.

➤ The Human Genome Project is also studying the DNA of mice, some bacteria, yeast, and other organisms.

➤ A researcher named J. Craig Venter announced in 1998 that his company plans to privately achieve the goals of the Human Genome Project before government agencies can.

➤ In 1998, U.S. Human Genome Project directors announced that their goals for the next five years included completion of the project by the year 2003 or earlier.

Chapter 22

What Is Gene Therapy?

In This Chapter

➤ How scientists hope to fix faulty genes with gene therapy

➤ Why early attempts at gene therapy failed

➤ How W. French Anderson performed the first successful gene therapy procedures

➤ What is antisense therapy?

➤ How scientists are beginning to create artificial chromosomes

---

A lot has happened in the field of genetics since scientists first began identifying genes for several rare hereditary diseases in the 1970s. New advances in technology, along with the ambitious goals of the Human Genome Project, have given geneticists more and more information about the machinery behind heredity.

One of the reasons why the Human Genome Project was started in 1990 was the hope that a greater knowledge of genes would translate into ways for scientists to find cures for otherwise incurable diseases. Scientists know that certain genetic mutations can cause diseases. The traditional way of treating diseases is, of course, with medicine. However, some scientists wondered what would happen if they got to the root of the disease, instead of just treating the symptoms. Could scientists correct misspellings in DNA and thereby prevent the disease from occurring in the first place?

Read on to find out how scientists are experimenting with gene therapy, which involves correcting faulty genes or adding healthy genes that are missing. Like many aspects of the new genetic technologies, gene therapy also has its controversial side. You can read more about these issues in Chapter 28, "People Problems."

A Genetic Challenge

The possibility of mending faulty genes has always been an alluring concept to some scientists. Getting to the root cause of an inherited disease, the faulty gene or genes themselves, and fixing it at its most elementary level has a lot of appeal. However, the technical tools haven't always been available.

Many scientists argue that too little is known about the mechanisms that cause diseases. They caution that unless we understand more about the complex interaction of different genes, it will be difficult or impossible to achieve lasting success in the field of gene therapy. They worry that it will potentially do more harm than good.

To date, there haven't been many success stories, but researchers are nonetheless hopeful that the Human Genome Project will yield more knowledge about the intricate interplay of the stretches of DNA in our cells. This knowledge might shed light on new strategies for gene therapy.

Of Mice and Marrow

In 1980, a researcher named Martin Cline became the first person to experiment with gene therapy in humans. Cline was the chief of a research unit at UCLA that studied blood diseases and cancer. In 1976, he chose a biologist, Winston Salser, as his mentor because of Salser's expertise in genetic engineering, a growing field that Cline was interested in.

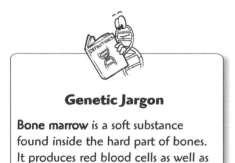

Genetic Jargon

Bone marrow is a soft substance found inside the hard part of bones. It produces red blood cells as well as certain types of white blood cells.

After Cline learned some genetic engineering techniques from Salser, the two scientists decided to see whether they could insert new genes into mice and get these new genes to function. They took some *bone marrow* cells from lab mice and put new genes into them.

This experiment was not 100 percent successful. Only about one-third of the mice seemed to take on the new gene in their cells. Even when the new gene was taken on, it happened in only a small amount of cells. In addition, even when the new gene had been incorporated into the mouse DNA, it often didn't continue to show expression after a few months. In other words, it ceased to function after a short time.

Despite these setbacks, Cline was encouraged by the results of his work with Salser and wanted to take things one step further. Because Cline was a specialist in blood diseases, he was familiar with thalassemia (thall-a-SEEM-ee-a), an inherited blood disorder that you read about in Chapter 9, "Irregular Genes." The disease is caused by a faulty gene. In one form of this illness, called beta thalassemia, the red blood cells can't survive for very long. Because red blood cells are responsible for carrying oxygen to various parts of the body, this disorder can lead to severe problems, such as deformation of the bones. Thalassemia victims often do not live beyond their 20s. Cline wanted to try his gene therapy experiments on patients with beta thalassemia to see whether he could bring about a cure by inserting the healthy gene that they lacked.

Seeking Approval

Before Cline could try his proposed gene therapy on human subjects, he had to ask for permission from UCLA's institutional review board. Because of the highly experimental nature of Cline's plans, the review board spent months discussing the pros and cons of the researcher's proposed projects.

In the meantime, Cline looked for prospective patients. Because beta thalassemia is a rare disease, he wasn't able to locate patients in the Los Angeles area, where he worked. Instead, he decided to try recruiting prospective patients from Mediterranean countries, because the blood disease occurs more frequently in people from places like Italy and Greece.

While the review board continued to discuss whether to give Cline the permission he requested, the researcher took another step. He contacted other researchers in Italy, Greece, and Israel and asked them if they were interested in joining him in the proposed project. Two researchers shared Cline's enthusiasm to try the experiment. They were Cesare Peschle, a researcher at the University of Naples, Italy, and Eliezer Rachmilewitz from the Mount Scopus Hospital in Jerusalem, Israel. Both were familiar with beta thalassemia and had seen a number of patients with the disease. They were open to the idea of trying gene therapy to find a possible cure for the disorder.

In 1980, Cline asked for a go-ahead from the Israeli and the Italian authorities to carry out his plans. He wouldn't need approval from UCLA to do these experiments, since they would be performed abroad. However, he did inform the university of his intentions.

Cell Mate

The results of a poll reported in the January 11, 1999 issue of *Time* magazine showed that 62 percent questioned think that the government should regulate gene therapy.

Experiments Abroad

On July 10, 1980, the Israeli authorities gave Cline permission to try out his experimental therapy in their country. Together with Eliezer Rachmilewitz in Jerusalem, Cline informed the patient, Ora Morduch, a woman in her early 20s, of the possible risks associated with the experiment. He also explained that this was not a miracle cure, even if the genes did get transferred into her cells.

Bone marrow cells were taken from the young woman. Using genetic engineering techniques, Cline mixed her bone marrow cells with the healthy cells that she lacked. He had brought these from his lab in the United States. Later that day, the engineered bone marrow cells were intravenously introduced back into the woman's body. Cline and Rachmilewitz observed her carefully to detect any complications as the altered cells were dripped into Morduch's veins from a vinyl bag. No problems developed. The next day, the same procedure was repeated.

A few days later, Cline headed for Naples, Italy, to perform the same procedure on another beta thalassemia patient. After getting approval to carry out his experiment in Italy, Cline repeated the procedure on a 16-year-old girl named Maria Addolorata. Like Morduch, she also showed no signs of complications from the experiment.

All seemed to be going well, but the very next day, Cline received word that the review board at UCLA had rejected his proposal. They concluded that before attempting such work on human patients, more experiments should be done on animals to test the safety of the procedure.

When word got out that Cline had already performed his experiments abroad, many people in the scientific community denounced his actions. The National Institutes of Health censured Cline, and as a result, he lost most of his federal funding for projects. He eventually resigned from his post at UCLA. In addition, Cline was banned from doing gene therapy experiments in the future because of his hasty decision to carry out his procedures abroad.

DNA Data

There are two major categories of gene therapy. The kind that Cline carried out is called somatic (so-MA-tic) cell gene therapy. A somatic cell is any cell in the body other than the reproductive cells. Germline gene therapy is performed on germ cells, another word for the sex cells, sperm and eggs.

Cline's gene therapy had little effect on his two experimental patients. Although they were not harmed by the treatment, the attempted gene therapy didn't help either. But one definite result of Cline's experiments abroad was to make the scientific community and laypersons alike more wary of this type of therapy. The general impression was: Why should anyone bother to pursue this type of treatment when it didn't work for Cline?

Ashanthi and the Misspelled Gene

Ten years after Cline's ill-fated attempt at gene therapy, another researcher achieved much happier results. W. French Anderson and other researchers at the National Institutes of Health began treating a sick four-year-old, and her life eventually changed for the better.

The little girl's name was Ashanthi DeSilva, and she suffered from a rare inherited disease called *ADA deficiency*. You may have heard of this disease, which forced its most famous victim, David, "the bubble boy," to spend most of his short life inside a germ-free tent environment. People with this disease lack the healthy gene that codes for an enzyme called ADA, which plays a critical role for the body's immune system.

Due to a genetic error that is inherited from both parents, a person with this disease cannot fend off germs. Even something as simple as a cold or a flu can be life-threatening. Sufferers must try to keep away from contact with germs, so it's nearly impossible to lead a normal life. Until the 1980s, almost all people with this rare disease died before the age of two.

To help defeat this serious and incurable condition, doctors gave Ashanthi a drug called PEG-ADA when she was two years old. This drug is made up of the ADA enzyme that Ashanthi's body does not produce, together with some chemicals that make it possible for the enzyme to function within Ashanti's body. However, it is by no means a permanent cure. The enzyme functions for just a few days before it has to be replaced, and this treatment is very expensive. In addition, it doesn't work equally well on every person with ADA deficiency. Things did not look hopeful for Ashanthi.

In the 1980s, researchers identified the healthy ADA gene that most people carry in their white blood cells. Ashanthi's copy of this gene was mutated. Dr. Anderson reasoned that if he could get copies of the normal gene and insert these into Ashanthi's cells, then her disorder might be cured.

Genetic Jargon

ADA deficiency is a rare genetic disease. In order for the disease to be present in a person, that person must inherit two copies of the faulty gene—one from the mother and one from the father. If only one gene is present, the other healthy gene will make up for this problem, and the disease will not occur. ADA is an enzyme that plays an important role for a person's immune system. Without ADA, the body cannot defend itself against even the simplest common sicknesses, such as colds and the flu.

Taking a Risk

Critics pointed to the fact that not enough is known about genes and how they function in concert with one another to attempt gene therapy. What would happen if a gene landed in the wrong place and disrupted the function of other genes? Would there be unexpected side effects? These thoughts must have gone through the minds of Ashanthi's family. But because the condition is so rare and so severe, Ashanthi's parents gave Dr. Anderson their consent to perform the new procedure on their ailing daughter.

Unlike Cline, Anderson waited until he received the go-ahead from the necessary authorities to perform his experiment on a human subject. One of the last agencies to give Anderson permission was the FDA, the United States Food and Drug

Administration. On September 14, 1990, just two days after Ashanthi's fourth birthday, W. French Anderson, R. Michael Blaese, Ken Culver, and a team of researchers called Ashanthi's parents and told them to bring their daughter in for the procedure.

New Genes for Old

The gene therapy that Anderson and his team used on Ashanthi is based on a simple principle. In order to get the new, healthy genes into Ashanthi's immune cells, the researchers needed something to deliver the gene to its target. They used what's called a *vector*, which is just a fancy word for a gene delivery system. This is usually a virus with a natural capacity to get inside cells.

Viruses invade cells in an effort to take them over. Once a virus worms its way into the nucleus of a cell, it tricks the cell into making more copies of the virus. Eventually, the copies of this tiny invader destroy the cell and go on to invade other cells. Researchers like Anderson wanted to use the virus's ability to get inside the cell, but in order to stop it from doing any harm, they had to disarm the virus by taking out its harmful parts.

Genetic Jargon

A **vector** is a delivery system used in gene therapy. It's what gets the new, healthy genes into a person's cells. Scientists often use organisms such as viruses to carry the new genes into a patient's cells. But first they take the disease-causing genes out of the virus or other invading organism.

This system is like a delivery service run by terrorists. They ring your bell and tell you there's a package for you, but once you open the door, they come in, point a gun at you, take over your house, and turn it into a terrorist training ground. After they've trained enough followers, they destroy your house as they leave.

What researchers are doing in gene therapy is like working with the terrorist delivery service, and giving them a package to bring to your house. Only this time, the scientists take away the terrorists' weapons, so they can't threaten you with harm while they're visiting.

Anderson took a virus that normally causes cancer in mice and disarmed it. Then he and his team used the disarmed viruses to insert the healthy gene into Ashanthi's immune cells. They removed some of her immune cells, genetically altered them so the healthy ADA gene would be part of their DNA, and then put these cells back into Ashanti's body intravenously.

It took only half an hour for the researchers to empty the contents of the vinyl IV bag into Ashanthi's bloodstream. The bag contained a billion engineered cells. Everyone hoped they would be taken up by Ashanti's body without complications. Although no one could tell immediately whether the therapy would have the desired results, Dr. Anderson was at least relieved that there were no negative effects.

More Therapy Needed

Ashanthi continued to receive these treatments every few months. It seemed that her immune system was responding well to this therapy. Because of these encouraging results, another child who suffered from ADA deficiency underwent the same gene therapy with Anderson's team four months after Ashanthi's first therapy session. The child was Cindy Cutshall, who was nine years old at the time, and she also responded well to the treatment.

For the next few years, both girls needed more therapy. But eventually, Anderson's team realized that the altered cells were alive and well. Even though both girls have less ADA in their bodies than the average healthy person who was not born with the disease, they still have enough of the enzyme to protect them. They can now lead normal lives without the life-threatening conditions of the past.

Both Ashanthi and Cindy are still taking reduced dosages of the PEG-ADA drug, but apparently the genetically altered cells are still functioning. Although the treatment is not a total cure for the disease, gene therapy in this case was a huge success. It removed the symptoms of a potentially fatal disease and showed that with more research in the future, the hope of successful gene therapy is still a possibility.

Special DNA Deliveries

The type of gene therapy used to help Ashanthi DeSilva and Cindy Cutshall involved introducing a healthy copy of a gene that the girls lacked. But other possible approaches to gene therapy are on the horizon. These different ideas include stopping a faulty gene's function and even trying to get cells to create artificial chromosomes.

Does Antisense Make Sense?

One approach to gene therapy other than the one used by Anderson and his team is called *antisense therapy*. It's meant to stop the function of a faulty gene so the wrong protein won't be manufactured in the cells.

Suppose that a person is born with a misspelled gene that causes a disease. You know that each gene produces a different protein. This particular misspelled gene produces a defective protein that causes an illness. Instead of adding a gene with the correct spelling of the gene, which was how Anderson's gene therapy worked, the antisense therapy is meant to stop the production of the unwanted protein.

Suppose the misspelled gene produces a protein that interferes with the healthy function of yet another gene. In this case, researchers might want to try the antisense approach. They would insert a new, engineered gene that would stop the faulty gene from working altogether. If they succeed, then the defective protein won't be produced, and the healthy gene that was inhibited will go back to functioning properly.

Genetic Jargon

Antisense therapy is a type of gene therapy that stops a faulty gene from producing its defective protein.

A Direct Approach

You learned that some of Ashanthi and Cindy's blood cells were removed and then put back after they were genetically altered, but not every inherited disease can be remedied by altering the blood cells. When scientists want to genetically modify mutated genes in other parts of the body, they have to depend on other approaches.

Suppose a researcher wants to fix a mutation in an organ like the kidneys. One way would be to take out some of the cells from that organ and genetically alter them. Then the modified cells could be transplanted back into the kidneys. But there's another way.

Dr. Clifford Steer and a team of researchers at the University of Minnesota Medical School used a novel approach on the liver cells of laboratory rats. Instead of using a virus as a delivery system, Steer's team created a special molecule made up of DNA and RNA that targets the rats' liver cells and gets inside their nuclei. As you've learned, the nucleus of a cell contains the hereditary material. Once inside a liver cell nucleus, Steer's synthetic molecule gets the cell to fix its own mutations.

The molecule that Dr. Steer's team designed can be compared to the correction key on an electric typewriter. In the same way that you can press this key to get the wrong letter deleted so the correct one can be put in, this special molecule lets the cell know that there's a mutation and prompts it to "type in" the correct letter.

Genetic Jargon

A **chimera** (kih-MEER-a) **molecule** is a synthetic molecule made up of DNA and RNA. It has been used in gene therapy experiments in laboratory rats to target and deliver genes to specific types of cells.

Although this technique is still in its infancy, some scientists are hopeful that it can be used more in the future. One of the advantages that it has over the use of viruses as delivery systems is the fact that it goes directly into specific cells. Also, this method can focus on one particular letter in the cell's DNA instead of having to replace an entire gene.

The special molecule is called a *chimera molecule*, and the official name of this method of gene therapy is called *chimeraplasty* (kih-MEER-a-plast-ee). So far, experiments using chimera molecules have been done only on laboratory animals. In the years to come, scientists will have to see whether this therapy works just as well in humans and whether there are any immediate or long-term side effects.

Creating Chromosomes

Sometimes gene therapists using viruses or similar delivery systems come up against a space problem. They can't always cram everything they want to into one virus. For instance, if they want to deliver a large gene plus all the DNA that acts as the on switch for that gene, that material might not all fit inside the virus.

To remedy this problem, some researchers are working on artificial chromosomes. You've already learned that regular chromosomes in humans contain several thousand genes. So if scientists could put together a whole chromosome, they could put a tremendous number of genes into it to deliver large stretches of DNA to needy patients.

In 1997, after more than 10 years of research, Dr. Huntington Willard, a geneticist at the Case Western Reserve University School of Medicine, reported progress in this area in a prestigious journal called *Nature Genetics*. Working with a team at the university and researchers from Athersys, a biotech company in Cleveland, Ohio, he created an artificial chromosome.

What the team did was something like giving an experienced cook all the ingredients needed for making a dessert and then sitting back and waiting until the cook puts everything together. The team put some human cells in test tubes. Then they provided these cells with some of the essentials needed for a chromosome. They created the very ends of a human chromosome, a stretch of DNA that belongs in the center of a chromosome, and another long stretch of DNA. When these basic ingredients were put together in the test tube, the repair systems in the cells went to work and put together the different pieces of DNA. In addition, the cells covered the DNA with the special proteins that cover chromosomes.

Dr. Willard's team said that this was an important step in the right direction. The team hopes that scientists can build upon its research and construct whole human chromosomes with genes that are needed to cure diseases. Artificial chromosomes also might be used in the future to become part of a cell's permanent genetic makeup. Theoretically, every time the cells would reproduce, the artificial chromosomes would be duplicated and passed along to new cells. These chromosomes might also be used to give scientists clues about how real chromosomes work.

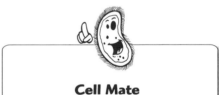

Cell Mate

Chromosomes are made up of *chromatin* (CRO-ma-tin). Chromatin consists of nucleic acids and proteins.

DNA Data

In 1987, researchers created artificial chromosomes in yeast, but the first attempts at creating human chromosomes failed. In 1997, Dr. Huntington Willard and a team of researchers at the Case Western Reserve University School of Medicine were responsible for the first step in creating a human chromosome.

The Least You Need to Know

➤ In 1980, researcher Martin Cline performed the first gene therapy experiments on two foreign patients, but the results were not promising.

➤ On September 14, 1990, Dr. W. French Anderson performed the first successful gene therapy on a four-year-old girl named Ashanthi DeSilva, who suffered from ADA deficiency, a rare inherited disorder that compromises the immune system.

➤ Antisense therapy is a type of gene therapy that is meant to stop the function of a faulty gene.

➤ In the future, researchers hope to deliver genes directly into people's organs.

➤ Scientists have taken the first steps in creating artificial chromosomes by giving cells the basic ingredients to make them.

Chapter 23

Molecular Medicine

In This Chapter

➤ The "magic bullets" that scientists hope will conquer disease

➤ Using genetic engineering to make life-saving medicines

➤ How interferon may help combat cancer and AIDS

➤ The history and controversy of Human Growth Hormone

➤ Fighting malaria and other diseases with DNA vaccines

With the recent discoveries about DNA and genes, scientists are hoping that they will gain a new understanding of disease. They look forward to the day when this knowledge will revolutionize the field of medicine.

Using genetic engineering and other types of biotechnology, researchers hope to develop new medicines specifically tailored to prevent diseases before they even start. Today's vaccines are made from weakened disease-causing bacteria or viruses. Doctors inject these substances into a person in the hopes that they will stimulate the immune system to fight off the disease. But tomorrow's vaccines may be engineered to carry small bits of the intruder's DNA.

Read on to find out how our expanding knowledge of genes and DNA is causing a rapid shift in the way medicine will be practiced in the years to come. Learn about the medicines that researchers are planning to have ready in the future.

Magic Bullets

Scientists have determined that there may be a better way to treat diseases than they have in the past. Using the new genetic technologies, they can try to develop molecular medicines to zero in on disease-causing organisms. The idea is that these molecular medicines might avoid some of the side effects that go along with modern-day medicines. One new development along these lines is called monoclonal antibody (MA-na-CLONE-ull ANT-ih-body) technology. It works on the principle that the body normally fights off invading viruses, bacteria, or other intruders using certain cells in the immune system.

Genetic Jargon

Antibodies are proteins produced by the immune system. They circulate in the blood to guard against invaders. Antibodies bind to specific invaders, which are then destroyed. An **antigen** (ANT-ih-jen) is a foreign substance, such as bacteria or viruses, that invades the body and is recognized by the body as foreign. **B-lymphocytes** (BEE LIM-fo-sites) are special cells that are called on to combat invaders. These cells are produced in the blood, lymph (LIMF) nodes, and the spleen.

Any invading substance that the body recognizes as foreign is called an *antigen*. The presence of an antigen triggers the immune system to activate special cells called *B-lymphocytes*. These cells are produced in the blood, lymph nodes, and the spleen. B-lymphocytes produce what are called *antibodies*; these are proteins that stop the invaders. Antibodies work by binding to the intruders, which prevents them from causing disease.

One thing that's special about antibodies is that they seek out and bind to one particular antigen, or type of body invader. Some of these helpful antibodies can keep on protecting you from a disease, even if it's been years since your body was invaded by that type of bacteria or virus. The antibodies that combat childhood diseases like chicken pox and the measles are just two examples of such antibodies. Once you've had these diseases, you won't get them a second time because your immune system has been activated and the antibodies remember to attack the invaders, even if the next time they come around is years and years later. The antibodies are still on the lookout for their antigens.

Fighting Diseases

Researchers wondered if they could get mice to act like small factories in the laboratory to produce specific antibodies, which could help people who need them to fend off diseases. To achieve this, researchers injected mice with a specific type of bacteria or other antigen. Then they waited until the immune systems of the mice went to work to produce the antibodies that go to war with these invaders.

It seemed like a good idea, but there were some problems to be overcome. To begin with, only a small amount of the antibodies were produced. After all, mice just aren't that big. In addition, it was hard to separate other substances that weren't needed from the antibodies.

So scientists tried a different approach. What if they got the cells to live outside of the mice? Maybe they could be coaxed into reproducing and making lots of antibodies. But there was a problem with this approach, too. B-lymphocytes, the cells that produce antibodies, just don't live for a long time in the laboratory.

So scientists came up with yet another idea. What if they took these cells and fused them to other cells that could continue reproducing in the laboratory? This is the idea behind monoclonal antibody technology.

A Winning Combination

In the monoclonal antibody procedure, a lab mouse is injected with the specific invader that researchers want to zero in on. Then the immune system of the mouse begins producing the antibody that will combat that specific intruder. Then the mouse is injected one more time just to make sure that the right form of antibody will definitely be produced.

At this point, the researchers take some of the mouse's B-lymphocytes from the mouse's spleen cells, for example. Then they fuse these cells to myeloma (MY-a-LOW-ma) cells. A *myeloma* is a cancerous tumor that forms in bone marrow. Scientists can get these cells to grow continuously in a culture dish or a test tube in the lab. The cells that result from the fusion of B-lymphocytes and myeloma cells are called *hybridomas* (HI-brid-O-mas). These cells grow continuously and produce large quantities of the antibody that the researchers want. The antibodies that are produced are called monoclonal antibodies.

Some scientists think that these monoclonal antibodies will be better at fighting disease than conventional medicines, because they are like "magic bullets" that only target specific invaders. Although conventional medicines also attack microscopic invaders, there are sometimes side effects such as allergic reactions or nausea because sometimes these medicines attack normal body cells, too. Because monoclonal antibodies are so specific, they produce few negative side effects.

Monoclonal antibodies have a number of potential uses:

➤ Scientists hope to eventually use them to conquer diseases such as cancer.

➤ They can help scientists to understand the structure of specific antibodies.

➤ They can be used to screen blood for the HIV virus, which causes AIDS.

➤ They can be used to diagnose certain types of cancer, even during the early stages of the disease.

Cell Mate

The word *monoclonal* comes from the Greek word *monos*, meaning sole or alone, as in *monologue*, a speech delivered alone by an actor, or *monopoly*, when a sole company controls an entire industry. Monoclonal antibodies produce just one specific disease–fighting protein. This protein combats one particular type of bacteria or virus and only that one type.

Monoclonal antibodies are made by fusing B-lymphocyte cells from mice with cancerous myeloma cells. This produces hybridoma cells that continue multiplying and producing antibodies in the lab.

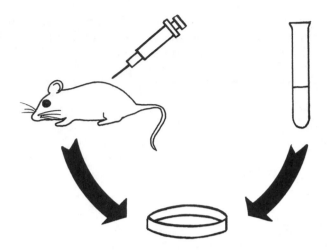

Approved Therapies

In 1997, the United States Food and Drug Administration (FDA) approved the first therapy for cancer using monoclonal antibodies. It is used on patients with a type of cancer called non-Hodgkin's lymphoma, also referred to as NHL. Researchers feel that using monoclonal antibodies to combat diseases like NHL will have less severe side effects than chemotherapy. They also feel that the treatment will not take as long as conventional chemotherapy treatment.

DNA Data

Non-Hodgkin's lymphoma, also called B-cell lymphoma or NHL, is the most common form of lymphoma, or cancer of the lymphatic system. The lymphatic system is a network of different vessels that circulates lymph around the body. Lymph is a colorless fluid that contains the white blood cells, which help the body fight off foreign invaders.

In this form of cancer, the B-lymphocytes, which are the cells that fight off bacterial and viral invasions, multiply too quickly. The result is cancerous tumors in the lymph glands. Often, these cancerous cells spread to organs such as the spleen or the liver.

In 1998, another type of monoclonal antibody treatment was approved by the Food and Drug Administration. This one was developed by Genentech, a biotech company, and its product will be used on patients with one form of breast cancer.

The latest development to watch for is genetically engineered plants such as corn, which can produce monoclonal antibodies. Because plants don't carry disease-causing organisms that can affect people, some scientists think they could be engineered to make *plantibodies* that would be safer than those made in animal cells. They'd be a lot cheaper, too.

In May 1998, researchers from a hospital in London reported that they had engineered plantibodies in tobacco. When applied to human volunteers' teeth, the plantibodies prevented a type of bacterial infection that can lead to tooth decay. In December 1998, researchers from Johns Hopkins University in Maryland wrote a report explaining that they had engineered plantibodies in soy plants that prevented infection by the herpes virus in lab mice.

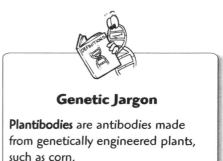

Genetic Jargon

Plantibodies are antibodies made from genetically engineered plants, such as corn.

Designer Drugs

Researchers are using new methods to try to conquer disease based on their improved understanding of the workings of DNA. They've also used genetic engineering to make large amounts of proteins that are in short supply. These engineered substances can be used to restore good health to patients who can't make these proteins for themselves or who are in need of large amounts of these proteins to combat disease.

Engineered Insulin

As you read in Chapter 13, "Cut-Down Genes," scientists now make insulin for diabetes patients in the lab. Instead of taking this substance from animals such as pigs, a method that was used in the past, scientists at several biotech companies have inserted the human gene for insulin into E. coli bacteria. When the bacteria are given a good home and plenty to eat, they multiply rapidly, and every time they do this, they also produce insulin according to the instructions contained in the human gene that's been added to their DNA.

Timely tPA

A protein called *tPA* is made by the cells that line the walls of blood vessels. It naturally dissolves blood clots, allowing blood to flow smoothly through the vessels of the body. A blood clot that isn't taken care of can result in the blockage of a blood vessel. In some cases, this blockage can lead to a heart attack or a stroke.

Scientists knew that if they gave tPA to a heart attack victim, it would dissolve the blockage. But

Genetic Jargon

Cells that line the walls of blood vessels in the human body use the **tPA** protein to dissolve blood clots so that blood can flow smoothly without blockage.

the problem was that this naturally occurring protein is produced in only small quantities in the human body. How could they get enough of it to use for needy patients?

Using genetic engineering, researchers found a way to produce large amounts of this life-saving substance. In 1987, the United States Food and Drug Administration (FDA), which oversees the legal aspects of drugs, approved recombinant (genetically engineered) tPA. However, doctors must be careful when using tPA on their patients, because tPA can interfere with normal blood clotting and lead to severe bleeding like that exhibited by hemophiliacs if used incorrectly. But when this drug is used properly, researchers have confidence in its effectiveness.

The Interfering Drug

In 1957, scientists discovered a class of proteins that they named *interferons* (in-ter-FEAR-ons) because the substances interfere with the reproduction of viruses that invade the body. In addition, interferons can stop infections caused by bacteria and parasites. What these proteins do is very important for the body's immune response.

The body produces interferons when it detects invaders such as viruses or bacteria. These protective proteins don't directly stop the viruses from multiplying. Instead, they find cells that are near cells that are already infected and make them aware of the virus or other threatening invader that's roaming around. Then these alerted cells begin producing special proteins to stop the viruses from being able to multiply inside them.

This stops the invasion, because viruses don't just multiply by themselves. They need to invade a cell and trick that cell's DNA machinery into making more copies of the virus. When this happens and lots and lots of these tiny invaders are made, they burst out of the cell, destroying it as they leave. Then they're free to inject their instructions for making copies of themselves into even more unsuspecting cells.

Scientists recognized the promise of curing many diseases by using interferons. But there was very little they could do to administer this protein to ailing patients, because it's naturally produced in such small amounts in the human body. In 1978, just one small dose of interferon cost approximately $50,000.

But then genetic engineering came on the scene. In 1980, Swiss scientists were the first to insert the gene for human interferon into bacteria. The bacteria churned out this substance in large quantities, and the price of a dose of the drug dropped to one dollar.

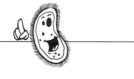

Cell Mate

In 1957, the three types of interferons were discovered by scientists Alick Isaacs and Jean Lindemann. These three types were named using the first three letters of the Greek alphabet: alpha interferon, beta interferon, and gamma interferon. Each type has its own way of reacting to viral intruders.

Alpha Advances

Researchers are focusing on interferons, especially alpha interferon, in their search for cures for types of cancer and also AIDS. Interferons might be used together with other forms of treatment to stop the progression of these and other diseases.

Although some view interferon as a miracle drug, it's not without its down side. Serious side effects can occur, such as fever, fatigue, and even inhibition of blood cell production. But despite these side effects, interferon therapy has had some success.

For instance, alpha interferon has been used in low doses to treat patients with a rare type of blood cancer called *hairy cell leukemia*. This type of blood cancer usually occurs in older men. In the 1980s, alpha interferon emerged as the preferred treatment for this disease. Since then, other drugs have replaced it.

In addition, alpha interferon is used to treat patients with *Kaposi sarcoma*, a type of cancer that affects the skin and often develops in AIDS patients. Researchers have found that alpha interferon can shrink tumors in some patients with this disease.

Alpha interferon is also used to treat patients with chronic hepatitis B and hepatitis C. These are both diseases of the liver. Before the use of alpha interferon, there was no treatment for these two diseases.

Genetic Jargon

Hairy cell leukemia is a cancer that occurs in certain cells in the blood and the bone marrow. It's called hairy because the cancerous cells in this disease look like they have hairs sticking out of them when viewed under a microscope. **Kaposi's sarcoma** is a type of cancer that affects the skin and often strikes people with AIDS.

Beta and MS

Beta interferon, another form of the protective protein, has been approved by the United States Food and Drug Administration (FDA) as a treatment for *multiple sclerosis,* also known as MS. When a person develops this disease, his own immune system turns against him. A disease with this effect is called an *autoimmune disease.*

Inside the brain and the spinal cord, there are nerve fibers. Something called a myelin (MY-a-lin) sheath, which is a fatty substance, acts as a protective cover for the nerve fibers. Think of this substance as something like the insulation on the electrical wires in your house. Without them, your electrical appliances would be exposed and would most likely

Genetic Jargon

Multiple sclerosis (skla-RO-sis), also known as MS, is a disease in which the body's immune system turns against the fatty substance that lines the nerve fibers in the brain and the spinal cord. Without this protection, the person with the disease becomes fatigued, has slurred speech and vision problems, and can suffer paralysis.

become damaged quickly. In multiple sclerosis, the myelin sheathes are attacked by the body's own immune system. People with this disease become fatigued, have slurred speech, suffer vision problems, and can even become paralyzed.

A Tall Tale

Another human substance that researchers have been producing through genetic engineering is called Human Growth Hormone (HGH). This hormone is what makes the difference in height between someone like you and Shaquille O'Neal. Human Growth Hormone is normally produced by the pituitary (pit-OO-it-er-ee) gland at the base of the brain, and it helps people grow normally. If there is a deficiency of this hormone, a person will not grow to more than three or four feet tall. This condition is known as *dwarfism*.

Until 1985, the only source of Human Growth Hormone for those who needed it was from cadavers. But unfortunately, some of the hormone that was obtained from this source was contaminated and caused a degenerative disease of the brain called Creutzfeldt-Jacob (KROYTZ-felt YAH-kob) disease. After this fact was discovered, the United States Food and Drug Administration (FDA) would not allow the use of Human Growth Hormone from human cadavers.

Then genetically engineered Human Growth Hormone came on the market. It did not carry the threat of the brain disease, because it was produced in harmless bacteria. However, there has been a considerable amount of controversy about the use of this engineered hormone. Because only a minority of people suffer from a severe deficiency in Human Growth Hormone, author Andrew Kimbrell called the drug "a cure in search of a disease" in his book, *The Human Body Shop*. Although the only legal use is for people who suffer from dwarfism or people with chronic kidney failure, others may be using this substance improperly.

In a report published in 1996 by scientists at Case Western Reserve University in Cleveland, Ohio, only 4 out of 10 children getting HGH treatment were deficient in the hormone or suffered from chronic kidney failure. An editorial written in the *Journal of the American Medical Association* by Dr. Barry B. Bercu of All Children's Hospital in Florida stated that some parents decide that their children who have short stature should receive treatments with HGH, even though their bodies produce it in normal amounts. Dr. Bercu said that this reflects the "cultural heightism" found in American society today.

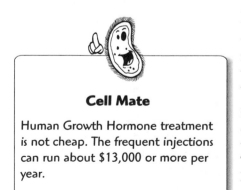

Cell Mate

Human Growth Hormone treatment is not cheap. The frequent injections can run about $13,000 or more per year.

In addition, a large-scale study of the hormone was carried out in France from 1973 to 1993 and reported in the *Journal of the American Medical Association* in 1997. This study left some people wondering whether the hormone is effective at all. The French study was carried out at the University of Paris, and it checked the progress of over 3,000 children who were using the hormone. The conclusion of the University of Paris study was that the hormone might have caused a decrease in the full adult height of the children who used it. However, some scientists feel that the study may have been flawed.

Another controversial issue goes beyond whether the children's height is increased. Some say that increased height doesn't necessarily translate into increased self-esteem. For instance, researchers at the Southampton University Hospitals in England conducted a study on girls with short stature. Some received the hormone, and others did not. Their conclusion was that the girls who were shorter were just as emotionally well-balanced and happy as those who had gained in height.

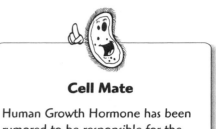

Cell Mate

Human Growth Hormone has been rumored to be responsible for the enormous stature of a few professional wrestlers.

Brave New Vaccines

Scientists are also looking at the possibility of developing new vaccines through the use of the new genetic technologies. In Chapter 14, "Seedy Science: Engineering Plants," you read that some researchers were developing foods such as bananas that can carry genetically engineered vaccines. In the future, people might eat their vaccines instead of being injected with them.

Traditionally, vaccines are made from disease-causing organisms that have been killed or weakened. They're injected into people in the hopes that the person's immune system will react to them by producing antibodies, substances that will attack the invaders and remain in the body for many years or sometimes permanently, rendering it immune to the disease. However, there are some drawbacks to conventional vaccination. Some people are allergic to certain vaccines, and there is also a small risk that the weakened organisms can cause the very disease that they were meant to prevent.

So some scientists are excited at the possibility of developing vaccines in the future through genetic engineering. For instance, instead of using a weakened form of a disease to provoke a protective reaction from the body, bits of the disease-causing organisms' DNA might be substituted for this same purpose. Researchers hope that these vaccines will give people longer-lasting immunity to certain diseases than conventional vaccines, which sometimes require booster shots. They're also considering developing a genetically engineered vaccine that could guard against a number of diseases simultaneously, instead of just one. But as with any new technology, some problems still need to be worked out.

One disease that's being targeted by researchers is malaria. The World Health Organization has estimated that between 300 million and 500 million people contract this disease annually, and about three million people die from it. Researchers at the Centers for Disease Control (CDC) in Atlanta, Georgia, decided to create a genetically engineered vaccine to protect people from this disease, which is caused by parasites—organisms that live off other organisms.

The malaria parasite goes through four stages during the course of this disease. During each stage, the parasites use different genes to produce different proteins to surround themselves. People who live in malaria-infested areas sometimes develop immunity to these parasites. Their bodies develop antibodies, so they can withstand attack by the parasites, which are spread via mosquito bites. The CDC researchers created a synthetic gene by putting together many small fragments of DNA that are present in the parasite during all of its four stages. The researchers tried out this vaccine on rabbits, and the rabbits produced antibodies that fought off the parasite's invasion.

However, some scientists are not sure that this vaccine will be a miracle cure for the disease. Years ago, people tried to combat the disease by using pesticides on the mosquitoes, and then later by using a drug that seemed to help at the time. But the mosquitoes have developed a resistance to these measures, and the incidence of the disease continues to increase. Some scientists think that maybe a combination of different techniques, including the new vaccine, may be needed to end the epidemic.

The Least You Need to Know

➤ Scientists have developed something called monoclonal antibody technology to produce disease-fighting substances in the lab.

➤ More and more drugs will be developed in the future using genetic engineering.

➤ Interferon is a naturally occurring disease-fighting substance, and scientists can now produce large amounts of it using genetic engineering.

➤ Human Growth Hormone is another naturally occurring substance that has been produced in large quantities with genetic engineering.

➤ Scientists are designing new vaccines using their knowledge of DNA.

Chapter 24

Sex Techs: Reproductive Technologies

In his classic 1932 novel, *Brave New World*, Aldous Huxley tells the story of a future society where technology has replaced parenthood. Virtually every child is a test tube baby, and droves of clones are produced according to a set plan. Some developing test tube embryos are given the required nutrients to thrive, and others are purposely deprived so they will grow up to become laborers in a society that has distinct classes. The concept of the family has disappeared, and the word *mother* is not uttered in polite company.

Huxley's unpleasant world has not materialized, but some people think that the future he described is taking root in today's assisted reproductive technologies. Will Huxley's world of high-tech births and the corresponding societal repercussions become a reality in years to come? Some people think so, and worry that its beginnings are already here. Yet others think that reproductive technologies are merely beneficial and not threatening. They point out that today's reproductive technologies have helped countless infertile couples who could not have otherwise experienced the great gift of parenthood.

Read on to find out how childless couples have been blessed with children through in vitro fertilization, in which sperm and eggs are mixed in a dish in the lab. You'll read about prenatal testing that can screen an unborn child for genetic diseases. You'll also learn how some parents are opting to choose the sex of their children, sometimes even before conception.

Genetic Jargon

Anorexia (a-nor-EX-ee-a), also called anorexia nervosa (a-nor-EX-ee-a ner-VO-sa), is an eating disorder. People who are anorexic stop eating because they think that they are overweight, although in reality they are underweight. The word *anorexia* comes from the Greek words meaning not and appetite.

Genetic Jargon

In vitro (in VEE-tro) **fertilization**, or IVF, is a type of assisted reproduction technology. According to this procedure, sperm and eggs are mixed in a culture dish or test tube. When the eggs are fertilized and begin to grow into embryos, some are implanted into the mother. The name *in vitro* comes from the Latin words meaning "in glass," and children produced using this technique have been dubbed "test tube babies" by the media.

How Babies Are Made

You read in Chapter 1, "Join the Genetic Revolution," that people like the ancient Romans and others took quite a while to figure out just how babies are made. From the idea of spontaneous generation, the belief that flies, for example, were formed spontaneously from meat, to the incorrect concept of the homunculus, a fully formed human inside the sperm, the science of reproduction has come a long way. Now we know that children are conceived when a man and a woman mate and a woman's egg is fertilized by a man's sperm.

But sometimes complications can occur. Some people are infertile, which means that they cannot have children. Infertility can occur for a number of reasons. For instance, a man might have a low sperm count. This can happen due to health reasons, drug use, or exposure to certain toxic chemicals. In women, certain hormonal disturbances can make conception difficult or impossible. Certain health problems such as *anorexia*, an eating disorder, can cause a woman to stop menstruating. Other women repeatedly have miscarriages when they try to carry a child to term.

Scientists have developed ways to help couples who are infertile. One of the major advances has been the use of in vitro fertilization.

A Baby in a Bottle

On July 25, 1978, a baby girl named Louise Brown was born in a small hospital in the north of England. This quiet, happy birth was later hailed as the result of a scientific miracle, and Louise was called "The Baby of the Century." What was it that made this birth so special?

The answer is that Louise was not conceived in her mother's womb from the start. Although she developed from her mother's egg, which was fertilized by her father's sperm, Louise was conceived in a Petri dish, a small dish used in laboratories. She was the product of what is called *in vitro fertilization*.

This method was originally used to produce animals, but when researchers became convinced of its success, they applied the procedure to humans. Although this method is much less than 100 percent effective, it has helped many infertile couples to become parents.

Excess Eggs

Here's how in vitro fertilization works. First, the mother is given superovulation drugs that cause her ovaries, the female reproductive organs that produce mature eggs, to develop many mature eggs. Normally, just one mature egg is released about once per month in the menstrual cycle. The reason why the fertility specialists want to have many eggs is so they can work with more than one embryo later in the procedure. Having more embryos increases the chances that at least one of them will develop successfully inside the mother.

The eggs are taken from the mother's ovaries and put into a special solution that will help keep them alive and healthy in the laboratory. Sperm from the father is then added to the eggs, and this reproductive mix is put into an incubator under strictly controlled conditions to ensure that the sperm and eggs will remain alive. Usually, after about one day, fertilization takes place.

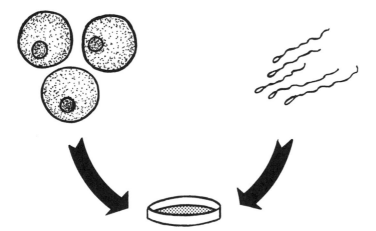

In vitro fertilization: Eggs and sperm are mixed in a lab dish. After the eggs are fertilized and grow into embryos, they are implanted into the mother.

After two or three days, the fertilized eggs—now developing into embryos—are put into the mother's uterus, where hopefully they will continue to develop until a child is born. A fertility specialist will usually implant about four embryos into the mother to increase the chances of a pregnancy. Sometimes this results in multiple pregnancies. If more embryos are created than the fertility specialist thinks are necessary, they can be frozen and stored. Frozen embryos can last for years, and then be thawed and implanted into a woman who can then give birth normally.

Techno-Turmoil

Although in vitro fertilization has helped many couples who were previously unable to become parents, this procedure has led to many ethical and legal complications. For instance, in 1995, a California couple paid $10,000 to have in vitro fertilization performed. Out of 46 eggs that were removed from the mother and mixed with her husband's sperm, 21 developed into embryos. Five of them were implanted, and she later gave birth to a son. The rest of the embryos were frozen and put in storage.

Some time later, a reporter from a local newspaper contacted them. To their shock, he explained that he learned that three of the couple's frozen embryos had been implanted into another woman, who later gave birth to twins. Biologically, the original couple were the mother and father of those twins. After a lawsuit, the fertility specialist gave up his practice.

Embryos on Ice

The existence of frozen embryos has led to thorny ethical problems. For instance, if a couple decides to undergo in vitro fertilization, and they have some embryos stored in liquid nitrogen, what will happen if the couple later gets divorced? This issue has come up in court, leading to battles over the custody of unborn, frozen embryos.

In a happier case, a married couple found out that they had some long-forgotten frozen embryos in storage. In 1989, the California couple paid $7,000 for in vitro fertilization. The woman was 36 years old at the time, and the procedure resulted in the birth of her first child. The couple apparently thought that this would be their only child, but in 1997, the fertility clinic contacted them with a happy surprise: unknown to them, other embryos from the woman and her husband had been frozen and put in storage.

Although the woman was 44 at the time, she and her husband considered having another child, because these embryos came from them when they were younger. So one of the embryos was implanted in the woman, and she gave birth to a son. Although the embryo they used had been frozen for over seven years, the baby was in perfect health.

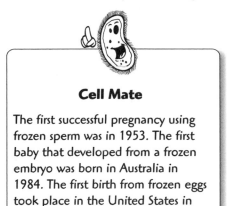

Cell Mate

The first successful pregnancy using frozen sperm was in 1953. The first baby that developed from a frozen embryo was born in Australia in 1984. The first birth from frozen eggs took place in the United States in 1997.

A Child Without Parents

In another case in 1995, the possible legal ramifications of fertility technology became painfully clear. A San Francisco couple tried in vitro fertilization four times. The cost was $10,000 for each try, but none of the attempts worked. Eventually, they decided to pay a woman to act as a surrogate mother for a child they would adopt. They signed a contract, and the woman who would bear the child was paid $10,000.

The embryo that was implanted in the surrogate mother came from an anonymous egg donor and an anonymous sperm donor, so there were no genetic ties to either of the adoptive parents. The adoptive mother-to-be was thrilled that she and her husband would finally be parents, but that happiness was short-lived. Just two weeks after they signed the surrogacy contract, the couple broke up.

After the baby girl was born, the adoptive mother sought child support from her estranged husband. But in an interesting twist, the judge who presided over the case declared that the woman's divorced husband was not the father of the child. The judge said that in the eyes of the law, the baby girl had no parents! Although the adoptive mother was allowed to keep the child, technically she was not considered the legal mother. As of 1998, the woman's lawyers were still appealing the court's decision.

Selecting Sexes

Some of the new reproductive technology procedures answer the question: Will the new baby be a boy or a girl? There are several methods for learning the gender of an embryo, but the controversy surrounding these methods comes from the fact that people feel that parents might use this information to determine the sex of their child by aborting or otherwise discarding embryos of the unwanted sex. Now there are even methods of identifying gender-causing chromosomes in sperm, which means the gender of the child can be determined before conception.

These advances in reproductive technology bring up an interesting question: If couples can choose the sex of their child, will there be a gender imbalance down the road? No one knows the answer to this question, but one thing is certain. Gender selection is available, and the accuracy rates are fairly high.

Post-Conception Procedures

The earliest methods for determining gender are also used to learn other information about an embryo or fetus. (During the early weeks of a fertilized egg's development, it is called an embryo. After the eighth week, it is called a fetus.) Originally, the point of determining gender was to help parents at risk for carrying sex-linked diseases, such as hemophilia or Duchenne muscular dystrophy, that only occur in males. However, some people think that this technology will be used by people who just have a preference for children of one gender or the other.

One genetic technology that can determine gender—and more—is called *PGD* or Pre-implantation Genetic Diagnosis. This procedure is used only in connection with in vitro fertilization. When the fertilized eggs grow in the Petri dish to about 16 cells in size, one cell is removed, and DNA analysis is performed on that cell. Not only gender, but also a lot of other genetic information can be gleaned from the DNA of this one cell. Remember, all of the DNA in the other cells is identical to the DNA in the one cell that's analyzed. Few clinics in the United States can perform this procedure, and it is reserved for prospective parents who are at risk for producing a child that can inherit a sex-linked genetic disease, such as hemophilia.

Ultrasound is a technology that was first developed in 1958. During an ultrasound procedure, a special instrument is placed across a pregnant mother's abdomen, and then sound waves create a visual picture, called a sonogram, of the fetus. Dr. John Stephens, a physician in San Jose, California, uses ultrasound to determine the sex of a fetus.

By taking a sonogram in the 12th week of pregnancy, Dr. Stephens claims he can accurately assess the gender of the fetus. He uses a method called FASA, or *Fetal Anatomic Sex Assignment.* This basically means that he carefully examines the sonogram to try to identify male or female genitalia. Dr. Stephens has patented his FASA procedure.

A procedure called amniocentesis is used to discover the gender of a fetus, among other information, but it generally is done weeks later than the FASA technique. Chorionic villus sampling (CVS) can find out the sex of the fetus as well. You'll read more about these two procedures in the section "Examinations Before Birth."

Selecting Sperm

Newer procedures can determine whether sperm will produce boys or girls. For instance, The Genetics & IVF Institute in Fairfax, Virginia, uses a procedure that separates sperm that carry X chromosomes (to make girls) from those that carry Y chromosomes (to make boys). The name of this procedure is *flow cytometric separation* or *micro sort.* The technique is based on the fact that a sperm cell that carries a Y chromosome has about 2.8 percent less DNA than a sperm cell that carries an X chromosome.

The first step in this procedure is to apply a special fluorescent dye to the sperm's DNA. Then a laser is shined on the sperm. Then the cells are sorted according to whether they will produce boys or girls. It takes close to an entire day to sort the sperm in just one sample.

Another gender-determining method is called *gametrics* (gah-MET-rics). It's also called the Ericsson method, because it was invented by Dr. Ronald Ericsson. This procedure involves putting sperm in a test tube with a special liquid that causes them to move slower than usual. The sperm cells that carry Y chromosomes are smaller and swim to the bottom faster than those with X chromosomes, which move to the top.

Genetic Jargon

PGD, or Pre-implantation Genetic Diagnosis, is a method used to test an early-stage embryo for inherited diseases and gender. When the embryo has grown to about sixteen cells in size, one cell is removed, and its DNA is tested.

Genetic Jargon

Fetal Anatomic Sex Assignment (FASA) is a method invented by Dr. John Stephens, a California physician, to determine the gender of a 12-week-old fetus. It uses ultrasound, an imaging process that employs sound waves, to create a sonogram, a visual image of the fetus. The sonogram is studied to determine whether the genitalia of the fetus are male or female.

Examinations Before Birth

PGD isn't the only type of genetic examination for prenatal patients. There are many others, including amniocentesis and CVS (chorionic villus sampling). Different procedures are carried out at different stages in the development of an embryo or fetus, and each method carries some risks.

Since the 1970s, more and more prenatal tests have become available for pregnant women who are concerned about the condition of their unborn child. As the Human Genome Project and independent research identifies genes that can cause inherited diseases, tests can be created to identify the presence of these genes in the embryo or fetus.

Fetal Fluids

Amniocentesis is a procedure that involves studying fluid from the amniotic sac, which surrounds the unborn child in its mother's womb. This procedure is usually performed between the 14th and 17th weeks of pregnancy.

To obtain the amniotic fluid, a doctor must insert a syringe into the mother's abdomen, and then into the uterus. Ultrasound helps the doctor to guide the needle so that it does not touch the fetus. A small amount of the amniotic fluid is then removed. The amniotic fluid contains some fetal cells, and these cells can be grown in the laboratory and later screened for inherited diseases such as Down syndrome and other abnormalities that can occur in the chromosomes.

Amniocentesis is recommended for women over the age of 35. It is also administered when one or both of the parents have a strong family history of genetic diseases. Women who have previously given birth to a child with Down syndrome may also wish to undergo amniocentesis.

Although amniocentesis can give results in a few weeks, there are associated risks. For every 200 procedures performed, there can be one miscarriage as a result. Less than one percent of women who undergo this procedure experience cramping, infection, or vaginal bleeding. There is also a small risk of harm to the fetus from the needle, and in a small percentage of cases, the amniotic sac can break.

Genetic Jargon

The **micro sort** (MY-cro sort) technique, also called flow cytometric (sy-toe-MET-rick) separation, is used at The Genetics & IVF Institute in Fairfax, Virginia. This procedure sorts sperm cells according to whether they carry an X chromosome (for girls) or a Y chromosome (for boys).

Genetic Jargon

Amniocentesis (AM-nee-o-sen-TEE-sis) is a procedure that involves studying fluid from the amniotic (AM-nee-OT-ic) sac, which surrounds the unborn child in its mother's womb. This fluid contains cells from the fetus. These cells can then be grown in the laboratory and screened for some inherited diseases.

The associated risks may increase if the procedure is carried out during the first trimester, the first three months of pregnancy. The January 24, 1998 issue of *The Lancet*, a respected British scientific journal, carried an article that debated the use of amniocentesis for disease screening during the first trimester, when the fetus is still in an early stage of development.

It cited a Canadian study carried out at British Columbia's Women's Hospital in Vancouver, Canada. The study involved over 4,000 women who had had an amniocentesis between the 11th and 13th weeks of their pregnancies. (Usually, an amniocentesis is performed later.) The researchers found that out of 4,374 women in the study, 29 of them gave birth to babies with a clubfoot. The journal suggested that other methods of fetal testing besides amniocentesis be used for disease screening during this early stage of pregnancy.

DNA Samples

Another method used to screen for disease in unborn children is called CVS, or *chorionic villus sampling*. This method involves taking a small sample of cells from the placenta (pla-SEN-ta). The placenta is an organ that develops in pregnant women. The unborn child is connected to it through the umbilical cord.

CVS is performed in either of two ways. In both types of procedure, the patient first undergoes ultrasound to help the doctor locate the placenta and the baby. In one procedure, a thin plastic tube called a catheter (CATH-a-ter) is inserted into the mother's vagina until it reaches the placenta. Then a small amount of placental tissue is suctioned out with a syringe. In an alternative form of the procedure, the doctor inserts a needle through the mother's abdomen and into the uterus until it reaches the placenta, and then a small amount of tissue is removed.

Genetic Jargon

Chorionic villus (KOR-ee-ON-ic VILL-us) **sampling**, or CVS, is a procedure used to screen a developing fetus for genetic diseases. CVS involves taking a small amount of cells from the placenta. Researchers can learn about the DNA in the fetus by studying these cells because these cells are genetically identical to the cells in the fetus.

The main advantage of CVS over amniocentesis is the fact that it can be performed when the fetus is 10 to 12 weeks old. Amniocentesis is usually performed after the 14th week of pregnancy. CVS can tell the parents whether their unborn child will have Down syndrome, Tay-Sachs disease, and other inherited disorders.

Like amniocentesis, there are associated risks with CVS. Minor complications such as cramping and vaginal bleeding may occur more frequently following CVS than amniocentesis. The greatest risk associated with CVS is the possibility of a miscarriage. About 1 in every 100 CVS procedures may result in the death of the fetus. This is a higher risk than with amniocentesis, where about one in every two hundred procedures might result in fetal death. There have also been reports of limb defects in some infants whose mothers underwent CVS. The estimated risk of limb defects is approximately 1 in every 3,000 births.

In 1996, the American College of Obstetricians and Gynecologists (ACOG) issued an opinion on CVS. It stated that the procedure is relatively safe when performed between the 10th and 12th weeks of pregnancy, but recommended that the procedure not be performed earlier than the 10th week. It also recommended genetic counseling before CVS to make the parents aware of the risks and benefits of the procedure. It stated that although further studies need to be done to determine whether CVS can lead to limb defects, patients should be counseled about this matter.

Designer Babies?

As scientists learn more and more about human genes, it's possible that in the future, parents will be able to learn a great deal more about their unborn child's health. If a fetus has defective genes or faulty chromosomes that will lead to life-threatening illnesses, parents may choose abortion.

DNA Data

The January 11, 1999 issue of *Time* magazine reported the results of a telephone poll of over 1,000 adult Americans. Among the questions was whether they thought the government should regulate genetic testing that identified traits in unborn children; 46 percent said yes, and 49 percent said no. When asked what traits they would pick for their unborn child, if they could, 33 percent said they would choose greater intelligence, and 11 percent said they would choose the gender.

The new genetic tests may further complicate the ethical debate that surrounds abortion. In addition, some people worry that in the years to come, parents may choose to abort a healthy fetus that does not meet the parents' standards for the type of child they want. Will some parents keep aborting until they have a boy who's 6'4", blond-haired, blue-eyed, and extremely intelligent? In a play called *The Twilight of the Golds*, a couple decide to abort their fetus when they learn that it carries genes that predispose it towards homosexuality. This is fiction, but in a recent real-life study, three-quarters of people interviewed said they would abort a fetus if it had a 50 percent chance of growing up to be obese.

In the 1997 sci-fi movie *Gattaca*, starring Ethan Hawke and Uma Thurman, prospective parents can flip through catalogues to choose the very best genes that money can buy for their unborn children. Is this just a fictional scenario, or could it come to pass? Will the new genetic technologies, which help so many childless couples, be misused to provide demanding parents with "designer babies"?

Some sperm banks already offer male sex cells from intellectually superior donors. For example, the Repository for Germinal Choice, which opened up in 1980 in California, stores sperm from men of above-average intelligence and who are in good health. For a fee, married couples may use this sperm to impregnate the wife if the husband is infertile.

In 1999, an anonymous couple ran an advertisement in the newspapers of outstanding universities. They offered $50,000 for eggs from a young Caucasian woman who had attained a high score of at least 1400 on her SATs (the college entrance exam).

Many scientists would point out that there is no scientific foundation for this concept. There are highly intelligent parents whose children are not of about average intelligence. We're back to the nature versus nurture debate. We can only hope that the future brings a deeper understanding of how to use our new knowledge of DNA to relieve human suffering and not to view unborn children as commodities and status symbols.

The Least You Need to Know

➤ In vitro fertilization is a procedure that helps childless couples become parents by fertilizing eggs in the laboratory, and then implanting the resulting embryo into the mother.

➤ New assisted reproduction technologies raise legal and ethical questions about the identity of a child's parents.

➤ Using new techniques, prospective parents can choose the sex of their unborn child.

➤ Currently available genetic tests can identify whether an unborn child will develop certain diseases.

➤ One ethical concern about genetic testing of embryos is whether the future will bring made-to-order designer babies.

Part 7
Playing God? Ethical Issues

The new technologies bring some very touchy issues. Are genetically engineered foods safe to eat? Is it OK to mix and match genes in animals? Will DNA testing infringe on our privacy?

This section covers the pros and cons of the many issues that have been raised along with the development of new genetic technologies. Most of these questions don't have simple answers.

Chapter 25

Patents on Life

> ## In This Chapter
>
> ➤ How the U.S. Patent and Trademark Office began granting patents on living things in 1980
>
> ➤ How patents have been taken out on whole species of genetically altered plants
>
> ➤ How researchers have obtained patents on genetically modified mammals
>
> ➤ How researchers have applied for patents on human cells

You've heard of patent leather and patent lies. But how about patents on living things? Since the 1980s, the United States Patent and Trademark Office (USPTO) has granted patents on several hundred plants, animals, and microscopic organisms.

Maybe you thought that patents, which grant an inventor or innovator the sole rights to produce, sell, and use a product or a process, were just for ingenious inventions like self-inflating cars and computer-assisted psychotherapy. But the definition of an invention has become broader these days. Biotech companies have applied for and received patents on lots of genetically engineered products.

But along with the new patents comes a new controversy. Some people think that if living things can be patented, that makes natural organisms commodities. But some companies feel that if they can't obtain patents on their products, then there's no point in spending millions on research when they might not get a return on their investments. What if competitors took the idea and marketed it? Read on to find out more about ethical questions that have been raised as a result of the patenting of new biotechnologies.

Building Better Bacteria

In 1971, a General Electric scientist named Ananda Chakrabarty experimented by putting together genetic material from four different strains of a family of bacteria. His goal was to create a super-oil-eating strain of microbes.

Since some forms of bacteria naturally like their food greasy, Chakrabarty reasoned that if he combined some genetic material from four such strains, he could produce an extra-hungry, oil-consuming microbe. This microbe could come in handy in case of potentially devastating oil spills, because it could gobble up the greasy meal before the oil could cause harm to the environment.

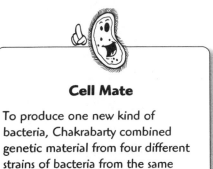

Cell Mate

To produce one new kind of bacteria, Chakrabarty combined genetic material from four different strains of bacteria from the same family, Pseudomonas (SUE-do-MO-nas) bacteria. This rod-shaped bacteria is found in soil and water.

As a result of combining genetic material from the different strains of bacteria, Chakrabarty created a type of bacteria that never existed before. He wanted to protect the ownership of his new bacteria, so in 1971, he applied for a patent in the United States.

When a person or a company gets a patent, that person or company owns the sole rights to that product or process. If other people or companies want to use that product or process, or if they want to manufacture it or sell it, they must obtain permission from the patent holder, and often they must pay a fee. In much the same way that writers protect their literary works by copyrighting them, inventors apply for patents to protect their inventions. In the United States, a patent is granted for a period of 17 years. After that time, anyone is free to use the product or process without asking the permission of the inventor.

DNA Data

Before Chakrabarty created his new bacteria, the United States Patent and Trademark Office (USPTO) refused to grant patents on several occasions because the "products" already occurred in nature. For instance, in 1928, the General Electric company wanted to patent pure tungsten, a hard metallic chemical element. The company was told by the Patent and Trademark Office that it could obtain a patent only for the special process that it had invented for purifying tungsten. But it could not get one for the element itself. The USPTO explained that because tungsten is a "product of nature," it is not patentable.

After reviewing Chakrabarty's patent application, the United States Patent and Trademark Office told him that a new life form is not an invention. No one had ever received a patent on a living thing. Inventions, according to the USPTO, have to be, among other things, "novel" and "useful."

A Supreme Decision

Although Ananda Chakrabarty's request wasn't granted by the U.S. Patent and Trademark Office, the scientist refused to take no for an answer. He decided to appeal the decision. The case was shuttled to higher and higher courts. Finally, years later, it reached the Supreme Court of the United States. In 1980, the much-debated issue was resolved, and the case called Diamond versus Chakrabarty came to an end.

The Supreme Court voted in a narrow five-to-four decision that the microbe in question was considered an invention created by Ananda Chakrabarty. They came to this landmark conclusion because they reasoned that the bacteria had never existed in nature before Chakrabarty created it. Therefore, they agreed that the General Electric scientist's "human ingenuity and research" could be rewarded with a patent.

Ironically, after the lengthy battle to get Chakrabarty's new bacteria patented, there was an unexpected anticlimax. The tiny critters were never marketed to eat up oil spills. The scientist discovered that even though the bacteria enjoyed a feast of oily eats in the laboratory, their culinary tastes weren't always the same out in the field.

The Joint Appeal

The 1980 Supreme Court decision in Diamond versus Chakrabarty fueled an ongoing debate about the patenting of living things. Some biotech companies applauded the decision, because they felt that without the protection of a patent, why should they bother to do any research? Without a patent, they felt, all the millions of dollars that might go into discovering a new drug or novel organism would go down the drain, because other companies would have the legal right to usurp their ideas and make money without doing the research.

DNA Data

Jeremy Rifkin is an activist who has long fought against some controversial aspects of biotechnology. As president of the Foundation on Economic Trends, an organization based in Washington, D.C., Rifkin organized The Joint Appeal Against Human and Animal Patenting. Probably the most outspoken opponent of biotechnology, he is the author of a number of books detailing his concerns. Among these books are *Algeny* and *The Biotech Century: Harnessing The Gene and Remaking the World.*

On the other hand, some people think that life is sacred and that granting a patent for a living organism transforms the way we think about it, and makes life a mere commodity. In May 1995, nearly 200 religious leaders representing almost every major religion came together in what was called The Joint Appeal Against Human and Animal Patenting. This coalition was organized by Jeremy Rifkin, an activist and long-time opponent of biotechnology.

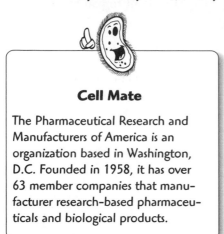

Cell Mate

The Pharmaceutical Research and Manufacturers of America is an organization based in Washington, D.C. Founded in 1958, it has over 63 member companies that manufacturer research-based pharmaceuticals and biological products.

A Roman Catholic bishop named William Friend, one of the religious leaders in the Joint Appeal, explained in a *New York Times* article that he feels the biotechnology industry is thinking only of science and business when they patent life forms. However, he stated that "we are dealing in a larger context—the sovereignty of God." Another religious leader, Rabbi David Saperstein, an attorney and the director of the Religious Action Center of Reform Judaism, said that no one should be allowed to lay claim to an entire living species, such as genetically engineered animals.

Protestants, Muslims, Buddhists, and representatives of other religious denominations joined in denouncing the patenting of life forms. Richard Land, director of the Southern Baptists' Christian Life Commission, spoke of the "commodifying of human materials" and described it as "grotesque."

In a quick response to the Joint Appeal, Gerald J. Mossinghoff, the president of the Pharmaceutical Research and Manufacturers of America in Washington, D.C., issued a press release entitled, "Changing Genetic Patent Practices Could Hurt Medical Progress." In the release, Mossinghoff commented that scientists need patents to protect themselves.

Without patents, anyone can reap the financial benefits of a scientist's research. He noted that companies might not be able to raise funds for their research without the assurance that they have the legal right to exclusively profit from what they develop, in order to recoup their investments. He concluded that although the religious leaders in the Joint Appeal do not want to deprive people of future cures for diseases, this could well be the result of their call for a ban on the patenting of living organisms.

Seeds of Discontent

Since Ananda Chakrabarty was granted a patent on a new form of bacteria, the U.S. Patent and Trademark Office has been called upon to consider granting other patents on plants, animals, and even human cells. Agriculture is one area that appears to be especially fertile ground for patent applications.

Biotech companies have applied for and received patents on genetically engineered plants that have been modified to give them more nutritional value, to create built-in resistance to insect pests, or to help them to withstand herbicides intended to kill weeds that grow around them. Again, when a company has spent a great deal of money conducting research to successfully modify a plant, it wants to make sure that it can benefit from the results of that research. Thus, companies want to patent their genetically engineered crops.

One problem that opponents of biotech have with the patenting of plants is the fact that farmers will no longer be able to save seeds at the end of the season. The practice of seed-saving allows farmers to plant another year's crop without having to spend more money on seeds. But if a company holds a patent on an altered variety of crop, then it legally has the right to forbid farmers to save seeds.

In 1993, concerns like this one led 500,000 farmers in India to organize a rally protesting proposed seed patenting by American agricultural companies. The Indian farmers saw seed patenting as a real threat to their livelihoods. Not only would they have to pay higher prices for patented seeds, they said, but they would also be forced to pay royalties if they wanted to save seeds for subsequent years.

A Tree for All Seasons

Indian farmers are also concerned that patenting a certain type of tree that grows in India might prevent them from reaping its benefits unless they pay for them. The tree in question is the neem tree, a fast-growing evergreen that can reach heights of up to 60 feet. It grows in India and other countries in Asia.

For centuries, Indians have relied on the neem tree for a wide variety of benefits. Vandana Shiva, an Indian scientist, explained some of these benefits in an article she authored for *Third World Resurgence* magazine. The bark, flowers, seeds, and leaves are used as medicine to treat a number of disorders. Neem oil is even used as a type of contraceptive, because it is a powerful spermicide. In addition, the oil is used in lamps. The wood itself is naturally resistant to termite infestation, so the neem tree has long been regarded as a good choice in construction.

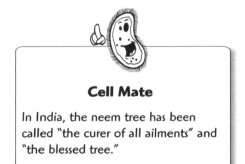

Cell Mate

In India, the neem tree has been called "the curer of all ailments" and "the blessed tree."

But the benefits don't stop there. The neem tree also has important agricultural applications. For example, its leaves are used to make the soil more fertile. In addition, the tree has insecticidal properties and can ward off locusts, nematodes, mosquito larvae, Colorado beetles, and boll weevils, just to name some of the approximately 200 insect pests that can be kept under control using the neem tree's natural properties.

In a country where there are about 14 million of these trees, these benefits are available for free or, at most, for a small price when, for example, villagers are paid to extract the oils from other components of the tree. What Vandana Shiva and others in India are deeply concerned about is the fact that since 1985, dozens of patents on components of the neem tree have been filed, and over a dozen patents have been granted to companies in the United States and Japan. They call the foreign patents a case of "biopiracy," because the neem tree's virtues were common knowledge among Indian farmers for as long as anyone can remember. They worry that the foreign patents may eventually lead to the end of free or low-cost access to the benefits of the tree, which has grown in India for centuries.

In response to these concerns, in April 1993, the Congressional Research Service issued a report to outline why they feel that patenting of neem tree components should be allowed. They explained that the active component does occur naturally and therefore cannot be patented. But a synthetic version of this natural compound could be patented, because it wouldn't be considered a "product of nature."

To this, Vandana Shiva comments that the patents that were issued have nothing to do with synthesizing the active ingredient. Rather, they were issued for the specific processes used to extract the chemical. She adds that these methods are "simply an extension of the traditional processes used for millennia for making neem-based products."

Broad Patents

The Rural Advancement Foundation International (RAFI), an agency based in Ontario, Canada, and North Carolina, keeps a watchful eye on patents that are granted to big biotech businesses. RAFI worries that if the food supply winds up in the hands of a few multinational corporations, this situation could lead to what they call "bioserfdom." Farmers would be dependent on these corporations, they say, and would have no choice but to purchase expensive seeds each year.

DNA Data

RAFI, an agency that keeps an eye on biotech developments, is challenging a species–wide patent issued by the European Patent Office that grants the Monsanto company rights to all genetically modified soybeans. The patent was originally granted to Agracetus, Inc., a subsidiary of W.R. Grace. However, when Monsanto bought Agracetus, they gained the rights to the patent.

But some corporations see things differently. They feel that because the world's population is growing at such a rapid rate, there may not be enough food to go around in the years to come. By using genetic engineering techniques, they hope to increase the world's food supply to avert a disaster. They contend that they are simply protecting their claims to these new food crops.

But RAFI and other critics, such as Jeremy Rifkin, don't agree. They are especially worried about what are called species-wide patents. For example, in 1992, Agracetus, Inc., a subsidiary of the W.R. Grace Corporation, was given a species-wide patent for genetically engineered cotton. This means that they were given the patent for any and all genetically modified varieties of cotton. So if another company wants to put a new bioengineered type of cotton on the market, it isn't allowed to do so unless it pays royalties to the patent holder. (Because of a corporate buyout, the patent is now owned by Monsanto.)

In the years to come, there will be more and more applications for patents on engineered plants and new methods of extracting existing natural substances. The debate is sure to continue as businesses hope to recoup their investments and farmers and others assert their right to their traditional approach to agriculture, one that is based on seed saving and free access to naturally occurring benefits of plants.

Critter Controversies

Another area of biotechnology that has come under scrutiny is the patenting of animals. Since Chakrabarty created his teeny microbes, higher forms of animals have been genetically altered and then patented. This development has fueled a debate over whether scientists are claiming ownership of what some consider to be the sacred creations of God.

You read in Chapter 15, "Down on the Pharm: Engineering Animals," that scientists can splice genes from one species into another. They have put human genes into pigs, and genes from a moth into potatoes. New genetic technologies make this possible, whereas in the past, there could never be such an interchange outside of a species.

Modified Mammals

In 1988, the first genetically engineered mammal was patented. It was called the Oncomouse, and it was developed by Dr. Philip Leder at Harvard University. The patent was issued to Harvard and the Du Pont Chemical Company, which provided funding for the research. The Oncomouse was genetically modified to develop breast cancer and other types of tumors. Researchers hope to use this type of animal as a tool to come up with treatments for cancer in humans.

The Oncomouse was the first, but certainly not the last, mammal to be modified and patented. In 1992, the same Harvard team that developed the cancer-prone mouse received a patent on another type of designer mouse, one that develops enlargement of the prostate gland. Two Dutch scientists, together with GenPharm International in California, designed and patented mice that lack immune systems.

The researchers who create these genetic modifications usually do so because they want to have unique animals to serve as models for experimental treatments for

Mutant Misconceptions

Don't think that if a person or company obtains a patent for an invention in the United States, that the patent applies all over the world. The inventor must apply for foreign patents to receive protection outside of the United States. But foreign patent offices may or may not agree with USPTO decisions. For instance, when the Upjohn Company applied for a patent in Europe for its hairless mouse, it was denied. The mice were genetically altered to be completely bald and were intended for use as laboratory tools to find a cure for baldness. The European patent office apparently didn't think that baldness constituted a major public health problem. (But some bald men might consider it worse than some fatal diseases!)

Cell Mate

The Council for Responsible Genetics is an organization based in Cambridge, Massachusetts. Among the distinguished scientists serving on the CRG's Board of Directors are Dr. Ruth Hubbard of Harvard University and Dr. Jonathan King of the Massachusetts Institute of Technology.

diseases. These models can also help scientists to understand the mechanisms behind certain diseases. Because it often takes many years and a great deal of funding to design these animals, the researchers or biotech companies that developed the altered species want to protect their investment, just as seed companies want to patent their new types of plants.

But there is a raging controversy over the issue of patenting animals. As you read in this chapter, the Joint Appeal Against Human and Animal Patenting is against such patents. Some people feel that living things cannot be owned, and this, of course, extends to animals.

Are Animals Inventions?

People challenge the U.S. Patent and Trademark Office on granting these patents for different reasons. A group called the Council for Responsible Genetics (CRG) in Cambridge, Massachusetts, carried an article challenging such patents in their newsletter. The article was written by Dr. Stuart Newman of New York Medical College and Nachama Wilker, the former executive director of the CRG.

In this article, Dr. Newman and Wilker questioned the notion that animals can be classified as inventions. They consider it erroneous to think that DNA is the blueprint responsible for a precisely defined set of characteristics in an organism. They said that scientists cannot accurately predict which traits will be modified by manipulating genes.

They point out that the use of the gene for Human Growth Hormone in animals produced different results in different species. When one researcher spliced this human gene into mice, he got extra-large rodents, but when another researcher tried this same approach with pigs, he obtained debilitated, arthritic animals instead of the robust porkers he hoped for.

Because there are sometimes unanticipated developments when genes are introduced from one species to another, Dr. Newman and Wilker conclude that the idea of a genetically modified animal as an invention is "based on an extreme oversimplification of the relation of genes to traits."

Half-Human Patents?

Dr. Stuart Newman joined with activist Jeremy Rifkin in 1998 to make a bold move. They applied for a patent on a method of making chimeras (kih-MIH-ras), animals that are the result of splicing DNA from two different species. In the case of Rifkin and Newman's patent application, they specified the combination of human embryo cells with animal embryo cells. The patent application is a broad one and includes many possibilities, ranging from early chimeric embryos that consist of a small number of human cells to entire organisms that would carry mostly human DNA.

Why would two opponents of biotech's patents apply for one themselves? The answer is that they are hoping to make the public aware of the type of patenting that they consider unacceptable. If they get the patent, they can prevent others from employing this technology. If the patent is denied, it will set a precedent for others who want to employ human/animal combinations for their research.

Owning the Rights to Life?

Some people are uncomfortable with the concept of a company owning the rights to higher life forms, especially mammals. But the Biotechnology Industry Organization (Bio) in Washington, D.C., sees this issue in a different light. In its pamphlet, "Animals, People, and Biotechnology," it states its view that patenting animals is akin to people owning them. The organization notes that "Owning animals is legitimate and traditional in our culture." Bio feels that in the same way that people own pets and farm animals, genetically modified mammals can also be owned.

DNA Data

In 1988, Harvard University was granted a patent in the United States for its development of the Oncomouse, a genetically engineered type of mouse that is susceptible to cancers. The Harvard researchers also applied for a patent in Canada. This was the first time that anyone had asked the Canadian patent office for the rights to a mammal. It refused, saying that the Oncomouse was not an invention as defined in the Canadian Patent Act. The Canadian patent office added that the Harvard researchers had invented a process, but not a life form.

Yet another issue that comes up in relation to animal patenting is the concern that there will be more and more uniform varieties of animals. Because it is easier to take out a patent on a specific type of animal, researchers are likely to stick to developing uniform varieties of animals to make their patent claims easier to enforce. Some scientists and laypersons alike worry that this could lead to a possible loss of biodiversity, the great diversity of life that helps animals and plants to successfully adapt to changes in the environment.

What would happen if a strong climate change occurred or a new type of disease developed among a species of animal? If there were only one uniform variety of cow or pig or sheep, and it was struck by this disease or environmental shift, how could that species adapt and defend itself against the threat? The more diversity there is in a species, the better the chances are that it can successfully adapt to challenging conditions. Genetic uniformity creates vulnerability to outside threats.

Patenting People

Probably the most heated debates have been over the patenting of human substances. You read in Chapter 21, "The Human Genome Project," about researcher J. Craig Venter's great success with mapping and sequencing fragments of genes. In the late 1980s, his employer at the time, the National Institutes of Health (NIH), wanted to obtain patents on these fragments of human genes. This created quite a stir.

Another patenting controversy soon surfaced as the result of scientists discovering that certain groups of people may carry genes that give them natural resistance to certain diseases or unhealthy conditions. For instance, some people in Limone, a town in Italy, seem to have natural protection against high cholesterol. If scientists can study the natural immunities of certain ethnic peoples, they might be able to use that information to find cures for diseases such as cancer and AIDS. They might even be able to stop these diseases before they start if they can understand how these illnesses are caused on a genetic level.

A Line of Cells

In 1990, the United States National Institutes of Health (NIH) filed for a patent on the cell line of a Guaymi (GWAH-ee-MEE) woman. The Guaymi are an indigenous people in Panama. A cell line is made from cells, such as blood cells, that have been taken from a person or animal and are made to grow continuously in the laboratory.

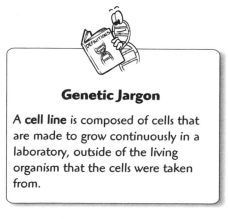

Genetic Jargon

A **cell line** is composed of cells that are made to grow continuously in a laboratory, outside of the living organism that the cells were taken from.

The researchers who applied for the patent said that the Guaymi woman gave her oral consent for them to take her blood sample. This event still sparked an international debate. Organizations such as RAFI led the opposition, and they were joined by Isidro Acosta, the president of the Guaymi General Congress. In a RAFI press release, Acosta was quoted as saying that he was shocked to learn that anyone would want to patent living things. He called these actions "immoral" and "contrary to the Guaymi view of nature and our place in it." He added that patenting the cell line of the Guaymi woman "violates the integrity of life itself and our deepest sense of morality."

Pat Mooney of RAFI-USA explained that the Guaymi were not trying to hold back medical research. But what they were objecting to was the "piracy and the immorality" of gaining the rights to human substances. Eventually, the patent application was withdrawn.

U.S. Patent Number 5,397,696

The issue resurfaced in August 1991, when the National Institutes of Health (NIH) again filed for a patent on a human cell line. This time it was from the cells of a man of the Hagahai (HOG-a-HI), an ethnic group from Papua New Guinea. In February 1992, the NIH sought a patent for the cell line of a man from the Solomon Islands, a small country in the Pacific.

In March 1996, the USPTO granted the NIH a patent on the cell line from the Hagahai man. This action angered groups like RAFI. In October 1996, the NIH filed paperwork with the USPTO to disclaim patent number 5,397,696 for the Hagahai cell line. The NIH explained that the patent was not commercially viable for its organization. However, RAFI is convinced that international pressure caused NIH to forfeit its patent claims.

Although some people, like the members of RAFI and other agencies, view the patenting of human cell lines as immoral and an example of what they term "biopiracy," the biotech industry has a distinctly different opinion. Biotech companies feel that to make it worth their while to spend the time and money to develop possible cures for the diseases that plague humanity, they need to ensure that they have the right to recoup their investments. But agencies such as RAFI contend that such companies have no intention of sharing their profits with the ethnic people whose genetic makeup might make future cures possible.

The Least You Need to Know

➤ In 1980, the United States Patent and Trademark Office granted the first patent for a living organism, a genetically modified type of bacteria.

➤ After the 1980 precedent, which ruled that a living thing could be considered an invention, researchers have obtained patents on genetically altered plants, including entire species.

➤ In 1988, Harvard University received the first patent on a mammal: the Oncomouse, which is a genetically altered mouse that is susceptible to cancers.

➤ In 1998, activist Jeremy Rifkin and biologist Dr. Stuart Newman applied for a patent on mixes of human embryos and animal embryos, in order to prevent other researchers from carrying out this type of work.

➤ In 1995, the National Institutes of Health was issued a patent for the cell line of a man in Papua New Guinea, but in 1996, the NIH disclaimed the patent.

Chapter 26

Food for Thought: Are Biofoods Safe?

In This Chapter

➤ How will altering one gene in a plant affect the plant's other genes?

➤ How will biocrops affect the economies of Third World countries?

➤ How will engineered crops affect the environment?

➤ A coalition sues the Food and Drug Administration for not labeling biofoods

➤ Should seed companies be allowed to make seeds that produce sterile plants?

You read in Chapter 14, "Seedy Science: Engineering Plants," about some of the new genetically altered biofoods. There are potatoes with chicken genes, soybeans with built-in resistance to herbicides, and lots more. Many biotech companies feel that creating genetically engineered foods is a dependable way to save the ever-expanding world population from starvation in the future. By providing higher yields and lowering the costs of food crops, genetically engineered crops are touted as having the potential to feed the world's hungry.

Others disagree. Organizations such as the Campaign for Food Safety, the Council for Responsible Genetics, and the Union of Concerned Scientists, to name a few, have serious doubts about some aspects of engineered food crops. Read on to find out how questions about food labeling, environmental impact, and other issues have sprouted up along with the new bioharvests.

Tasty Cuisine or Frankenfoods?

In Chapter 14, "Seedy Science: Engineering Plants," you learned that in 1994, the FLAVR SAVR™ tomato went on the market. The first genetically altered plant to be approved by the U.S. Food and Drug Administration (FDA), and the result of $25 million worth of research, the tomato was engineered to retain its freshness without rotting before hitting your dinner table. The FLAVR SAVR™, which was developed by Calgene, a California biotech company, wasn't on the market for very long. But it will go down in history as the first of many plants that had their genetic makeup rearranged to improve upon nature.

Genetic Jargon

Blood serum is the clear, yellowish liquid part of blood. It separates out from the clot in coagulated blood, and it's about 90 percent water. More than half of blood's volume is liquid.

Calgene voluntarily labeled their altered tomatoes, but there was no legal requirement to do so. The United States Food and Drug Administration (FDA) took the stand that because all DNA consists of different arrangements of the same As, Cs, Ts, and Gs, then any spliced-in genes are just more DNA. In other words, the FDA feels that because nothing unnatural is being added to the fruits or vegetables, such as chemical additives or preservatives, genetically engineered foods don't require special labels.

But not everyone agrees with the FDA's position that biofoods are safe. They say that the new combinations are not like anything that could occur in nature and worry that we might be in store for some unpleasant surprises in the future.

Nothing to Sneeze at

For instance, what will happen to people with allergies? According to the Food Allergy Network in Fairfax, Virginia, about 5.2 million people in the United States have food allergies. How would eating biofoods affect them? Suppose a person is allergic to tomatoes. If a gene from a tomato is spliced into a potato, could that one little tomato gene cause an allergic reaction when the allergic person eats the potato?

For a while, no one knew the answer to that question. But in 1994, Dr. Steve Taylor, a food scientist at the University of Nebraska at Lincoln, conducted tests to find out whether some genetically engineered soybeans could produce an allergic reaction in sensitive individuals.

The soybeans he tested were developed by Pioneer Hi-Bred International, a giant seed company. Pioneer's researchers added a gene from Brazil nuts to the soybeans. The extra gene codes for a protein that has high amounts of a specific amino acid. They did this to produce soybeans that would be more nutritious than average. These soybeans were intended to be used to feed cattle, but some could also be used as additives in food for people, too. Before these soybeans were put on the market, Pioneer asked Dr. Taylor to find out whether the addition of the one Brazil nut gene could cause a reaction in people who were allergic to these nuts.

Because Dr. Taylor is an expert on food allergies, he just happened to have a laboratory freezer chock-full of samples of *blood serum*, the liquid part of blood, from people with this allergy. When Dr. Taylor tested the blood serum samples with extracts of the altered soybeans, he found antibodies in seven out of the nine samples tested. You'll remember from Chapter 23, "Molecular Medicine," that antibodies form when the body is attacked by invaders. Allergic people's immune systems sometimes perceive nonthreatening substances, such as those found in Brazil nuts, as invaders and produce antibodies to combat them.

Dr. Taylor's experiment suggests that the one Brazil nut gene could cause a full-blown allergic reaction in sensitive individuals who ate these altered soybeans. As a result of Dr. Taylor's study, Pioneer Hi-Bred decided not to market the soybeans.

DNA Data

When a person is allergic to a food, typical reactions include rashes or distress in the stomach and intestines. But some people suffer more severe reactions, such as anaphylactic shock (AN-a-fill-ACK-tic). Anaphylatic shock is characterized by a drop in blood pressure, difficulty in breathing, and even a loss of consciousness, and it can sometimes be fatal.

Religious Questions

One touchy issue that surrounds genetically engineered foods is the question of religious laws and ethical beliefs. Some religious people, such as devout Muslims and Orthodox Jews, follow specific laws concerning which types of food they may or may not eat. For instance, Orthodox Jewish dietary law does not allow followers to eat meat and dairy products at the same meal. Both Muslims and Orthodox Jews refrain from eating pork. Vegetarians, while not following religious laws, follow an ethical belief that they should not eat meat. Vegans (VEE-guns) are vegetarians who will not eat or drink any animal product, including all dairy products and eggs.

Because the FDA does not require special labeling on genetically altered foods, a difficult question arises. If a person's religion or ethical belief forbids them from eating a certain food, does that include one gene from that food that's been spliced into another food?

Suppose a person's religion forbids them to eat pork, and a gene from a pig is spliced into a potato. If this person goes to the supermarket to buy a potato, they won't know that pig genes are spliced into it, because there is no special labeling. When that person cooks and eats french fries or potato salad, will he be transgressing his religious laws?

The FDA's view is that splicing one gene from an animal does not change the nature of a fruit or vegetable. You won't see a potato that oinks and rolls around in the mud or a french fry that smells suspiciously like bacon. The FDA points out that sometimes different species share the same type of genes. For instance, some genes in humans are also found in rice.

A Well-Pressed Suit

On May 27, 1998, a coalition of scientists, religious leaders, health professionals, consumers, and chefs filed a suit against the U.S. Food and Drug Administration. They are asking for mandatory safety testing and labeling of all genetically engineered foods in the country. The suit was coordinated by the Alliance for Bio-Integrity in collaboration with the International Center for Technology Assessment (CTA) in Washington, D.C., which is doing all the legal work in the case. The two groups are nonprofit organizations dedicated to protecting human health and the environment through sustainable technologies.

The coalition is concerned because some of these genetically engineered foods are used as ingredients in processed foods such as soy-based infant formulas and corn chips. One of the coalition's focal points is their contention that consumers have a right to know whether they are buying and eating genetically altered produce.

The suit lists 33 genetically engineered foods that are currently being sold without special labeling or required safety testing beyond the tests that the biotech companies opt to perform themselves. Insect- and virus-resistant potatoes, herbicide-tolerant corn, virus-resistant squash, and many other foods are included in the suit. Andrew Kimbrell, executive director of the International Center for Technology Assessment (CTA) and co-counsel in the case, said, "By failing to require testing and labeling of genetically engineered foods, the…[FDA]…has made consumers unknowing guinea pigs for potentially harmful, unregulated food substances."

The suit, which was filed in federal district court, also mentions the religious concerns that surround this issue. The suit alleges that the FDA policy of not requiring special labels is a violation of religious freedom.

DNA Data

According to the International Service for the Acquisition of Agri-Biotech Applications (ISAAA), genetically altered soybeans and corn accounted for more than 80 percent of all the transgenic crops grown in the world (not including China, where data was not available) in 1998.

Andrew Kimbrell, executive director of the International Center for Technology Assessment, is an attorney in the lawsuit against the FDA for not labeling genetically engineered foods.

Bio-Surprises

Another concern about genetically engineered foods is the question of unpredictability. Although scientists are learning more and more about the way genes function, surprises can still happen. What if inserting a gene into a food crop disrupts the function of another gene in that plant? Genetically altered foods have been on the market only since 1994, so there haven't been any studies on their long-term effects.

Author and environmentalist Jack Doyle discusses some of these questions in his book, *Altered Harvest*. He gives an example of a classically bred potato that yielded some unpredictable results when it was harvested. This potato wasn't genetically engineered; it was developed by selective breeding. In other words, researchers took different types of potatoes and crossbred them in the hopes that they would get a new type of potato with the best qualities of each type.

The Lenape potato, as it was called, didn't work out too well. Some plants like potatoes and tomatoes naturally contain small amounts of toxins. The amounts aren't enough to get sick from, but they're there. The Lenape potato produced higher levels of toxins than normal. As a result, some people who ate the potato got sick, and the spuds had to be pulled off the market.

If this could happen with selective breeding, might the same sort of thing happen with genetically altered foods? Doyle concluded, "You can elevate a

Cell Mate

It's hard to believe, but many years ago, some people were afraid to eat tomatoes because they thought that they were poisonous. If that idea survived, just imagine all the Italian restaurants that would've never existed.

naturally occurring toxin. Lots of them occur in very low background levels in lots of food, and you might inadvertently amplify these traits. There are lots of interactions that we don't really understand."

Built-In Bug Killer

Doyle is also worried about the splicing of genes from B.t., a type of naturally occurring soil bacteria, into altered plants. For years, organic farmers used B.t. as a harmless biopesticide, a natural bug killer, because it doesn't harm animals or beneficial insects. A certain substance gets into the digestive systems of pesky insects when they eat plants sprayed with B.t. A toxin is released, and the bug dies. Animals or people who eat the plant are not harmed.

By splicing a gene that causes the demise of these bugs into food crops, scientists have essentially created plants with built-in pesticides. Some people see this as a positive development, because it could reduce the use of other pesticides that can be toxic to animals and the environment.

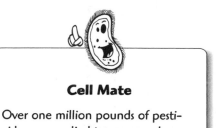

Cell Mate

Over one million pounds of pesticides are applied to crops each year in the United States. Some biotech fans hope that genetically engineered crops can reduce the need for pesticides.

But Doyle cautions against embracing this high-tech idea too soon. He worries that if the genes in B.t. are spliced into too many food crops, insects could develop a resistance to B.t. In organic gardening, B.t. is used on a small scale, but in genetically altered plants, the B.t. gene that fights insects is present in every one of the plants' cells.

Doyle makes a comparison to what's going on with antibiotics today to illustrate how organisms can become resistant. Some scientists feel that we have overused antibiotics and that this overuse has led to a resistance to antibiotics in some microscopic disease-causing agents. In the same way, Doyle is concerned that insect pests may develop a resistance to the B.t. toxin, and then organic farmers will lose a precious natural resource.

The Dirt on B.t.

The toxins in naturally occurring B.t. exist in what's known as an inactive form. For the toxins to be liberated and do their job of killing insect pests, they need to be eaten by the bugs. When the inactive toxin is digested, it lets the active toxins loose. These active toxins are what eventually does the bugs in. But genetically altered crops that have genes from B.t. spliced into them carry the active form of the toxin.

In 1998, Guenther Stotzky and C. Crecchio of New York University performed laboratory experiments that showed these active B.t. toxins might accumulate in the soil. In the New York University experiments, the active toxins became bound to certain particles in the soil. As opposed to the naturally occurring B.t. toxins, these bound toxins cannot be broken down by tiny microbes in the soil. These bound toxins also keep their ability to kill insects. This study suggests that some beneficial insects that are normally not affected by the inactive B.t. toxin might be susceptible to the buildup of the active B.t. toxin in the soil.

Beyond Appearances

Scientists know that genes often act in conjunction with each other. When genes are added or deleted in a fruit or vegetable, the added or missing genes might cause a reaction with the other genes. What if this reaction led to a decrease in the nutritional value of some biofoods?

The FLAVR SAVR tomato that came out in 1994 was engineered to stay fresh for a longer period of time than the average tomato. The gene that is responsible for the softening that occurs during ripening was taken out and then reinserted backwards to cancel its effects. Scientists from groups like the Council for Responsible Genetics voiced their concerns about whether these tomatoes could have the appearance of being fresh and nutritious, but in fact be older and less nutritious than they looked, a phenomenon known as "counterfeit freshness."

DNA Data

What do most Americans think about genetically altered foods? The January 11, 1999 issue of *Time* magazine reported on their nationwide poll on various aspects of biotechnology. When people were asked whether they wanted labels on genetically engineered foods, 81 percent responded yes. The next question asked whether these people would buy foods that were labeled as genetically engineered. Fifty-nine percent answered no.

The Economics of Food

Another aspect of biofoods that raises concern is their possible affect on the economies of Third World countries. In his book *The Gene Hunters: Biotechnology and the Scramble for Seeds*, Dr. Calestous Juma of the African Centre for Technology Studies in Nairobi, Kenya makes the case that genetic engineering might bring about a decrease in crop sales in his country and elsewhere. Dr. Juma points out that biotechnology might provide consumers in the United States and other developed nations with products such as genetically engineered sweeteners that could eventually replace natural products such as sugar, which is traditionally grown in a number of Third World countries.

If so, what would happen to the millions of people whose livelihoods are based on the sale of sugar?

Some less-developed nations rely almost exclusively on a few food crop exports to sustain their economy. What would happen to their trade with larger nations if a biotech company came up with, for example, coffee that could be grown in Maine or New York? This hasn't happened yet, but it is conceivable that in the future, someone could splice genes into warm-weather crops that would protect them from the ravages of snow and ice. If that happened, a small country's economy might collapse from the loss of trade. In his book, Dr. Juma concludes, "...those who have competitive advantage in technological know-how are the ones who will define the future of global agriculture."

Biotech companies answer these concerns by saying that the new technologies will be available to people all over the globe. They point out their opinion that biotechnology could increase the yields and efficiency of traditional farm practices, and therefore might lead to higher profits for Third World farmers.

Growing Concerns

How will the environment be affected by biofoods? Some genetically altered plants can withstand herbicides that are sprayed on crops to keep pesky weeds down. As of 1998, there were more than a dozen herbicide-resistant plants in the United States that were approved for commercial use by the F.D.A. What would happen if pollen from a genetically altered plant got carried by the wind or hitched a ride on a bee? Could it fertilize a wild plant that's related to it? What if this wild plant were a weed that farmers were trying to keep from taking over large portions of their land and crowding out their food crops? If the pollen from the genetically modified plant carried a gene for resistance to an herbicide, would that gene transfer to the weed? If that happened, would we end up with superweeds that couldn't be stopped?

For a while, no one knew the answer to these questions. But there seems to be some growing evidence that modified plant genes may be capable of spreading into wild populations. A scientist named Joy Bergelson and others at the University of Chicago conducted experiments that cast some light on this issue. They reported their findings in the September 3, 1998 issue of *Nature* magazine, in an article provocatively titled, "Promiscuity in Transgenic Plants." But rest assured that this is not X-rated.

The scientists investigated mustard plants that were genetically altered to withstand herbicides. They also studied another variety of mustard plant that was naturally resistant to herbicides. They tested both

Cell Mate

According to the Union of Concerned Scientists (UCS) in Washington, D.C., most patents granted for genetically engineered food crops in the United States so far have been for herbicide-resistant varieties. The UCS is concerned that the existence of more herbicide-resistant crops might result in the increased use of herbicides.

varieties of mustard plants to see how likely they were to interbreed with relatives. (This was the racy part of the article.) The surprising result was that the genetically altered mustard plants were 20 times more likely to interbreed with relatives than the traditionally bred plants were. No one quite understands why this is so, but some people think that it could be a cause for concern.

Some fans of biocrops contend that even if the genes from altered plants spread to wild populations, the resulting plants would eventually die out. The plants would simply have too much competition from their wild relatives.

DNA Data

The Laboratory for Comparative Invertebrate Neurobiology (LNCI) in France conducted experiments to learn how altered plants could affect beneficial insects. The researchers wanted to know how honeybees were affected when they were given a diet of genetically engineered rape plants, which are related to turnips. When honeybees were fed these plants for two weeks, their memory for smell was affected. This sense is very important for bees because it helps them find their food. This study suggests that genetically altered plants may adversely affect the honeybees.

In the past, some people reasoned that altered plants would not fare well in the wild. They assumed that even if the herbicide-resistant gene could be transferred into wild populations, the plants with these genes wouldn't thrive. But a 1998 study showed that this might not be the case. Allison Snow, an ecologist at Ohio State University, conducted experiments using wild mustard plants in their natural state and other mustard plants that were genetically altered to be herbicide-resistant. Snow discovered that the altered plants produced the same amount of seeds as the natural, wild plants.

The Terminator

In March 1998, the Monsanto Corporation, one of the world's largest biotech companies, received a patent that led to a tremendous controversy that may continue for years to come. The patent covers a genetic technology that can make some plants sterile. In other words, they cannot produce healthy seeds for the next year's crop.

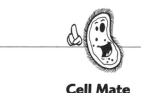

Cell Mate

The world's three largest seed companies are Pioneer Hi-Bred International and Monsanto in the United States and Novartis, headquartered in Switzerland.

Dr. Margaret Mellon wrote an editorial on this technology in the Fall/Winter 1998 issue of *The Gene Exchange*, the newsletter of the Union of Concerned Scientists in Washington, D.C. Mellon wrote that plants can be genetically engineered "to produce an exquisitely targeted dose of poison to its own seeds. Sounds like a nightmare. Not if you are an executive in an agricultural biotechnology company. To you, it's a celestial vision."

Killer Genes

Monsanto did not create this technology; it was developed by the U.S. Department of Agriculture (USDA) and Delta and Pine Land, a seed company in Mississippi. The patent was granted to both parties, but Monsanto later bought Delta and Pine Land, and the patent is now their property. Mellon calls the sterilization technology "one of the cleverest applications of biotechnology yet." But she and her colleagues hardly approve of it.

The system works by inserting three new genes into the plant. One codes for a toxin that can kill seeds, but another stretch of DNA is added to block the production of the toxin. Before the seeds go to market, they are treated with a special solution that removes the gene that blocks the toxin. When farmers plant the seeds, the seeds grow in the normal way, but as they mature, the toxin is produced. The toxin causes the plant's new seeds to die; therefore, the farmers cannot save seeds from one year's crop to the next.

The Delta and Pine Land Company describes the patented process as a "technology protection system," because the patent holder doesn't have to worry about farmers saving newly produced seeds at the end of the harvest. But the Rural Advancement Foundation International (RAFI), a group opposed to the process, has another name for it: "terminator technology."

Going Wild?

Some people are concerned that if pollen from crops like these spreads, it might cause other plants to become sterile. In an article in *Time* magazine on February 1, 1999, a Monsanto spokesperson said that there is no risk of this ever happening.

DNA Data

In December 1997, the United States Department of Agriculture (USDA) proposed national standards for organic foods. Among other things, it proposed allowing genetically engineered foods to be classified under this category. But when it solicited comments on the proposals, it received over 200,000 negative reactions. Most of the comments opposed the idea of genetically engineered foods being classified as organic foods.

Some people think that it's perfectly all right for companies to use technologies like this one, which help them reap profits from their products. If this technology catches on, farmers will have to buy seeds every year, instead of just once, allowing the company to make enough profit to cover its investment in creating new seeds.

But Mellon cautions, "the implications of the concentration of power in the food system, which the terminator will exacerbate, go deeper than high prices. Those who control the food system will have an inordinate say over our economy, our food production systems, and the use of our land."

Monsanto has estimated that the first crop to utilize this technology, which will be cotton, won't come to market until the year 2005. If opponents of the technology, like RAFI and the Union of Concerned Scientists, have their way, the technology will be delayed even further or, if possible, indefinitely.

The Least You Need to Know

➤ The FDA does not currently require that genetically engineered foods be labeled as such, but in 1998, a coalition of scientists, religious leaders, consumers, and others filed a lawsuit against the FDA for not requiring special labels for these foods.

➤ Groups like the Union of Concerned Scientists have doubts about the safety of genetically engineered foods.

➤ Some people have concerns that genetically engineered crops might adversely affect the economies of Third World countries.

➤ Some people worry that if biocrops interbreed with wild relatives, superweeds might be the result.

➤ A patent was granted in 1998 for a new genetic technology that causes plants to poison their own seeds, making it necessary for farmers to buy new seeds every year.

Chapter 27

Creature Concerns

In This Chapter

➤ Is it unsafe or unnatural to use pigs as organ donors?

➤ Is it ethical to genetically alter animals?

➤ Selective breeding causes genetic problems. Will genetic engineering cause even bigger problems?

➤ How will engineered animals affect the environment?

➤ Should Bovine Growth Hormone be used to increase milk production in cows?

You read in Chapter 15, "Down on the Pharm: Engineering Animals," about how biotechnology experts are redesigning animals. Some sheep and goats, for instance, have been engineered to produce human proteins in their milk, and genetically engineered mice are used as laboratory tools to help researchers understand human diseases. There are even pigs that have had human genes spliced into their DNA in the hope that they may some day alleviate the shortage of healthy organs that are needed for people with failing hearts, livers, and other organs.

But along with these breakthroughs in genetic technologies come complex ethical questions that are difficult to answer. Should scientists use animals as tools and biofactories? Are these actions ethical? Does using animals as models for human diseases cause the animals suffering? And if so, do the possibilities of future cures for people warrant this animal experimentation? Is it safe or ethically sound to use animals as organ donors? This chapter explains some of the different viewpoints held by scientists, ethicists, and other concerned parties on these complicated issues.

Transplanting Hopes

In Chapter 15, "Down on the Pharm: Engineering Animals," you read about xeno-transplantation (ZEE-no-trans-plant-AY-shun). This is the proposed use of animals as organ donors for people. Some biotech companies are doing research on the possibility of taking hearts, lungs, livers, and other needed organs from animals and transplanting them into humans who cannot survive for long without a healthier body part.

The animal of choice these days is the pig. One reason why the pig has gained in popularity for this purpose is the fact that its organs are approximately the same size and shape as those in humans. Plus, there are lots of pigs to go around. In addition, because pigs are routinely slaughtered for their meat, not to mention pigskin gloves and footballs, the controversy over using them for this purpose would presumably not be as heated as it would be if, for example, monkeys or other animals closer to humans were used.

Even when a human organ is transplanted into a patient, doctors must administer special drugs to keep the immune system from perceiving it as a foreign invader and promptly destroying it. Normally, a pig's organ would be detected and destroyed very quickly.

Some researchers hope that giving pigs some human genes will increase the chance that their organs will be accepted by the patient's immune system. DNX in Princeton, New Jersey, Baxter Healthcare in Illinois, and Imutran in Great Britain are just some of the companies investigating this possibility of using "humanized" pigs as organ donors. But the very idea of genetically engineering animals for xenotransplantation has caused a controversy that centers around the welfare of the animals, the safety of the procedure, and even the possibility of the creation of new epidemics.

DNA Data

In 1992, surgeons at Johns Hopkins Hospital in Baltimore used a pig's liver to help a woman survive until a human liver became available. The pig's liver was not transplanted into the woman's body. It was kept inside a plastic bag and connected to her body from the outside. The next day the woman was given a human liver. Later that same year, a patient at Cedars-Sinai Medical Center in California became the first person ever to receive a pig liver transplant. After surviving for 30 hours, the woman died before surgeons could give her a human liver.

Pig Parts

Some proponents of the use of pig organs for transplants point out that humans have used pig parts medically in the past. For instance, some heart patients have had their failing valves replaced by pigs' heart valves. However, these heart valves are first treated with a special chemical before they can safely be put into human recipients. Synthetic valves are also being used for human heart patients, which may eventually end the use of pig valves.

Pig insulin was also used at one time for diabetes patients, but in 1982, genetically engineered insulin became available. Many people preferred the biotech version because it is manufactured from human genes spliced into bacteria. It does not carry any threat of diseases carried by pigs, and there are supposedly fewer allergic reactions because the insulin is made by bacteria according to the human hormone's code.

Questions of Safety

Some people seriously question whether humanized pig organs can be safe for use in the human body. Animal organs may not necessarily work in exactly the same way as human organs, even though they may be similar.

Adding to these concerns was the evidence of new viruses in pigs discovered in a 1998 study. Among the new viruses were *retroviruses*. They are a type of virus that has genetic material consisting of RNA instead of the usual DNA.

Some scientists think that these viruses could possibly spread to humans. As a result, organizations such as the Campaign for Responsible Transplantation are calling for a delay in the use of pigs as organ donors, and others are calling for a complete ban on this procedure.

A Desperate Need

But supporters of xenotransplantation say that it is desperately needed due to the current lack of human organ donors. In 1993, 2,800 ailing patients died while waiting for organs. In 1994, the United States Department of Health and Human Services cited

Genetic Jargon

A **retrovirus** (REHT-ro-vy-rus) is a type of virus that has RNA instead of the usual DNA as its genetic material. It carries a special enzyme that converts their RNA into DNA, which can then combine with the DNA in the cell that the virus is invading. There are different types of retroviruses, including HIV, the virus that causes AIDS.

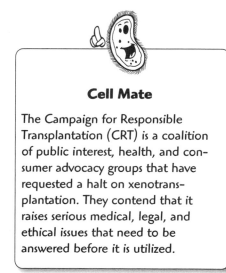

Cell Mate

The Campaign for Responsible Transplantation (CRT) is a coalition of public interest, health, and consumer advocacy groups that have requested a halt on xenotransplantation. They contend that it raises serious medical, legal, and ethical issues that need to be answered before it is utilized.

that even though more than 18,000 organ transplants were performed in the United States, more than 44,000 patients were left waiting for hearts, lungs, livers, and other organs. But that's not the whole story. Some people reason that there are more than just the patients noted in the official statistics. They say that some patients who receive transplants require new organs after about five years.

Second transplants are sometimes necessary because about half of all the successfully transplanted organs begin to lose their capacity to function properly after some time because the organs are slowly attacked by the body's immune system. Although drugs keep the immune system at bay so the organs aren't completely destroyed, they do tend to become scarred or show other signs of wear and tear. Some patients whose first and second transplants have been successful live long enough to require a third organ replacement.

But if pigs carry viruses, could placing pig organs in humans result in harmful side effects? Could using infected pig organs unleash a new epidemic? Alix Fano, in an article in *GeneWatch*, the newsletter of the Council for Responsible Genetics, points out that there are 25 known diseases carried by pigs that can be passed on to humans. Fano mentions that the Asian flu virus in 1957 and the Hong Kong flu virus in 1967 mutated in pigs. In February 1998, Peter Kirkland, a virologist in Australia, found that a type of virus that causes brain deformities in pigs caused flu-like symptoms in two of his co-workers.

Proponents of xenotransplantation counter that if animal organ donors are screened properly, mishaps can be avoided. But thorough screening of these animals may be difficult, and some diseases may not develop immediately. If this were the case, could an infected person silently carry the disease and infect others over a long period of time?

DNA Data

One theory for the origin of the HIV virus, which causes AIDS, is that it originated in monkeys in Africa. The theory states that people who had contact with the blood of infected monkeys picked up the virus. But they did not develop AIDS immediately and unknowingly allowed it to spread.

Animal Welfare

One touchy topic that arises from the use of genetically engineered animals is the question of animal rights. Similar to this is the question of whether it is ethical or natural to mix the genes of one species with another. There are many different views on these complex topics, and even scientists remain divided in their opinions.

Should scientists use animals to obtain organs or other biological material? Critics say that this technology exploits animals by reducing them to biological spare parts. They point out that engineering animals to have traits that they would never naturally possess causes them needless suffering. For example, when mice or other animals are modified to develop lethal human diseases, such as cancer or AIDS, they endure pain that they would not normally have to suffer.

DNA Data

Dr. Vernon Jennings of SustainAbility, an environmental consulting company in London, England, wrote in the January 1994 issue of *GeneWatch*, "Reducing animals to mere machines...raises deep ethical concerns when we consider the effect of genetic engineering on the welfare and dignity of nonhuman species."

But proponents of xenotransplantation explain that animals have been used experimentally for many years to develop cures that have prevented human suffering and saved countless human lives. Lab animals enable researchers to study the courses of serious diseases in ways that would be considered unethical if researchers studied these diseases in humans. Many life-saving drugs have been developed due to experimentation on laboratory animals such as mice, sheep, and dogs. Scientists tested the safety and efficacy of these drugs on animals before making them available to humans. Every person who has received a heart, liver, or other organ transplant owes his or her life to the experimental animal testing that helped develop the technology and make it safe for humans.

In a brochure that the Biotechnology Industry Organization put out on the subject of engineered animals, it explained its view that laboratory animals, transgenic or otherwise, must be treated properly. Under proper conditions, the organization concludes, it is ethically sound to engineer animals to find cures for human diseases or to utilize transgenic animals as living factories to produce human substances that are in short supply.

Those who are in favor of the use of engineered animals point out that in the long run, researchers might not have to use as many laboratory animals if they use genetically altered animals for their research. If mice, for example, are engineered to lack a specific gene or to develop a certain disease, then researchers will probably need fewer of these mice to obtain successful results in the quest for cures or a better understanding of diseases than they would if they were using regular mice.

Critics of the use of animals for experiments or biotech purposes assert that we should learn to see ourselves as a part of nature. Traditional cultures have always seen humanity as one segment of the interconnected web of life. Both of these groups hold the opinion that humans share the earth with animals and should live in harmony with them, instead of trying to control them and use them for people's wants and needs. These questions go way beyond science. They bring us into the realm of philosophy and to the very question of what it means to be a human being. Understandably, the debate continues because such questions have no simple answers.

Genetic Jargon

A **bioethicist** (BY-o-ETH-a-sist) is a person who specializes in discussing the controversial issues of biology and the medical field, such as genetic engineering topics and the care of the terminally ill. Bioethicists tend to have an in-depth background in philosophy and ethics.

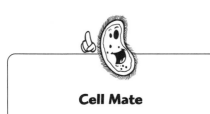

Cell Mate

Dr. Michael Fox, a veterinarian concerned about animal welfare, calls the use of altered animals as disease models "genetic parasitism." In 1988, shortly after the Oncomouse was patented by Harvard University, Dr. Fox organized a conference on the ethics of genetically altering and patenting animals.

A Conference on Ethics

Although some people view the use of animals as a necessary aspect in the development of biotechnology, others want to draw the line at mixing human genes with those of animals. In 1988, shortly after the Oncomouse was patented by researchers at Harvard University with funding from Du Pont, a veterinarian named Dr. Michael Fox decided to do something about it. As you read in Chapter 25, "Patents on Life," the Oncomouse is a transgenic mouse that was genetically manipulated to be susceptible to human breast cancer.

Dr. Fox organized a conference to discuss ethical issues related to the engineering of animals. Among the attendees at the conference, which was held at the Airlie House Conference Center in Virginia, were religious leaders and *bioethicists*. At the conference center, there were Catholics, United Methodists, Presbyterians and representatives of many other religions. Fox asked them to address two main issues: the creation of transgenic animals and the patenting of life.

After much discussion, the conference participants concluded that decisions about genetically altering animals to fulfill human needs shouldn't be made by corporations and scientists alone. They felt that there should also be public input, because the technologies affect everyone, not just those who are working in the field of biotechnology. They also came to the conclusion that new genetic technologies should not cause animals any suffering.

Standardized Animals?

Opponents of the use of animals in genetic technology caution that one result could be a decrease in biodiversity, the wide range of different forms of life that exist or will develop in the future. You read in Chapter 25, "Patents on Life," that it might be easier for researchers to obtain patents if they develop more uniform varieties of plants or animals. If this were to happen, then there might be herds of cattle or goats that are genetically similar—or if cloning becomes more widespread, genetically identical.

When there is less biodiversity, a species is a lot more susceptible to changes in the environment, such as climate fluctuations or new diseases. When there are natural variations within a species, there are usually some individuals that have a natural immunity to one disease or another. Other individuals continue to thrive even if the weather gets much hotter or colder. But uniformity within a species undercuts its natural resiliency to such environmental changes.

If a disease were to develop within a normal herd of cattle, for example, many of the herd would die out. But some would naturally be able to survive, and their offspring would carry that natural immunity to the disease. But theoretically, if an entire herd had the same or very similar genetic makeup, then a disease could conceivably wipe out the whole herd.

A Breed Apart

No one knows what will happen to genetically engineered animals in the long term, because they haven't been around for hundreds of years yet. But some people point out that if we look at what's happened to animals that have been selectively bred for a long time, that might give us a clue to what could happen with bioengineered animals in the future.

DNA Data

Many species of canine have been bred for generations to have a certain look or to perform a certain task. But sometimes breeding for these characteristics hasn't been genetically the best for the dogs. As a result of breeding, Labrador retrievers have a genetic susceptibility to dwarfism and hemophilia. German shepherds and Akitas often develop hip dysplasia (dis-PLAYZ-a), a hereditary weakness of the joints. Some opponents of the genetic engineering of animals caution that similar or worse hereditary conditions could result from adding foreign genes to animal species.

Even with selective breeding, unfortunate mistakes sometimes go along with the search for the perfectly bred animal. For instance, out of the approximately 20 million purebred dogs in the United States, one out of four has some sort of serious genetic problem. Dalmations, for example, are often deaf. About 70 percent of all purebred collies have genetic eye trouble, and female English bulldogs often need to undergo cesarean sections to deliver their litters because their pups' heads are too large to fit through the birth canal.

The reason for these genetic problems is the fact that these breeds have been developed over thousands of years and continue to be bred for a certain desired look or to perform a desired task. But what looks good may not be genetically the best for the animal's well-being. What happens if an animal is genetically altered in ways that harm its health or the overall hardiness of its species?

Because genetic engineering can be used to alter one specific gene, its supporters contend that there may be fewer problems than those that occur with selective breeding. Instead of randomly mixing the DNA of two animals that have desired characteristics, they say, biotech methods pick out a specific trait or two that are desired in the next generation.

Opponents answer that unpredictable results frequently occur when genes are mixed and matched. This would appear to be especially true for genetic engineering, which can add genes from one species to a completely unrelated one. These interspecies mixes could never occur in nature. If diseases and poor health have been the result in dogs that have been bred by selective breeding, then will the same or worse happen when researchers genetically cross unrelated animals or even animals with plants?

Those in favor of the genetic technology explain that all life is thought to have evolved from a common ancestor. Chapter 11, "Gene Gangs," pointed out that even yeast cells have some genes that are similar to human genes. All DNA is made up of the same building blocks—the same As, Cs, Ts, and Gs—although their order can vary tremendously from one species to the next. This view led the United States Food and Drug Administration (FDA) to decide not to require special labeling of genetically engineered foods.

But there is a concern that what works for one species will not necessarily work for another. Genes that work in a specific, predictable way in one type of plant or animal may not act as predictably when transplanted into the DNA of another species.

For instance, you read in Chapter 15, "Down on the Pharm: Engineering Animals," about the supermice that were developed by Dr. Ralph Brinster at the School of Veterinary Medicine at the University of Pennsylvania. When Human Growth Hormone was added to their DNA, they grew to nearly twice their normal size. But when researchers at the United States Department of Agriculture (USDA) put the Human Growth Hormone gene into pigs in an attempt to produce superpigs, they were in for an unpleasant surprise. Some of these altered pigs were sickly, impotent, cross-eyed, and arthritic. So it may be difficult or impossible to accurately predict how genes will work in a foreign species.

DNA Data

In 1992, the United States Department of Agriculture (USDA) decided to find out how the general public felt about the genetic manipulation of animals, so they funded a public opinion survey on some of these issues. The study found out that 53 percent of those polled thought it was "morally wrong" to genetically alter animals. Seventy percent said that putting animal genes into plants was "unacceptable." Close to 90 percent said that putting human genes into animals was "unacceptable."

Could mixing species cause health hazards for humans as well? You read in Chapter 15, "Down on the Pharm: Engineering Animals," about the AIDS mouse that was developed in 1987 by Dr. Malcolm Martin and a team of researchers at the National Institutes of Health in Maryland. In 1990, Dr. Robert Gallo, the co-discoverer of the HIV virus, wrote a paper warning of the possibility that the human AIDS virus could potentially combine with viruses found in mice.

A Great Escape?

What would happen if some of these genetically altered animals escaped from the laboratories where they were developed? What if they bred with naturally occurring species? Proponents of the technology state that engineered animals are restricted to laboratories under controlled conditions, but mistakes can and often do happen.

Consider fish that are raised in fish farms. These fish are not transgenic fish; they're regular fish raised primarily for the food market. An article in the *New York Times* on October l, 1996, explained that between 200,000 and 650,000 salmon escaped from fish farms in Norway in 1995. There are several ways that fish find their way from fish farms into the wide world of Norwegian rivers. During severe storms, it's hard to keep a good fish down. During several storms in Norway in 1990, about four million salmon made their escape.

Even though these fish are not genetically altered, their escape still spells problems for the naturally occurring fish in the area. Wild fish from different rivers are genetically diverse from one another. But when the farm-raised escapees come to town, they have love on their minds. With each mating between the farm fish and the wild fish, the wild gene pool could eventually be diluted.

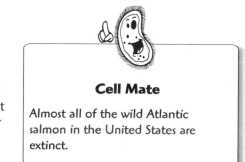

Cell Mate

Almost all of the wild Atlantic salmon in the United States are extinct.

Some people caution that the same thing could happen if genetically altered species were to mate with their wild counterparts. Because so many types of animals are being modified these days, could this mean major changes in future generations of sheep, goats, mice, fish, and other animals? Will this upset the environmental balance of nature? And if unexpected and negative results happen, how will researchers be able to retract their mistakes once they've spread to wild populations? Will this upset the environment balance of nature?

Holy Cow!

One of the greatest controversies over a genetically engineered product happened in November 1993, when the Monsanto company was granted permission by the U.S. Food and Drug Administration (FDA) to market *Bovine Growth Hormone* (BGH), also called Bovine Somatotropin or BST.

The naturally-occurring gene for Bovine Growth Hormone (BGH) is taken from the cow and inserted into E. coli bacteria. Then the bacteria produce large amounts of the hormone. This hormone can then be injected into cows to increase milk production.

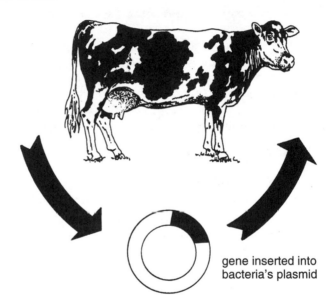

gene inserted into
bacteria's plasmid

Cows have a naturally-occurring hormone that causes their bodies to produce milk to feed their calves. Researchers took the gene that naturally codes for this substance and spliced it into bacteria. As with many of the genetically engineered substances you've read about, the process is simple. The bacteria are given a warm, cozy environment with lots of food. When they reproduce, they also reproduce what the extra gene codes for, in this case BGH. These mini-factories will keep making the hormone until the cows come home.

After the bacteria have produced a large amount of the hormone, they are killed off, and the substance is harvested and purified. The hormone is sold to dairy farmers, who inject it into their cows. This hormone increases milk production by about 25 percent.

When the FDA approved the sale of this engineered hormone, it had already been tested, and the FDA was confident that it was safe. But many consumers and some scientists are not so sure. They claim that there are consequences that need to be considered, including human health issues, animal welfare, and the economy of the dairy industry.

Is It Safe?

Those in favor of the use of the hormone point out that because it occurs naturally in cows, it is safe. However, opponents counter that when cows are injected with extra amounts of the hormone that it requires their bodies to use more energy than they normally would. Because they produce so much more milk, many of these treated cows develop mastitis, or udder infections.

Genetic Jargon

Bovine (BO-vine) **Growth Hormone** is also called BGH or rBGH (the *r* stands for recombinant). Others call it Bovine Somatotropin (SO-ma-toe-TROPE-in) or BST. This genetically engineered substance is injected into cows to increase their milk production.

To care for their cows with udder infections, dairy farmers must administer antibiotics. This worries many people. They fear that antibiotic residues may remain in the milk and dairy products that people eat and drink. Monsanto, one of the companies that now produces the engineered hormone, insists that federal and state regulations are adequate to monitor whether there are antibiotic residues in the milk of treated cows.

Dr. Michael Hansen, a research associate at the Consumer Policy Institute, conducted studies on the engineered hormone. He found a higher somatic cell count (SCC) in milk from treated cows. This is a measure of pus in milk.

Dr. Hansen also has concerns about a substance called IGF-1, also called insulin-like growth factor. He and Jean M. Halloran, director of the Consumer Policy Institute, wrote a letter to the commissioner of the FDA in 1993 to warn of a possible connection between IGF-1 and cancer. Hansen and Halloran cited studies that show an increase in IGF-1 in milk obtained from cows that are treated with BGH. They also cite a study that suggested that IGF-1 might survive human digestion and possibly be absorbed by the body.

Spilt Milk

In 1994, activist and biotech opponent Jeremy Rifkin and his Foundation on Economic Trends engaged in what they described as "Milk Wars." Through their Pure Food Campaign, they organized milk dumps in close to 200 cities. They poured containers of milk from hormone-treated cows onto the streets.

But despite the opposition, there are still some people who are in favor of the engineered hormone. Dr. M. Roy Schwartz of the American Medical Association said that use of the hormone could ensure that we have an adequate food supply, and Sara C. Parks, R.D., of the American Dietetic Association stated that the hormone does not change the composition of the cows' milk.

But critics point out that the United States already has a surplus of milk. They conclude that an engineered hormone that induces cows to produce more milk is not needed. What will happen, they ask, if more and more milk appears on the market? When prices are driven down, will the small dairy farmers be able to survive?

DNA Data

Bovine Growth Hormone, also called Bovine Somatotropin, has been banned in a number of countries. It cannot be used in Australia, Greece, Austria, Germany, Italy, Denmark, New Zealand, France, Belgium, Luxembourg, Ireland, Great Britain, the Netherlands, Spain, Portugal, or Israel.

The debate continues. The United States Food and Drug Administration is one of the few agencies in the world that has given its approval for the use of this hormone. In 1999, Canada formally decided not to allow the use of Bovine Growth Hormone to treat their dairy cows. Their decision came after considering the many pros and cons of the engineered substance. Among the points that they considered were animal welfare issues. Canadian officials decided against the hormone when they considered that there is a higher incidence of udder infections in treated cows and that these animals can also have a shorter life span.

The Least You Need to Know

➤ Although some scientists are considering using genetically engineered pigs as organ donors in the future, other scientists caution that pigs carry viruses that might infect humans.

➤ Some people think that engineering animals to produce human substances or to act as organ donors constitutes cruel and unethical treatment.

➤ Selective breeding has caused genetic problems in dogs, and some people worry that mixing genetic material from foreign species could cause greater problems.

➤ Another concern is that engineered animals will escape and interbreed with wild populations, depleting the variety of animals.

➤ A genetically engineered hormone called BGH or BST, which increases milk production in cows, has been welcomed by some and denounced by others.

Chapter 28

People Problems

In This Chapter

➤ Is DNA testing reliable?

➤ Should people be able to get DNA tested for incurable diseases?

➤ Will the future bring genetic discrimination?

➤ Should gene therapy be performed on unborn children?

➤ Will people order designer babies in the future?

The study of genes and DNA promises to help us understand more about how our bodies function in health and in disease. It could possibly lead to cures that will eradicate once-dreaded diseases. But along with new knowledge about genetics come some very thorny issues. The technologies have developed far ahead of the ethical issues that they raise.

For instance, who will have access to information about our DNA? If someone in your family decides to get tested for a genetic disease, and some negative results surface, what will that mean for you? Suppose you get a DNA test and discover that even though you're healthy, you carry a gene for a disease, or you have a susceptibility for a disease. How will that affect your job opportunities? Will employers deny jobs to someone who is genetically "flawed"? Will it decrease your chances of getting insurance?

If you or your spouse gets a prenatal test to see how your unborn child is doing, what will you do if there's bad news? If your fetus carries genes for a debilitating illness, will you choose to abort? If not, will your insurance company refuse to pay for your child's treatment?

Questions like these have no easy answers. There are many sides to the issues of DNA testing, genetic privacy, and other aspects of the Genetic Revolution that affect people's lives and the way they view themselves and others. Read on to find out how promising genetic technologies bring along a host of difficult questions that need to be answered.

How Important Are Genes?

To read the newspapers and watch TV these days, you'd think that all you need to do to understand who you are and what you'll become is take a DNA test. But some of these ideas are just the result of media hype. DNA has become a popular topic in our society, and genes have been elevated to an almost mythical status.

Not everyone thinks that genes are that important. Dr. Ruth Hubbard of Harvard University co-authored a book with Elijah Wald entitled *Exploding the Gene Myth*. In this book, Dr. Hubbard explains that we are not simply bits of DNA; we are complex organisms that are much more difficult to understand than some would lead you to believe. She writes, "The myth of the all-powerful gene is based on flawed science that discounts the environmental context in which we and our genes exist."

Dr. Hubbard does not agree with the popular view that our genes are entirely responsible for our future mental and physical health, who we are, or what we become. Rather, she worries that the recent announcements of discoveries of genes for this and for that may cloud the real issues that control our well-being. Hubbard says that the popular view holds genes responsible for conditions that may result from environmental or societal factors. She thinks that if we continue to consider genes to be the basic cause of things that go wrong in our lives and in the world, it will prevent us from addressing the actual causes of our problems.

For instance, if a company in the future learns that it has toxic chemicals in its workplace, will it look for workers with a genetic constitution that allows them to tolerate these chemicals for a longer period of time than most? If this scenario happened, the company might not hire workers who couldn't tolerate the chemicals. So rather than addressing the real issue of using toxins in the workplace, this company would be shifting the focus, and implicitly the blame, to their workers' genes.

If we were to accept as fact the notion that some people are genetically disinclined to succeed because they are programmed to have lower-than-average intelligence, then why would we try to help people in impoverished areas? Instead of trying to help them rise above their surroundings by instituting social programs to improve their schools and neighborhoods, we might just throw up our hands and say that no matter what we could do to help them, they would still be prisoners of their genes.

Cell Mate

In a poll published in *Time* magazine on January 11, 1999, participants were asked if they agreed or disagreed with the fact that the government is spending three billion dollars to map all human DNA. Fifty–five percent said they disagreed with this use of taxpayer money.

The same genetic blame game can be played with diseases, if we follow that line of thinking. Instead of cleaning up the environment to prevent the occurrence of certain diseases, we could "blame the victims" by claiming that they had a genetic predisposition to those diseases. But Dr. Richard Lewontin, a geneticist at Harvard University, gave another opinion in the Bullfrog Films video *Gene Blues—DNA Testing Dilemmas.* He said, "Most people in the world are not dying of genetic diseases. Most people are dying of overwork and undereating."

Difficult Predictions

Chapter 20, "Did Your Genes Make You Do It?" covered the nature versus nurture debate. This topic still fuels some of the most heated arguments today. The most logical view seems to be that we are shaped by interactions between nature and nurture. The importance of this interaction becomes evident when you consider the fact that if twins have the same gene for the same disease, one may develop the disease at an early age, and the other might get sick at a later age or not at all.

Even in the case of a serious hereditary illness such as Huntington's disease (which you read about in Chapter 19, "DNA Diagnosis"), no one can predict exactly when the disease will strike or how severe the symptoms will be. It's also impossible to pinpoint how long the course of the disease will run before a person dies from it.

The Role of Genes

The Council for Responsible Genetics (CRG) in Cambridge, Massachusetts is an organization that has a number of distinguished scientists on its board. In the Council's "Position Paper on the Human Genome Initiative," CRG states its view that the role of genes has been overemphasized in our culture.

DNA Data

In a book called *The DNA Mystique—The Gene as Cultural Icon,* authors Dorothy Nelkin and M. Susan Lindee explain how genes have taken on mythic proportions in our culture. From comic books to sitcoms to science fiction and beyond, the Almighty Gene has inspired awe in the eyes of the public.

The paper states, "The most that the complete sequence of an organism's genes can tell us is what proteins that organism can make. [It] cannot tell us how they will interact and operate together." It makes a comparison with a recipe that can tell you what goes into a certain type of meal, but doesn't guarantee that you'll get the same meal every time.

The paper explains that "Genes have their part to play in the ways people function, but they are always only part of the story." They worry that an overemphasis on the importance of genes might take funding away from what they consider to be more pressing problems in biology. They also caution that the view of genes as reliable predictors of the future could radically alter our concepts of normal health. Will people's DNA be examined routinely in the future? If so, will we view people with genetic "flaws" as a "genetic underclass?"

DNA Data

The Americans with Disabilities Act (ADA) protects people with physical disabilities from unfair treatment. The Equal Employment Opportunity Commission (EEOC) enforces the ADA. The ADA forbids prejudicial treatment or refusal to hire someone who has a disability or if the potential employer has knowledge that the person may become disabled in the future. Some people think that genetic "disabilities"—that is, having genes that might cause diseases that a healthy person may carry—should fall under this act.

Testing Tests

How reliable are DNA tests? If a person has a gene that is associated with a disease, does this mean that that person will definitely develop that disease? Some scientists debate this point. For instance, the Council for Responsible Genetics put out a position paper that comments on the Human Genome Project. It stated that although genes are important, the way that they function is affected by many factors, including a person's environment and what they eat.

Sometimes statistics may lead people to overestimate risks. For instance, Dr. Barbara Weber, Professor of Medicine and Genetics at the University of Pennsylvania Cancer Center, was quoted on this topic in *Time* magazine on January 25, 1999. Commenting on the statistic that one in nine women will develop breast cancer, Dr. Weber noted that this statistic doesn't mean that a woman has a one in nine chance of developing the disease tomorrow. She said that this statistic means that "over a lifetime of 85 years, one out of nine women will develop breast cancer." She added that the number one killer of women in the United States is in fact heart disease and, after that, lung cancer.

Dr. Ruth Hubbard also takes up this point in her book, *Exploding the Gene Myth*. She asserts that this much-quoted statistic refers to the *average* probability that a woman will get breast cancer and concludes that it "has little predictive power." She comments that the probability refers to a woman's chance over her whole lifetime, not just at one specific time in her life. Dr. Hubbard concludes that the probability of a woman dying from cancer by the time that she is 55 is only about 1 in 180.

She adds that another way of looking at this one-in-nine statistic would be to say that "all the women will die some time, and eight of them will die without ever having breast cancer." The one who does develop this disease, Dr. Hubbard says, still has a two-out-of-three chance of surviving.

Incurable Diseases

If you could take a test that would tell you whether you carry any genes for lethal diseases that have no cure, would you want to know the answer? And if the test results showed that you would definitely get sick and later die of an incurable condition, would that knowledge help you to accept your situation, or would it merely plunge you into a deep depression? These are the types of thorny questions some people are facing now. As more and more genetic tests become available, the number of people who will receive the bad news about their genes may increase.

An example is Nancy Wexler, who was profiled in Chapter 19, "DNA Disease Diagnosis." Wexler's mother died of Huntington's disease, a degenerative illness that is inherited. Presently, there is no cure for this fatal disease, even though the gene that causes the condition was found in 1993. Although Wexler has brought her compassion and boundless energy to people who have the disease in Venezuela and elsewhere, she herself does not discuss whether she has been tested for the disease.

Although DNA testing for Huntington's can tell a person whether they will eventually get the disease, no one can tell just when it will occur or how quickly or slowly the fatal disease will progress. Some people want to be tested because they are hoping to hear good news. If they find out that they don't carry the gene, then they know that they will not develop the disease later in life.

But what happens if these people find out that they do carry the gene? They will know for sure that they will develop Huntington's disease at an unpredictable time in the future. How will they respond to this news? Will they be gripped by panic every time they accidentally drop something, and think that this is the beginning of the slow progression of the dreaded disease?

Because of the emotional nature of this knowledge, genetic counselors should be consulted before a person decides to take such a test. But will all insurance companies pay for this counseling? What if a person who considers testing denies the possibility that they might receive bad news and goes through with the test, even though they are not psychologically prepared for the worst?

Mutant Misconceptions

Don't think that just because scientists have discovered the gene for a disease that this knowledge automatically leads to a cure. The gene for cystic fibrosis was discovered in 1989, and there is still no cure. There's been a test for sickle cell anemia since the 1970s, but no one has come up with a cure for this disease, either.

Genetic Privacy

One unsettling issue related to genetic testing is the question of genetic privacy. If a person takes a test that reveals that they have a gene for a condition that may or may not develop in the future, will their employers or insurance companies gain access to this information?

On October 25, 1996, the journal *Science* published the results of a study conducted by the American Association for the Advancement of Science (AAAS) on some of the ethical issues that revolve around genetic testing. They studied the attitudes of 332 members of genetic support groups. These participants came from families with about 100 different genetic disorders.

Twenty-five percent of the people surveyed with family members with diseases felt that they were denied life insurance because genetic diseases exist in their families. Twenty-two percent of the participants in the survey said that they were denied health insurance for the same reason. Thirteen percent thought that they were either denied jobs or lost their existing jobs because of genetic issues in their family.

DNA Data

If you think we have possible problems with genetic privacy in the United States, then consider this: An entire country is having its genetic information sold to a corporation. In December 1998, Iceland's parliament passed a law that gave a biotech company, deCODE Genetics, the right to compile a massive database from the genealogical records, health records, and DNA profiles of every single person in the country. Medical records will be gathered on the basis of presumed consent—that is, from everyone who does not specifically deny their consent. Blood samples will only be gathered from those who give their consent. The data from the 270,000 Icelanders will be used to try to understand the genetic basis of disease and find the answers to other questions. Because there has been little immigration to the island, most of the population's DNA has largely been unchanged for a thousand years.

In addition, nine percent of the participants or their family members refused to undergo genetic testing because they feared that the resulting information might lead to genetic discrimination. Eighteen percent refused to reveal genetic information to insurance companies due to worries about discrimination. Seventeen percent did not offer any genetic information to their employers for the same reason.

A Call for Legislation

Although the goal of genetic testing is to help people understand which diseases they are at risk for or to put their minds at ease by informing them that they do not carry a mutant gene that runs in the family, that information could get into the wrong hands and be used to put people into unfair categories. Will there be a genetic underclass in the future? Will people who carry just one copy of a mutant gene be labelled as unfit for jobs or insurance policies?

To respond to these and other questions, on January 20, 1998, Vice President Al Gore released a government report, "Genetic Information and the Workplace," which explained the current and future problems that could arise from the misuse of genetic information. Gore also called for federal legislation to stop employers from treating their employees unfairly on the basis of their genetic tests.

DNA Data

A Harris poll conducted in 1995 showed that more than 85 percent of Americans are concerned that employers or insurance companies could have access to information about their DNA.

Banking on Genetics

In 1992, the United States Army started collecting blood samples from new recruits as part of their genetic dog tag program. The Pentagon's goal is to collect DNA samples from all of the two million active servicemen. They're aiming to complete this task by the year 2000.

The Army initiated this program so there would never be another unknown soldier. (Chapter 17, "Stories Genes Can Tell," told about the DNA testing and identification of the remains of one unknown soldier.) But not everyone supports the Army's plan. Organizations such as the American Civil Liberties Union (ACLU) considers DNA data banks to be "intrusive government surveillance" and a violation of people's right to privacy.

Cell Mate

In a *Time* magazine poll reported on January 11, 1999, people were asked whether they thought insurance companies should be allowed to access their genetic records without their permission. Ninety-four percent responded no.

A Germ of an Idea

Another touchy issue is the question of how much scientists should tinker with human genes. Chapter 22, "What Is Gene Therapy?" explained how Dr. W. French Anderson used gene therapy to successfully treat Ashanti DeSilva and Cindy Cutshall for ADA deficiency in 1990. ADA deficiency is a rare inherited disorder, which left the girls virtually without immune systems. Although Dr. Anderson did not totally cure the disease, the therapy has at least partially improved the quality of the two girls' lives.

If these girls decide to become mothers when they grow up, their modified genes will not be passed on to their children. This is because the cells that were genetically altered in these two girls are *somatic cells*, or body cells. Somatic cells are all the cells in the body except for the reproductive cells. The sperm and eggs are called *germ cells*. If gene therapy is ever performed on the germ cells, the altered genes will be passed on to the children, grandchildren, and all subsequent descendants of the original patient.

Genetic Jargon

Somatic cells are all the body cells except for the reproductive cells. **Germ cells** are the reproductive cells, the sperm and eggs.

When modifications are made to the germ cells, this is known as *germ line gene therapy*. So far, no one has performed this kind of therapy. But because it would affect all of the generations that follow the patient who receives the genetic alteration, it raises some very complex questions.

Prenatal Patients

In 1998, Dr. Anderson said that he would eventually like to perform gene therapy on embryos and fetuses that have been diagnosed with ADA deficiency or alpha-thalassemia, a blood disease. (You read about thalassemia in Chapter 9, "Irregular Genes," and Chapter 22, "What Is Gene Therapy?".) Dr. Anderson and others have submitted their preliminary ideas on a new type of gene therapy to the Recombinant DNA Advisory Committee (RAC) at the National Institutes of Health (NIH).

What Anderson and his team hope to do is to use viruses as vectors (delivery systems) to get the new, correct copies of the genes into the cells. They would like to alter the genes of affected fetuses that have been developing for between 13 and 15 weeks. Anderson and his colleagues chose this time frame because experiments with animals suggest that when genes are manipulated during the early development of a fetus, the procedure might be more successful than if the same procedure were performed later on. But this type of gene therapy could lead to irreversible changes in the reproductive cells of the fetus, and this possibility raises many ethical questions, because such changes would affect all of the descendants of that person-to-be.

DNA Data

The Recombinant DNA Advisory Committee (RAC) was established in 1974 to address public concerns about the safety of using recombinant DNA (genetic engineering) techniques. The RAC advises the director of the National Institutes of Health on matters that concern genetic technologies. The committee has 25 members. Two-thirds of them are scientists. The remaining one-third of the members represent the interests of the public.

Evolving Errors?

If researchers can develop therapies that could help reverse the negative effects of genetic errors, then why would anyone argue against it? Wouldn't getting rid of severe diseases benefit humankind? Why would anyone oppose a procedure that could reduce the level of human suffering?

The answer is that human germ line gene therapy might have unexpected repercussions for many generations to come. The Human Genetics Committee of the Council for Responsible Genetics (CRG) wrote a position paper on this topic. It states that even when researchers believe that a gene is responsible for a specific disease, the way that gene affects that person remains unpredictable.

The paper stated that scientists do not understand enough about how genes will affect a person to be able to guarantee that modifying genes will result in the successful elimination of a specific disease. It also points out that even if researchers could begin such a program, it would take thousands of years to get rid of the genes implicated in some diseases and would entail conducting a widespread, mandatory program of gene therapy to get these results. And what if new diseases arose?

The CRG paper also points out that if parents carry certain genes that are associated with severe disorders, options other than germ line gene therapy can be used. For instance, prenatal tests already exist for some diseases, such as cystic fibrosis. (Chapter 24, "Sex Techs: Reproductive Technologies," covered these tests.) Of course, the results of these tests might lead some people to

Cell Mate

In a *Time* magazine poll published on January 11, 1999, 62 percent of respondents said that they believe the government should regulate gene therapy.

choose abortion. Other methods, such Pre-implantation Genetic Diagnosis (PGD), which is used with in vitro fertilization, allow researchers to select a developing embryo that does not carry the mutant gene or genes and implant it into the mother-to-be.

The Perfect Baby?

The Council for Responsible Genetics also cautions that germ line gene therapy could result in people wanting to pick desired traits for their children that have little to do with health. In the past, certain people have advocated "improving" humankind by eugenics. Sir Francis Galton, a cousin of Charles Darwin, coined the word *eugenics* from the Greek words meaning wellborn. The idea was that by encouraging "superior" people to have more children and discouraging "inferior" people from reproducing, human beings would become stronger and smarter over generations.

Mutant Misconceptions

If you think that the idea of a "superior race" started with the Nazis in Hitler's Germany, think again. Hitler got some of his disturbing ideas about genetic control from the United States' eugenics movement.

There was a strong eugenics movement in the United States in the beginning of the twentieth century, and by 1931, about 30 states had involuntary sterilization laws. The involuntary sterilization law in California wasn't repealed until 1979. The idea of genetic superiority and inferiority also led to the U.S. Immigration Restriction Act of 1924.

DNA Data

In 1904, the Eugenics Records Office (ERO) was established in Cold Spring Harbor in Long Island, New York. The purpose of the establishment was to study inherited disorders. As well as studying disorders such as Huntington's disease, the staff of the ERO also sought the genetic basis for traits such as vagrancy, needlework skills, and seafaring ability. These days, geneticists do not believe that there is a genetic basis for such behaviors.

In her book *Exploding the Gene Myth*, Dr. Hubbard explains her concerns about what she dubs "the new eugenics." She comments that prenatal testing is becoming more and more routine. She adds that if people place too much stress on the influence of heredity, this may keep them from dealing with and fixing environmental, social, economic, and other problems that play a large role in the development of diseases.

Selecting Sex and Other Traits

Chapter 24, "Sex Techs: Reproductive Technologies," explained there are already tests that can tell you whether your unborn child will be a boy or a girl. There are even ways to choose the gender of a child-to-be before the sperm fertilizes the egg. These technologies were originally developed to prevent the occurrence of severe sex-linked genetic diseases. But some parents might use this technology for what is called "family balancing"— or having both boys and girls in the familiy. Also in some countries, male children are explicitly preferred over females, so this technology might be to ensure that a mother gives birth to a boy.

In the future, will the selection of gender, eye color, hair color, height, and intelligence be available for the choosing? Will there be culturally dictated standards for what constitutes a more "valuable" child? Who will get to pick their children from a genetic "catalog" in the future? Will only the very rich be able to afford the "best" genes that money can buy for their unborn children?

A Genetic Underclass?

Prenatal DNA testing aims to screen unborn children for diseases that will develop later on. Adults can also be tested for existing diseases or for genetic predispositions for some illnesses. But when test results show that a severe or incurable disease will develop, what are the options?

For instance, if a pregnant woman learns that her unborn child carries the gene for a serious disease such as cystic fibrosis, then what are her choices? Because there is no cure for cystic fibrosis at this time, she might opt to have an abortion. But what would happen if she decided to go through with the pregnancy and to accept the fact that her child will be born with the disease? Some people feel that abortion is wrong under any circumstances, so they might decline genetic testing altogether. How would this decision affect the family's health insurance? Would the mother be stigmatized for making such a decision? Would some people argue that she made the wrong decision and should have aborted the fetus?

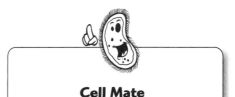

Cell Mate

According to a Gallup poll in England, 18 percent of respondents would use genetic enhancement, if it existed, to remove their unborn child's predisposition to alcoholism, 10 percent would be in favor of genetically preventing a child from becoming a homosexual, and 5 percent would use gene therapy to make their children better-looking.

Cell Mate

Who should control which genetic experiments get carried out and which do not? Although the government has some control over scientific experiments, it usually influences only those proposed by researchers who apply for government funding. If private institutes or individuals want to conduct experiments on their own, they might not have to confer with government agencies for approval.

Who has the right to decide whether an unborn child is "good enough" to live? Some critics fear that the new technologies might be misused in the future. What would happen if people decided that they would abort all fetuses that were not male, 6'3" or taller, blond-haired, and blue-eyed?

What happened in Hitler's Germany occurred long before the latest genetic technologies were developed. What might have happened, some critics wonder, if someone with Hitler's political aims had the new genetic technologies available to them?

A Life Worth Living

In the video *Gene Blues*, a woman named Lynne Morrow explained that she never wanted to give birth to a child who would suffer from sickle cell anemia. As an African American, Ms. Morrow was aware that she might pass this gene on to her child. So when she became pregnant, she had her unborn child tested for the disease. The results were negative, and Ms. Morrow happily carried her child to term.

But when Ms. Morrow's daughter Khalilah was born, doctors discovered that she did have a mild case of sickle cell anemia. Ms. Morrow subsequently sued the hospital that had conducted the prenatal tests. Nonetheless, she loves her daughter and is happy that she was born. "Having a genetic disease is not the tragedy that you might think," she commented. Her daughter said that it's true that she gets sick at times, but so do other people.

In the same video, Dr. Marsha Saxton of the World Institute of Disability comments that people with disabilities "…in the 1990s are being welcomed into mainstream society, and yet at the same time, the message we get from genetic screening technologies is that our lives are a mistake and should've been eliminated before we were even born."

All geneticists agree that every one of us carries some genes that may not be considered perfect, or could even lead to disease. But with the new tests, does this mean that some people's lives will be considered not worth living? When considering these issues, we should keep in mind what Dr. Mary-Claire King of the University of Washington said in the *Gene Blues* video. She reminds us that there is "an enormous number of different ways to be human," and we should "celebrate the genetic diversity of our species." (You can read more about Dr. Mary-Claire King in Chapter 17, "Stories Genes Can Tell.")

The Least You Need to Know

➤ Some people think that genes and gene testing are not reliable predictors of disease.

➤ Now that genetic testing is becoming more widespread, we must decide as individuals and as a society what the most ethical way to use this genetic information is.

➤ Some people worry about the possibility of genetic discrimination by insurance companies and employers.

➤ Dr. W. French Anderson wants to perform prenatal gene therapy, but some people worry about the repercussions of this therapy for future generations.

➤ Some people are concerned that genetic testing in the future may lead people to choose "perfect" sperm and eggs for their children before they are born.

Chapter 29

Should We Clone Around?

> ## In This Chapter
>
> ➤ Should humans be cloned?
>
> ➤ Why some people may want to clone themselves in the future
>
> ➤ Why some people want to ban human cloning
>
> ➤ Is it scientifically possible to clone humans?

If they made a clone of you, would you be beside yourself? Does the prospect of identical copies of human beings excite you or frighten you? Could copies of yourself be put to use to help you straighten out your files, to get all those jobs around the house finished, or to go out with two different people at the same time when you can't decide which one you want to date?

Although animals have been cloned for a long time, Dolly the sheep made history when her birth was announced to the world in the February 27, 1997 issue of *Nature* magazine. What made Dolly so special was that she was created from the DNA of an adult sheep. Some people think that this brings us one step closer to the day when humans can be cloned.

Although science may one day allow us to make an unlimited number of copies of ourselves, the ethical questions that surround this issue still have to be worked out. Some people consider cloning to be just an extension of reproductive technologies such as in vitro fertilization (test tube babies), which you read about in Chapter 24, "Sex Techs: Reproductive Technologies."

But others think that it's unnatural and unethical to raise someone in the exact likeness of another person. What would happen if this technology got into the hands of someone with political or even just egotistical motives? To find out what different people are saying about this issue, read on and decide for yourself whether cloning is a blessing-to-come or future double trouble.

Dividing to Multiply

You read in Chapter 16, "Send in the Clones," how scientists have taken animal embryos and multiplied them. In 1993, Jerry Hall and Robert Stillman, two researchers from George Washington University in Washington, D.C., took 17 human embryos and divided them into 48. (They deliberately used defective embryos, so they never grew into full-blown human clones.)

Then, of course, there was Dolly, the famous sheep that was cloned from DNA obtained from the mammary cells of an adult sheep. For years, scientists believed that this could never happen because most of the genes in our cells are "turned off." In other words, they can't be made to function, because you don't need genes that code for eyes to get turned on in your back so you could see where you've been already. But Dolly's birth proved that genes that have been turned off can be turned on again. Ian Wilmut, the embryologist at the Roslin Institute in Scotland who was responsible for Dolly, achieved what many considered to be an impossible feat.

The cloning experiments continued. Fifty mice were cloned in Hawaii. Researchers at the Kyunghee University Hospital in Seoul, South Korea, took DNA from a woman's cells and put it into her own eggs. Early embryos started to develop before the re- searchers stopped their growth to conform with their country's laws.

Here's a recap of the cloning that's gone on so far:

➤ In 1952, Robert Briggs and Thomas King cloned frogs from the cells of tadpoles.

➤ In 1962, John Gurdon repeated Briggs and King's experiment but used cells from older tadpoles.

➤ In 1993, Jerry Hall and Robert Stillman, researchers at George Washington University in Washington, D.C., experimented on 17 human embryos. By creating an artificial coating and teasing the cells apart, they divided them into 48 embryos.

➤ In 1996, Ian Wilmut, an embryologist, cloned Dolly the sheep from the mam- mary cells of an adult ewe at the Roslin Institute in Scotland.

➤ In 1997, Teruhiko Wakayama, a postdoctoral student at the University of Hawaii working with laboratory director Ryuzo Yanagimachi, cloned 50 mice by directly injecting their DNA into cells that had their own DNA removed. This becomes known as the Honolulu method.

➤ In 1998, researchers at Kinki University in Nara, Japan, cloned eight calves from the DNA of a cow.

➤ In 1998, researchers at the Kyunghee University Hospital in Seoul, South Korea, put DNA from a woman's cells into her eggs, and embryos started to grow. The experiment was terminated at an early stage because South Korean laws restrict experiments on human embryos.

Because of these advances in genetic technology, some people think successful human cloning might be carried out in the near future. An editorial in the January 9, 1999 issue of *The Lancet*, a prestigious British scientific journal, stated, "the creation of clones of human beings is inevitable."

The future procedure for human cloning would likely be similar to the one used to clone Dolly the sheep. You read about this method, called nuclear transfer, in Chapter 16, "Send in the Clones." Researchers would remove the nucleus from a woman's egg and discard it. (Recall that the nucleus contains the genetic material.) Next, the researchers would take DNA from the cells of the person who's going to be cloned and put that DNA into the "empty" egg. When an embryo starts to develop, it could be implanted into a woman, who could carry and later give birth to the child.

Reasons to Replicate

Why on earth would some people want to make copies of themselves? Bioethicists who study issues related to cloning feel that there may be some situations in which people might look to cloning to serve their physical or emotional needs.

Cloning for Kids

Some people consider cloning to be a natural extension of the assisted reproductive technologies that you read about in Chapter 24, "Sex Techs: Reproductive Technologies." John Robertson, who writes frequently about law and reproduction, thinks that cloning should be considered as an alternative for infertile couples. In his essay, "Cloning as a Reproductive Right," which appeared in *The Human Cloning Debate*, edited by Glen McGee, Robertson stated that cloning in the future may play an important role in providing infertile couples with children who are genetically related to one of them. He says that families who decide to clone should be allowed to do so, because they have already made a commitment to raise a child. He considers the decision about whether to have children to be one of our fundamental human rights.

Robertson adds that although some genetic selection techniques, such as sex selection, exclude certain traits, he believes cloning to be a more positive technology, because it picks out the genetic traits that parents want, rather than excluding traits they don't want. However, Robertson is opposed to human cloning when the DNA donor doesn't want to raise the child or wants to use the clones as workers like those in the nightmare scenario of Aldous Huxley's *Brave New World*. Robertson feels that this type of cloning should be banned.

Another situation where people might want to clone themselves would be in the case of homosexual couples who want to raise a child that is genetically related to one of them. If a lesbian couple, for example, wanted to raise a child that was biologically related to one or both of them, cloning might offer an option. Male homosexuals would need to have an egg donor and a surrogate mother to carry the child to term in order for cloning to work for them.

Clones for Cures

Another scenario that might make some people consider cloning would be in the case of parents who have a dying child. Suppose their daughter needed a bone marrow transplant or a new kidney. If they cloned her, the clone would have the identical DNA as the sick child. Then they could use the clone's bone marrow or kidney for a transplant. Because the new sister would have the same DNA as the sick child, there would be less concern that the transplanted organ or bone marrow would be rejected by the sick child's immune system.

But what would happen if someone wanted to raise a clone expressly for the purpose of using an organ that they couldn't live without, such as a replacement heart? To create a human clone with the intent of sacrificing that person later would certainly be unethical, and many people would want such a practice banned. Some people even consider the whole scenario a ludicrous one.

Cell Mate

In 1999, the Israeli Knesset, the Israeli parliament, passed a law placing a five-year moratorium on human cloning or the creation of a human through germ line gene therapy. However, the law does not ban research and development of cloning technologies.

DNA Data

Cyber Dialogue, a New York City online research firm, conducted an online cloning survey of 364 respondents in March 1998. Two-thirds of the participants were opposed to the cloning of humans. Thirty-eight percent were very opposed and said cloning should be banned. Twenty-nine percent thought that cloning efforts should not be banned, but should be kept under control. Sixteen percent were in favor of human cloning. Of the people who were opposed to cloning, 34 percent said they were concerned about government or business misuse of the technology, 18 percent had ethical reasons, and 16 percent cited religious reasons.

Some people envision a similar scenario where parents have a dying child, but in this case, the disease is incurable. The grieving parents might want to have a clone made to "replace" the child that they know they will inevitably lose to the sickness.

Clones for Continuity

Another possible lure of cloning could be the feeling that a person can in some sense be "immortal"—or at least get a lot of mileage out of their DNA—by cloning themselves. Some people might feel that this transfer of identical DNA would be something like living forever, although they will age and die themselves. Many people would see this as an egotistical reason to raise a child. This kind of motivation raises concerns that the clone would not be seen as an individual, but as a mere extension of the DNA donor.

A No Clone Zone

Many people think that cloning should be banned altogether. They say that it reduces human beings to their component parts, their genes and DNA, and makes them into standardized beings. They worry that eventually all individuality might be lost and point out that alternatives such as in vitro fertilization already exist for infertile couples, obviating the need for cloning technology for reproduction. Of course, infertile parents also have the option of adopting one of the many children who need families.

Leon Kass, a professor at the University of Chicago, writes about this topic in "The Wisdom of Repugnance—Why We Should Ban the Cloning of Humans," in *The Human Cloning Debate*. Kass says of the prospect of cloning in the near future, "I insist that we are faced with having to decide nothing less than whether human procreation is going to remain human . . ." He calls the possibility of cloning oneself "narcissistic recreation."

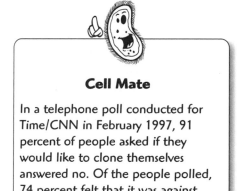

Cell Mate

In a telephone poll conducted for Time/CNN in February 1997, 91 percent of people asked if they would like to clone themselves answered no. Of the people polled, 74 percent felt that it was against God's will to clone humans, and 65 percent thought that the federal government should regulate the cloning of animals.

Others fear that cloning will change the way families function. Instead of a mother and a father, a clone would have just one parent who donated his or her DNA to create the next generation.

In addition, if cloning were to become a popular form of baby-making, there might be a new market for donor eggs to receive DNA in the cloning process. Today, some infertile couples use donor eggs that can be fertilized by the sperm of the husband, so the resulting child will be biologically related to one parent. Will cloning lead to a bigger market for eggs to be fertilized? Does this turn women's reproductive capabilities into marketable commodities?

Moaning About Cloning

Will cloning lead to the end of marriage? Will some women without partners, tired of looking for a compatible man to share their lives with, decide to clone themselves so

349

they can have children? Will this lead to a world where women's liberation takes on a radical new meaning, insofar as women will be literally liberated from needing men to reproduce? Will cloning usher in a world in which women feel that men are useless?

(Some women feel this way about men already.) One view of cloning paints the exaggerated scenario that if women only cloned themselves, then there would be no more men.

If human cloning catches on in the future, who will reap the benefits of this reproductive technology? Will having a child identical to oneself be a luxury reserved only for those who can afford to pay for this high-tech solution to infertility?

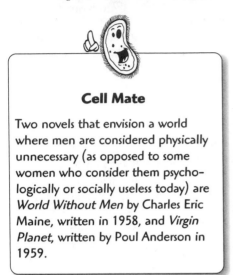

Cell Mate

Two novels that envision a world where men are considered physically unnecessary (as opposed to some women who consider them psychologically or socially useless today) are *World Without Men* by Charles Eric Maine, written in 1958, and *Virgin Planet*, written by Poul Anderson in 1959.

Genetically Blended Families

Would human cloning add further confusion to the convoluted relationships that already exist in some families? Many of us have lived in blended families, in which half-siblings share the same mother or father, but not both, and many adopted children long to find out about their biological mother and father.

If a single woman cloned herself, this "mother" would in fact be the clone's identical twin, because they have the same DNA. But the twin is raising the clone, so they have a different relationship than natural twins who are raised as siblings.

Suppose this clone wants to know about her biological roots. She knows who her mother/twin is. But who is her father? One simple answer is that she has none, because she was created from female donor DNA obtained from an adult cell and an egg that had its own DNA removed and replaced. (This is assuming she was cloned by nuclear transfer, the procedure used to create Dolly the sheep.)

You know that the father of a child is simply the male whose sperm fertilized the egg to give rise to the resulting embryo. Half of that developing embryo's chromosomes are supplied by the father's sperm, and the other half are supplied by the mother's egg. So even though it's true that the clone has no father in the traditional sense, she could still look to her mother's roots to find out more about her ancestry.

This is where it can start to get really confusing. The DNA donor, the "mother", who technically is an identical twin born many years before the clone, got half of her genetic material from her mother and the other half from her father. So in one sense, the "father" of the clone, the man whose sperm provided half of her DNA, would be her grandfather. So should she call him daddy or grandpa? In a sense, the woman she calls "Mom" could even be considered her aunt, because that woman is the daughter of the clone's grandfather. Things get bewildering very quickly in the family that clones.

Would this confusion take place if human cloning became popular? Or does the actual role that a person plays in the life of a clone mean more than the technical genetic relationship? These questions have no simple answers. In the future, if and when human cloning becomes legally and scientifically possible on a large scale, you will probably get some highly complicated family reunions.

DNA Data

On December 8, 1998, regulatory agencies in the United Kingdom published a report on their recommendations for human cloning in that country. The report was written by the Human Genetics Advisory Commission, which advises the British government on issues relating to genetics, and the Human Fertilisation and Embryology Authority, which regulates assisted reproductive technologies. They advised that all forms of human cloning be banned in the United Kingdom. But the report also recommended allowing researchers to experiment with therapies for mitochondrial diseases and diseased organs in humans.

Clone Control

Shortly after the February 27, 1997 issue of *Nature* magazine came out, so did Ian Wilmut's secret. Dolly had been born months earlier, but researchers at the Roslin Institute had kept things quiet, just in case Dolly didn't develop normally. As soon as the news was out, the cloning debate began in full force.

In response to the news, President Clinton called for a moratorium on using federal funds on human cloning research. He also asked for a report from the National Bioethics Advisory Commission within 90 days to give its recommendations for or against tighter control of the technology. The 107-page report, "Cloning Human Beings," came out a few weeks late. The committee recommended that human cloning should be banned. This recommendation came as no surprise, considering the controversial nature of the topic.

Since then, some states have passed laws to ban human cloning. California became the first to do so on January 1, 1998. Other states are considering doing the same. But legal analysts believe that there could be ways around these laws if someone really wanted to create human clones.

Identical, but Not the Same

One futuristic scenario that some people fear is that cloning could lead to egomaniacal dictators making genetic "photocopies" of themselves to carry on their evil plans or to raise whole armies of clones. Another fear is that whole herds of human clones could be created as organ donors to be sacrificed and harvested for privileged people who need their hearts or livers or lungs.

Some scientists say that these scenarios are just the stuff that pulp science fiction is made of. They point out that even though a clone's DNA is the same as that of the DNA donor, this doesn't mean that the clone will grow up to be exactly the same type of person. Environment is also an important factor in shaping personality.

You read about Thomas Bouchard, who conducted the famous 1988 Minnesota Study of Twins Reared Apart, in Chapter 20, "Did Your Genes Make You Do It?" Even Bouchard is convinced that cloning will not produce people with identical thoughts and behavior. He said that this is one reason why the thought of cloning does not bother him.

Twins Out of Time

So it looks like we're back to the old debate about nature versus nurture, but with a new twist. The clone would be an identical twin born in a different time and maybe a different place than the first one. They won't have the same sisters or brothers or other relatives around as they grow up, and they'll be raised in a world that will have more technological advances than the first "twin."

If Einstein were cloned, for example, the new baby would have to be born at the same time and in the same place to grow up to be just like him. In the 1976 novel *The Boys from Brazil* by Ira Levin (which was made into a movie in 1978), neo-Nazis try to raise a bunch of Hitler clones, but it just doesn't work. Because they aren't raised under the exact same circumstances as the original Adolf Hitler, the clones don't turn out as evil as he was.

Imagine if you were raised in the 1920s or the 1950s or the year 2200. You'd have the same genetic tendencies, but you'd be living in different types of houses, driving around in different types of cars, and in all probability eating different types of food. Of course, you'd also have different friends and would grow up in different societies.

Consider that different cultures have a variety of ways of looking at people's physical appearances and personalities. What is accepted in one society and era may be unacceptable in another. Just think of all the personal ads that describe a woman as "Rubenesque." In the sixteenth century, when the Flemish painter Peter Paul Rubens lived, hefty women were considered the pinnacle of sensuality and beauty. But they'd never get to walk down a runway and model new fashions in the twentieth century.

Cell Mate

The 1989 novel *The Cloning of Joanna May* by Fay Weldon was turned into a television miniseries for British viewers. The story centered around a man who leaves his wife when he discovers she has been unfaithful to him. But he clones her before divorcing her, because he hopes that he can get an identical replacement for her in the future.

A Fatherless Child

Glen McGee, Associate Director for Education at the Penn Center for Bioethics, writes in his book *The Human Cloning Debate*, "A clone is the product of an odd, high-tech configuration of various biological products, which produces offspring that is 'identical' to none of the donors."

Glen McGee of the Penn Center for Bioethics. Photo Credit: Tommy Leonardi Photo courtesy of the Penn Center for Bioethics

Consider the fact that even though the DNA taken from the donor cell's nucleus will be identical to the clone's, the egg that the DNA is put into will be different from the one that the donor was formed in. This brings us back to mitochondrial DNA, which we discussed in Chapter 4, "Me, My Cell, and I." The mitochondria are like tiny powerhouses in the cell. You'll remember that they have their own DNA, but it's not quite the same as the DNA in the cell's nucleus. Mitochondrial DNA is passed down only from the mother to her offspring. This may be a small difference, but it's a difference just the same. Within the context of cloning, if there are defects in the mitochondrial DNA in the egg cell that is used, a genetic disorder could develop in the clone, even if the donor DNA from the parent cell's nucleus was chosen for its "superior" qualities.

The pregnancy that leads to the clone's birth also needs to be considered. If the mother—or surrogate mother, depending on who is going to give birth to the clone—eats well and gets plenty of rest, the fetus will obviously develop in a much healthier way than if the mother of the clone drinks, takes drugs, and eats nothing but Danish pastries for the nine months that the child is developing inside her.

353

So all these factors come into play. It may be difficult or impossible to replicate the exact conditions that led to the birth of one unique individual.

DNA Data

Some critics worry that being a clone would be a psychological burden. They think the clone might feel like less of an individual. They might be unfairly compared to their DNA donor. And would it be psychologically damaging for clones to look at their parent's wrinkles and envision what they'd look like in the distant future? Would they worry about developing the same diseases, or growing up to be as neurotic as their parents later in life?

Is Human Cloning Possible?

One important aspect of the cloning debate is the fact that some scientists still question whether human cloning will ever be possible. Although Dolly the sheep is a successfully cloned mammal, skeptics point out that hers was no easy birth.

Out of 277 adult cells that gave their DNA to "empty" eggs, only 29 embryos lasted longer than six days. Only 13 pregnancies resulted, and out of these, Dolly was the only lamb to make it through to a live birth. Most people agree that with odds like that, the technological state of cloning isn't so great for sheep, so it could be even less effective for humans.

Another concern that some people have is the fact that if researchers embark upon the cloning of humans, that in the beginning stages at least, the technology might produce a lot of mishaps. This could mean not only miscarriages, which would be tragic enough, but it might also mean the creation of deformed children before the technology is perfected. Most people would consider it unethical to conduct cloning experiments while disregarding the possible negative effects.

In 1997, Richard Lewontin of Harvard University wrote in *The Confusion Over Cloning* that a cloned embryo could develop in a different way than an average embryo. He explains that when a regular embryo's cells divide, they're perfectly timed to produce complete,

Mutant Misconceptions

Don't think that by cloning Dolly the sheep, Ian Wilmut was trying to perfect a method for cloning humans. Wilmut is very much against human cloning. Dolly's birth was the result of Wilmut's search for an effective method to reproduce the same type of animal over and over again. After genetically engineering sheep or other animals to produce human substances in their milk, for example, cloning could be used to make copies of this type of valuable animal.

accurate copies of all the chromosomes. But because the nucleus of clone cells comes from another source, Lewontin cautions that things might work differently.

With cloning, an egg has its genetic material removed, and the "empty" egg is fused together with another cell that has a nucleus with all of its chromosomes. Lewontin thinks that the donated chromosomes might divide out of step with the egg cells in the embryo. If this were to happen, he says, there could be missing chromosomes or extra chromosomes, and this could lead to an abnormal embryo that could die. But in some cases, a fetus might survive and be born with abnormalities.

DNA Data

Some researchers worry that an all-out ban on human cloning will thwart their efforts to work with the possibility of cloning human organs. This would involve using something called stem cells, which are undifferentiated, meaning that the cells haven't decided what they want to be when they grow up. Researchers hope that by using cloning technologies, they might be able to create new organs for needy patients.

A Bad Seed?

In January 1998, a physicist with the unlikely name of Dr. Richard Seed went on National Public Radio and bared his intentions to set up a human cloning clinic in the near future. He said that if he could raise the necessary funds, he would open up a commercial cloning clinic in the Chicago area, where he lives.

The now notorious Dr. Seed previously owned a company that transferred embryos in cows. Then he and his brother Randolph, a surgeon, started a company that would use the same embryo technique in women. But the procedure never took off, because in vitro fertilization became popular. Dr. Seed lost his house and several million dollars.

Dr. Seed told interviewers that he had four anonymous clients who already expressed a desire to use his future cloning services. Dr. Seed stated that in his opinion, cloning is "the first serious step in becoming one with God."

Cell Mate

Many religious leaders are opposed to the cloning of humans on religious grounds. For instance, the Reverend Albert Moraczewski of the National Conference of Catholic Bishops called attempts at human cloning "playing God," and Dr. Abdulaziz Sachedina, a representative of the Islamic faith, said that human cloning is at odds with Islamic teachings.

But some experts doubted that he could raise the funds to start up his clinic. And if he did, would he be able to overcome all the scientific and legal barriers, they wondered? Dr. Seed's answer to this question was that if the U.S. government put an all-out ban on human cloning, he would seek another location that would allow it, perhaps in Mexico or the Caribbean.

In November 1998, just when most people thought that Dr. Seed had faded into oblivion, he announced his intention to set up a center in Japan. Although there are no specific laws that ban the cloning of humans in Japan, Dr. Seed will still be up against some restrictions. A subcommittee of the Council for Science and Technology in Japan has recommended a ban on the cloning of humans, so Dr. Seed's idea may never sprout.

The Least You Need to Know

➤ Some people believe that with advances in cloning technology, scientists will be able to clone humans in the near future.

➤ There are possible scenarios in which people would want to clone themselves, such as infertile couples, same sex couples, and couples who want another child with identical DNA to donate bone marrow to their first child, who is dying.

➤ Some people think that clones, although they have identical DNA as their parent-donors, would not be exactly the same because of environmental and other factors.

➤ Some people are seeking an all-out ban on human cloning, because they consider it unnatural and unethical.

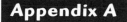

Resources

Here are some resources in different areas of genetic interest.

Paternity/Identity Testing

Cellmark Diagnostics
20271 Goldenrod Lane
Germantown, MD 20876
(800) 872-5227
www.cellmark-labs.com

DNA Diagnostics Center
205 Corporate Court
Fairfield, OH 45014
(800) 362-2368
www.dnacenter.com

LIFECODES Corporation
550 West Avenue
Stamford, CT 06902
(800) 543-3263
www.lifecodes.com

Education

Access Excellence
Genentech Incorporated
(800) 295-9881
www.gene.com/Company/
Responsibility/about_ae.html

Blazing a Genetic Trail
Howard Hughes Medical Institute
(301) 215-8855
www.hhmi.org/GeneticTrail

Centre for Applied Ethics
University of British Columbia
www.ethics.ubc.ca/brynw

CityLab
Boston University School of Medicine
715 Albany Street, S-408
Boston, MA 02118
(617) 638-562
www.bumc.bu.edu/Departments/
HomeMain.asp?DepartmentID=285

DNA Learning Center
Cold Spring Harbor Laboratory
1 Bungtown Road
Cold Spring Harbor, NY 11724
(516) 367-7240
www.vector.cshl.org

The Gene Letter
The Shriver Center, Inc.
www.geneletter.org

Primer on Molecular Genetics
Department of Energy
www.bis.med.jhmi.edu/Dan/DOE/
intro.html

Genetic Diseases

Alliance of Genetic Support Groups
4301 Connecticut Avenue N.W.
Suite 404
Washington, D.C. 20008
(800) 336-4363
www.geneticalliance.org

National Organization for Rare Disorders, Inc.
P.O. Box 8923
New Fairfield, CT 06812-8923
(800) 999-6673
www.rarediseases.org

National Society of Genetic Counselors, Inc.
233 Canterbury Drive
Wallingford, PA 190886-6617
(800) 872-7608
www.nsgc.org

Biotechnology and Bioethics

Biotechnology Industrial Organization
1625 K Street, N.W.
Suite 1100
Washington, D.C. 20006-1604
(202) 857-0244
www.bio.org

Bullfrog Films
P.O. Box 149
Oley, PA 19547
www.bullfrogfilms.com

Council for Responsible Genetics
5 Upland Road
Suite 3
Cambridge, MA 02140
(617) 868-0870

Food Bytes
www.geocities.com/Athens/1527

Genetic Engineering News
2 Madison Avenue
Larchmont, NY 10538
(914) 834-3880
www.genengnews.com

International Center for Technology Assessment
www.icta.org

Rural Advancement Foundation International-USA (RAFI)
P.O Box 655
Pittsboro, NC 27312
(919) 542-1396
www.rafi.org

Third World Network
www.twnside.org.sg

Union of Concerned Scientists
26 Church St.
Cambridge, MA 02238
(617) 547-5552
www.ucsusa.org

Virtual Library on Human Genome Project
www.ornl.gov/techresources/
HumanGenome/genetic.html

Genetic Jargon

ADA deficiency   A rare genetic disease that prevents the immune system from developing. Even the simplest infection like a cold or the flu can prove fatal for people with this disease.

allele   One of several forms that a gene for a trait can take.

amino acids   Chemicals that are the building blocks of proteins. There are 20 amino acids, and they are linked together in different combinations to form different proteins.

amniocentesis   A procedure that studies fluid from the amniotic sac, which surrounds the unborn child in its mother's womb. Cells from the fetus found in this fluid are grown in the laboratory and screened for some inherited diseases.

Angelman's syndrome   Like Prader-Willi syndrome, a disease caused by a mutation on chromosome 15. Both of these syndromes cause mental retardation. Angelman's is inherited from the mother, but Prader-Willi is inherited from the father. The two diseases are caused by the same mutation in the same gene.

antibody   A type of protective protein that is produced in humans and other higher mammals in order to help the body fight back when it is attacked by infections.

antigen   A foreign substance, such as bacteria or viruses, that invades the body.

antisense therapy   A type of gene therapy that stops a faulty gene from producing its defective protein.

atom   The smallest particle of an element. It is composed of a nucleus and electrons.

base   A sub-unit of the DNA molecule. There are four bases in DNA: adenine (A), thymine (T), cytosine (C), and guanine (G).

base pair   Two bases that bond together to form one rung on the twisted ladder shape of DNA. The base adenine (A) can pair only with thymine (T), and cytosine (C) can pair only with guanine (G).

B-lymphocytes   Special cells that are called on to combat invaders. These cells are produced in the blood, lymph nodes, and the spleen.

behavioral genetics   The branch of genetics that studies the relationship between the genes you're born with and the behaviors you exhibit.

biochip   A microchip about the size of a dime that is used in diagnosing genetic diseases or predispositions to diseases. Biochips have also been referred to as DNA arrays, chip arrays, DNA chips, and microarrays.

bioethicist A person who specializes in controversial ethical issues related to biology and medicine, such as genetic engineering topics and the care of the terminally ill.

blood serum The clear, yellowish liquid part of blood that separates out from a clot in coagulated blood.

bone marrow A soft substance found inside the hard part of bones. It produces red blood cells as well as various white blood cells.

carcinogens Cancer-causing substances.

carrier A person who carries one copy of a mutant gene for a recessively inherited genetic disease that can only develop if a person has two copies of the mutant gene. A carrier does not develop the disease but may or may not pass the gene for the disease on to their children.

chorionic villus sampling (CVS) A procedure used to screen a developing fetus for disease. CVS involves taking a small amount of cells from the placenta. These cells are genetically identical to the cells in the fetus. By studying the cells of the placenta, researchers can find out about the DNA in the fetus.

complementary DNA (cDNA) A synthetic form of DNA made from messenger RNA. Usually, the tiny ribosome "factories" in the cell use messenger RNA as the blueprint for manufacturing a specific protein. But cDNA is used by scientists for purposes such as locating specific genes and producing proteins in bacteria.

cell cycle The four phases a cell goes through when it divides to eventually produce two new cells. These phases are called the G1 phase, the S (for synthesis) phase, the G2 phase, and the M (for mitosis) phase.

cell fusion The fusion of two different cells to become one single cell. This can occur naturally during sexual reproduction, when a sperm and an egg fuse together to produce a fertilized egg that will grow into an embryo and eventually produce a full-grown human or animal. Cell fusion is used in the laboratory to bring together two different types of cells.

cell line Cells that are made to grow continuously in a laboratory, outside of the living organisms that they were taken from.

cell membrane The "skin" that covers a cell. The cell membrane acts like a security guard and checks what goes in and out of the cell.

chimera molecule A synthetic molecule made up of DNA and RNA. It has been used in gene therapy experiments in laboratory rats to target and deliver genes to specific types of cells in their bodies.

chloroplasts Structures in plant cells that do not exist in animal cells. They contain a green substance called chlorophyll, which absorbs light energy from the sun. This energy is later turned into food for the plant.

chromatin The stuff that chromosomes are made of. Chromatin is about one-half DNA and one-half protein.

chromosomes   The rod-like structures in the cell's nucleus that contain genes.

clones   Two or more organisms that have exactly the same DNA. Identical twins are naturally occurring human clones.

cloning   Making an exact copy of all the genetic material in a plant, animal, or human. Identical twins are technically clones because they have all the same genes because they form from the same egg, which splits into two separate embryos. Scientists have found ways to artificially clone different organisms.

codon   A grouping of three nucleotides that represents a specific amino acid. The way these codons signify different amino acids is known as the genetic code.

collagen   A protein that is found in skin. It doesn't get replaced, so as it gets older, it loses its elasticity, and this leads to wrinkles.

crossing-over   A process that can occur when chromosomes from the mother and chromosomes from the father line up in pairs during meiosis, when egg and sperm cells are produced. The two chromosomes swap some genes so that some of the genes originally on the father's chromosomes are now on the mother's chromosomes and vice versa.

cystic fibrosis   A genetic disease characterized by the buildup of large amounts of mucus in the lungs. Patients with this disease rarely live beyond their 20s or 30s.

cytoplasm   The jelly-like inside of a cell within the barrier of the cell membrane. All the smaller structures of a cell swim within the cytoplasm.

DNA (deoxyribonucleic acid)   A molecule containing hereditary information that is passed to the next generation.

DNA probes   Short sequences of As, Ts, Cs, and Gs that have been given radioactive tags. They join up with complementary sequences of DNA. For instance, AATC will latch onto TTAG, because As are attracted to Ts and Cs are attracted to Gs. DNA probes are used in DNA testing.

DNA profiling   Genetic testing used to compare DNA samples and determine identity. For instance, these tests can be used to prove maternity or paternity. Also called DNA typing or DNA fingerprinting.

dominant   A dominant form of a gene always appears in the next generation, even if only one copy of the gene is passed on. Compare *recessive*.

dominant disease   If a disease is dominantly inherited, only one copy of the mutant gene is needed to cause the disease in a child who receives the mutation. Compare *recessive disease*.

electrophoresis   One of the steps involved in DNA profiling. In this process, electricity is passed through a gel containing DNA fragments. The smallest pieces move through the gel quicker than the longer pieces, so the pieces are separated.

361

embryo twinning   A method used to clone animals. Researchers split a developing animal embryo into two at an early stage. This process usually results in two separate embryos, which develop into identical animals.

enzyme   A protein that can speed up certain chemical reactions in cells.

eukaryote   Any cell that has a nucleus.

exon   The part of a gene that codes for a protein. Compare *intron*.

expressed sequence tags (ESTs)   The name that researcher J. Craig Venter gave to parts of genes that are "fished out" using a synthetic form of DNA called complementary DNA.

factor VIII and **factor IX**   Proteins that help in the coagulation (clotting) of blood. A deficiency in either of these factors leads to different types of hemophilia, a disease in which the sufferer bleeds profusely, even from a small cut.

forensic DNA testing   The use of DNA testing as evidence in court cases.

gene gun   A device that shoots tiny metal spheres coated with foreign DNA into plant cells. Some of the foreign DNA becomes part of the plant's permanent genetic makeup.

gene machine   A machine that can construct a desired gene fragment from scratch. The researcher determines the sequence of the gene fragment, types it into the machine's computer, and then the machine produces that sequence of As, Cs, Ts, and Gs.

genes   The basic units of heredity that are found in living cells. Genes are stretches of DNA.

gene therapy   The manipulation of genes to add correct copies or delete incorrect copies to cure genetic diseases.

genetic locus   A specific area of DNA being studied. A genetic linkage map compares the position of one genetic locus with others.

genetics   The study of heredity.

genome   The totality of all the DNA in an organism. The human genome refers, of course, to all the DNA found in humans. Each type of plant and animal has its own genome.

genomics   The study of genomes, including genome mapping and gene sequencing.

genotype   The actual genetic constitution of an organism. A person's genotype might be one gene for brown eyes from his mother and one gene for blue eyes from his father. Compare *phenotype*.

germ cells   The reproductive cells, the sperm and eggs.

germ line gene therapy   Gene therapy performed on sex cells. Any genetic alterations to germ cells will be passed on to future generations.

ghost gene   A stretch of DNA that once functioned as a gene that coded for a protein, but has been changed and can no longer function that way. Also called pseudogenes.

hairy cell leukemia   A cancer that occurs in certain cells in the blood and bone marrow. The cancerous cells in this disease look like they have hairs sticking out of them when viewed under a microscope.

helix   Another word for a spiral. DNA is a double helix, or a double spiral.

heredity   The way traits are passed from parents to their offspring.

hormones   Substances formed in the tissues or glands of the body that usually circulate through the bloodstream. They influence the different functions and growth of cells and organs.

Human Growth Hormone (HGH)   The substance responsible for normal growth in humans. An excessive lack of the hormone results in dwarfism. HGH is normally secreted by the pituitary gland, which is found at the base of the brain.

hybrid   A plant or animal that results from the crossing of two different species.

imprinting   The way mutations in genes can cause one disease if inherited from the father, but cause another disease if the same mutation is inherited from the mother. Also called genomic imprinting. Compare *Angelman's syndrome* and *Prader-Willi syndrome.*

in vitro fertilization (IVF)   A type of assisted reproduction technology. Sperm and eggs are mixed in a culture dish or test tube. When the eggs are fertilized and begin to grow into embryos, some are implanted into the mother. The name *in vitro* comes from the Latin words meaning "in glass." Children produced by this technique have been dubbed test tube babies by the media.

incomplete penetrance   When a person has a mutated gene or genes, but it does not affect that person and the disease associated with that gene or genes does not develop, it's called incomplete penetrance. This may be due to the influence of other genes, the environment, or both.

insulin   A protein that regulates blood sugar. People with diabetes don't have enough of this protein in their bodies, and they usually need to get injections of insulin to make up for this deficiency.

interferons   Three proteins that help improve the body's natural defenses against disease. Interferons fight viruses and can slow the growth of tumors. The three types of interferons are alpha, beta, and gamma interferons.

intron   A segment of a gene that does not code for a protein. Compare *exon.*

Kaposi's sarcoma   A type of cancer that affects the skin and often strikes people with AIDS.

keratin   A protein found in your hair, fingernails, and toenails.

Klinefelter's syndrome   A condition that results from an inherited mutation that gives offspring two Xs and one Y chromosome. Males who inherit this mutation suffer from sterility and sometimes enlarged breast tissue and speech problems.

Lamarckism   The theory that adaptations to the environment cause changes in plants and animals that can be inherited by their offspring.

ligase   An enzyme that acts like molecular glue in genetic engineering. DNA ligase joins DNA segments, and RNA ligase joins RNA segments.

marker   A genetic variation on a chromosome. The marker is like a signpost that points the way to the general location of a specific gene.

meiosis   The process of duplicating all the chromosomes in sex cells, and then reducing their number by half, so that there are four new cells, and each one of them contains half the number of chromosomes in the original cell. Meiosis starts out like mitosis, but it has one more division.

meningitis   A serious disease characterized by an inflammation of the membranes that surround the brain and the spinal cord.

micro sort technique   A method used at The Genetics & IVF Institute in Fairfax, Virginia that sorts sperm cells according to whether they carry an X chromosome (for girls) or a Y chromosome (for boys).

mitosis   The process of duplicating all of a cell's chromosomes during cell division. When the cell finally divides, each of the two daughter cells, as they are called, has two copies of all the chromosomes that were in the original cell.

molecule   A group of two or more atoms that are bonded together.

monoclonal antibody technology   A technology that produces antibodies. Scientists take B-lymphocytes from lab animals and fuse them together with myeloma cells, which are cells from cancerous tumors in bone marrow. When fused together, these cells can grow continuously in a lab, so they can produce a large quantity of the desired antibody.

monogenic diseases   Diseases that are caused by a mutation in a single gene. Compare *polygenic diseases*.

multiple sclerosis (MS)   A disease in which the body's immune system turns against the fatty substance that lines the nerve fibers in the brain and the spinal cord. Without this protection, a person with the disease becomes fatigued, has slurred speech and vision problems, and can suffer paralysis.

mutation   A change in the DNA "letters" of a gene.

natural selection   Darwin's theory of natural selection, also known as survival of the fittest, states that some plants and animals within a species are more suited to their environment than others. These better-suited plants and animals tend to thrive and reproduce more of their own kind. In the same way, those that aren't especially suited to their surroundings tend to do poorly and reproduce less. Over time, this difference leads to species-wide changes.

nematode   A type of worm with a threadlike body. They are also called roundworms. The name *nematode* comes from the Latin word meaning "the threadlike one." Scientists have sequenced the entire genome of C. elegans—a type of nematode—and it's used to study development in other organisms.

non-Hodgkin's lymphoma (NHL)   A type of cancer in which the B-lymphocytes, the cells that combat bacterial and viral invasion, begin to multiply too much and too rapidly. This results in cancerous tumors that can spread to organs such as the spleen or the liver.

nuclear transfer   The process used to clone Dolly the sheep. The DNA from an adult cell was put into an egg cell that had its own nucleus removed. After this egg was stimulated by electricity to grow, it was implanted into another sheep, which later gave birth to Dolly.

nucleotide   One of the building blocks of the DNA molecule. The four nucleotides are A, C, T, and G. Because of their chemical structures, A can pair only with T, and C can pair only with G. These nucleotide pairs are known as base pairs.

oncogene   A gene that causes cancer. A tumor-suppressor gene, or anti-oncogene, normally stops the growth of cancer cells, but when it mutates, it causes cancer.

parasite   Any organism that lives off another organism, but does not contribute anything good to the host. In some cases, a parasite can make the host sick or eventually destroy it.

phages   Viruses that attack and infiltrate the bodies of living bacteria. Also called bacteriophages.

pharming   The genetic altering of animals to produce human proteins that can be used to treat people. Also called molecular farming.

phenotype   A living thing's observable display of a trait. Compare *genotype*.

phenylketonuria (PKU)   A genetic disease characterized by the lack of a liver enzyme that can break down an amino acid called phenylalanine. If the disease is not controlled by a special diet, the toxic buildup in the brain of the sufferer can lead to severe mental retardation.

pistil   The pistil of a plant is the seed-bearing reproductive organ of a flower. The pistil includes parts called the stigma, the style, and the ovary.

plantibodies   Antibodies made from genetically engineered plants, such as corn.

plasmid   A circular piece of DNA found in bacteria. A Ti plasmid is a tumor-inducing plasmid in a bacterium that infects plants. Genetic engineers sometimes splice a new gene into a Ti plasmid, because it can naturally invade plant cells.

pneumococcus   The name of the bacteria that causes pneumonia in mice.

point mutations   Mutations that occur when just one nucleotide, an A, T, C, or G, is copied incorrectly in a gene.

polygenic diseases   Diseases that are caused by several mutated genes acting together. Compare *monogenic diseases.*

polymerase   An enzyme that copies DNA.

Polymerase Chain Reaction (PCR)   A method that produces millions of copies of a specific sequence of DNA, even if it has broken down considerably. This technique was devised by Kary Mullis, a scientist in California, who was awarded the Nobel prize for this discovery in 1983.

polymorphism   A variation in short stretches of DNA.

population genetics   A branch of genetics that deals with the hereditary makeup of different groups within the population, such as by analyzing how frequently a version of a gene appears in a certain ethnic group.

Prader-Willi syndrome   One of two diseases caused by a mutation on chromosome 15. Both diseases cause mental retardation. Prader-Willi is inherited from the father. The two diseases are caused by the same mutation in the same gene. Compare *Angelman's syndrome.*

Pre-implantation Genetic Diagnosis (PGD)   A method used to test an early-stage embryo for inherited diseases and gender. When the embryo has developed to about 16 cells, one cell is removed, and its DNA is tested.

primates   The group of mammals that includes monkeys, apes, and homo sapiens, or humans.

primer   A short sequence of As, Ts, Cs, and Gs that starts the process of copying DNA. Polymerase can't do its job without a primer to begin, or "prime," the process.

prokaryote   The simplest type of cell, such as bacteria, which are one-celled life forms. This word is made up of some Greek words that mean "before nucleus." Prokaryotes don't have a nucleus, which is the small, dark body inside a eukaryotic cell that contains its genetic material.

promoter   A region of DNA in front of a gene that promotes the expression of that gene. It's something like the on/off switch on a lamp or an electrical appliance.

protein synthesis   When a cell manufactures protein, this process is called protein synthesis. Proteins are made by placing a number of amino acids in a specific order in a chain.

recessive   A recessive form of a gene only appears in the next generation when present in two copies. Compare *dominant.*

recessive disease If a disease is recessively inherited, a child needs to inherit two copies of the mutation, one from each parent, to develop the disease. Compare *dominant disease*.

restriction fragment length polymorphism (RFLP) analysis A method used in creating DNA profiles.

restriction enzymes Special enzymes that occur naturally in some bacteria. When the bacteria are attacked by invaders such as viruses, restriction enzymes cut the DNA of the attackers to protect the bacteria. Restriction enzymes are used in genetic engineering.

retinoblastoma A rare cancer of the retina in the eye.

retrovirus A type of virus that has genetic material that consists of RNA instead of the usual DNA.

schizophrenia A mental illness characterized by hallucinations, delusions, and disorganized thinking. In some cases, a schizophrenic cannot tell the difference between fantasy and reality. The disease usually appears in a patient's teenage years or early 20s.

selective breeding The process of crossing plants or animals that have desired characteristics with other plants that have other desired traits and continuing this breeding until the desired traits are mixed in a new generation. This traditional form of plant or animal breeding is also called artificial selection.

sickle-cell anemia An inherited disease that occurs when some of the red blood cells take on a long, thin sickle or crescent shape. These cells clog the blood vessels and damage the internal organs.

silent mutations "Misspellings" in DNA that go unnoticed because they have no effect on the outcome of a protein.

somatic cells The body cells, except for the reproductive cells.

spacer DNA Noncoding DNA that occurs in between genes.

SRY gene The gene that plays a crucial role in the sex determination of human males. SRY stands for "sex-determining region Y."

stamen The part of a plant that produces pollen, which is the plant's sperm cells.

stem cells Cells in an embryo that are still undifferentiated. This means that they can grow into any type of cell in the body, such as heart cells, blood cells, or brain cells.

stigma The top of a pistil of a plant, where pollen is deposited during pollination.

tandem repeats Specific sequences of DNA that repeat over and over again, one immediately after the other, like a genetic stutter. They are also called VNTRs (variable number of tandem repeats).

testosterone The male sex hormone, or chemical messenger, responsible for many sexual characteristics in men.

thalassemia A genetic disease characterized by abnormally small red blood cells. Also called Cooley's anemia.

tPA A protein naturally produced in the human body. Cells that line the walls of blood vessels use this substance to dissolve blood clots, so blood can flow smoothly throughout all the blood vessels.

Turner's syndrome A disorder that occurs in women due to a chromosomal mutation. The sufferer of this disorder has inherited only one X chromosome instead of two. Women with this mutation are short of stature and infertile and can experience problems with their hearts, kidneys, and thyroids.

transcription The creation of a messenger RNA strand by copying the letters that pair up with the gene in DNA.

transgenic animal An animal that has one or more genes from another organism added to its DNA or one or more genes removed from its DNA through genetic engineering.

transgenic plant A plant that has been genetically altered. It may have one or more genes from an animal, a human, or another plant spliced into its genome. It might also have a gene or two deleted from its genome.

transgenic organism A new living organism that scientists produce by mixing and matching genes. Using the new genetic technologies, scientists can mix different types of plants, animals, and even humans in ways that were never done before.

transposons Stretches of DNA that travel from one position on the chromosome to another. Barbara McClintock first discovered this phenomenon in the 1940s, but few scientists accepted her theory at the time. Also called jumping genes.

vector A delivery system used in gene therapy that gets new, healthy genes into a person's cells. Scientists often use organisms such as viruses to carry the new genes into a patient's cells. First they take the harmful genes out of the virus or other invading organism.

virus A microscopic bag of genes with a protein coat. A virus tricks cells into letting it in, and then the cell is instructed to use its resources to make more of the virus. This makes the cell sick or kills it. Scientists don't consider viruses living things because they cannot reproduce on their own and rely on their hosts to carry out functions for them.

xenotransplantation The use of animal organs to replace needed human parts.

Selected Bibliography

Here's a list of books for further reading on a variety of genetic topics.

Coleman, Howard and Eric Swenson. *DNA in the Courtroom: A Trial Watcher's Guide.* Seattle, WA: GeneLex Press, 1994.

Council for Responsible Genetics. *Genetic Engineering: Unresolved Issues.* Cambridge, MA: Council for Responsible Genetics, 1992.

Crick, Francis. *What Mad Pursuit? A Personal View of Scientific Discovery.* New York, NY: Basic Books, 1990.

DeSalle, Rob and David Lindley. *The Science of Jurassic Park and the Lost World: or How to Build a Dinosaur.* New York, NY: Basic Books, 1997.

Dodson, Bert and Mahlon Hoagland. *The Way Life Works.* New York, NY: Times Books, 1995.

Gonick, Larry, and Mark Wheelis. *The Cartoon Guide to Genetics.* New York, NY: Harper Collins Perennial, 1991.

Harris, John. *Wonderwoman and Superman: The Ethics of Human Biotechnology.* Oxford, United Kingdom: Oxford University Press, 1992.

Hubbard, Ruth and Elijah Wald. *Exploding the Gene Myth.* Boston, MA: Beacon Press, 1993.

Keller, Evelyn F. *A Feeling for the Organism: The Life and Work of Barbara McClintock.* New York, NY: W.H. Freeman and Company, 1983.

Kevles, Daniel J. *In the Name of Eugenics: Genetics and the Uses of Human Heredity.* Berkeley, CA: University of California Press, 1985.

Kevles, Daniel J., and Leroy Hood, eds. *The Code of Codes: Scientific and Social Issues in the Human Genome Project.* Cambridge, MA: Harvard University Press, 1992.

Kolata, Gina. *Clone: The Road to Dolly and the Path Ahead.* New York, NY: William Morrow and Company, 1998.

Levine, Joseph S. and David T Suzuki. *The Secret of Life: Redesigning the Living World.* New York, NY: W.H. Freeman and Company, 1998.

Nelkin, Dorothy and M. Susan Lindee. *The DNA Mystique: The Gene as a Cultural Icon.* New York, NY: W.H. Freeman and Company, 1995.

Sayre, Anne. *Rosalind Franklin & DNA*. New York, NY: W.W. Norton and Company, 1978.

Shapiro, Robert. *The Human Blueprint: The Race to Unlock the Secrets of Our Genetic Code*. New York, NY: Bantam Books, 1992.

Steen, R. Grant. *DNA and Destiny: Nature and Nurture in Human Behavior*. New York, NY: Plenum Press, 1996.

Suzuki, David and Peter Knudtson. *Genethics*. Cambridge, MA: Harvard University Press, 1989.

Tagliaferro, Linda. *Genetic Engineering: Progress or Peril?* Minneapolis, MN: Lerner Publications Company, 1997.

Watson, James D. *The Double Helix*. New York, NY: Penguin Books, 1969.

Index

Council for Responsible Genetics (CRG), 302, 313, 333–34, 339–40
Cows
  BGH (Bovine Growth Hormone) and, 328–30
  cloning, 191–92
Creationism, 138
Creutzfeldt-Jacob disease, 280
Crick, Francis, 9, 38, 62–65, 67–69
Cross-fertilization, 19
Crossing-over, 50
Culver, Ken, 268
Cutshall, Cindy, 269, 338
CVS (chronic villus sampling), 288, 290–91
Cyber Dialogue, 348
Cystic fibrosis, 107, 126, 229, 235
Cytoplasm, 42–43
Cytosine, 60

D

Darwin, Charles, 7, 136–38
Dating methods, 144–45
Daughter cells, 129
DeCODE Genetics, 336
Deletions, 106–7
Delta and Pine Land, 316
Deoxyribose, 60
"Designer babies," 291
DeSilva, Ashanti, 267–68, 338
De Vries, Hugo, 29–30
Diabetes, 80
Disease resistance, in plants, 168

Diseases. *See* DNA testing, for diseases; Gene therapy; Genetic disorders; *And specific diseases*
DNA cards, 234
DNA data banks, 220, 337
DNA (deoxyribonucleic acid), 3, 8–9, 35
  in bacteria, 40–41
  discovery of the presence of, 37
  Hershey and Chase experiments and, 36
  junk, 81–84
  mitochondrial, 143–44, 207–9
  of Neanderthals, 145
  in plasmids, 150–53
  rearrangement of DNA sequences, 75–76
  recombinant. *See* Genetic engineering
  repair system, 74
  spacer, 82–83
  strands of, 73
  structure of, 38, 59–65, 91
DNA polymerase, 203
DNA probes, 200, 228, 233–34
DNA replication, 68–72
  errors in, 73–74
DNA testing, 197–210, 332–35
  for diseases, 225–36, 335
    biochips, 232–34
    breast cancer, 229–30
    currently available tests, 235–36
    cystic fibrosis, 229

DNA probes, 228
  electrophoresis, 227–28
  RFLP analysis, 227
  sickle cell anemia, 227
identification of criminals by, 211–22
legislation on, 337
mitochondrial DNA analysis, 207–9
origin of homo sapiens and, 145–46
paternity, 199, 202, 205–7
PCR (Polymerase Chain Reaction) technique, 202–5
prenatal, 341
reliability of, 334–35
Southern blot technique, 199–200
STR technique, 204
Dogs, 325–26
Dolly, 175, 186–89, 345, 354
Dominantly inherited diseases, 105, 127
Dominant traits, 20
Double helix, 64–65
Down syndrome, 86, 109, 289–90
Doyle, Jack, 311–12
Drosophila melanogaster, 33
Drossard, Aurore, 205
Duchenne muscular dystrophy, 80, 107, 204
Du Pont Chemical Company, 177
Dwarfism, 280
Dysentery, 151

X

Y